PUBLICATIONS OF THE NEWTON INSTITUTE

Numerical Methods in Finance

Publications of the Newton Institute

Edited by H.P.F. Swinnerton-Dyer

Executive Director, Isaac Newton Institute for Mathematical Sciences

The Isaac Newton Institute of Mathematical Sciences of the University of Cambridge exists to stimulate research in all branches of the mathematical sciences, including pure mathematics, statistics, applied mathematics, theoretical physics, theoretical computer science, mathematical biology and economics. The four six-month long research programmes it runs each year bring together leading mathematical scientists from all over the world to exchange ideas through seminars, teaching and informal interaction.

Associated with the programmes are two types of publication. The first contains lecture courses, aimed at making the latest developments accessible to a wider audience and providing an entry to the area. The second contains proceedings of workshops and conferences focusing on the most topical aspects of the subjects.

NUMERICAL METHODS IN FINANCE

edited by

L.C.G. Rogers

University of Bath

and

D. Talay

INRIA, Sophia-Antipolis

PUBLISHED BY THE PRESS SYNDICATE OF THE UNIVERSITY OF CAMBRIDGE
The Pitt Building, Trumpington Street, Cambridge, United Kingdom

CAMBRIDGE UNIVERSITY PRESS
The Edinburgh Building, Cambridge CB2 2RU, UK http://www.cup.cam.ac.uk
40 West 20th Street, New York, NY 10011–4211, USA http://www.cup.org
10 Stamford Road, Oakleigh, Melbourne 3166, Australia

First published 1997
Reprinted 1999

Printed in the United Kingdom at the University Press, Cambridge

Typeset in Computer Modern

A catalogue record for this book is available from the British Library

ISBN 0 521 57354 8 hardback

CONTENTS

CONTRIBUTORS

F. AitSahlia, Hewlett-Packard Laboratories, 1501 Page Mill Road, MS 4U-1, Palo Alto, CA 94304, USA.
farid@hpl.hp.com

Renzo G. Avesani, University of Brescia, Department of Economical Science, Via Porcellaga 21, I 25121, Brescia, Italy.
rga@opoipi.opipi.it

G. Barles, Laboratoire de Mathématiques et Applications, Faculté des Sciences et Techniques, Université de Tours, Parc de Grandmont, 37200 Tours. France.
barles@univ-tours.fr

Pierre Bertrand, Université Blaise Pascal, Laboratoire de Mathématiques Appliquées, 63177 Aubière Cedex, France.
bertrand@ucfma.univ-bpclermont.fr

Peter Bossaerts, HSS 228-77, California Institute of Technology, Pasadena, CA 91125, USA.
pbs@rioja.caltech.edu

M. Broadie, Columbia University, Graduate School of Business, 3022 Broadway, 415 Uris Hall, New York, NY 10027, USA.
MBROADIE@research.gsb.columbia.edu

P. Carr, Johnson Graduate School of Management, Cornell University, Ithaca, NY 14853, USA
peter@johnson.cornell.edu
and Morgan Stanley, Equity Derivatives Research, 1585 Broadway Avenue, 6th floor, New York, NY 10036, USA.
carrp@morgan.com

D. Chevance, Université d'Orleans, Bâtiment de Mathématiques, B.P. 6759 45067 Orleans, France.
chevance@lagrange.univ-orleans.fr

J. Detemple, Faculty of Management, McGill University, 1001 Sherbrooke Street West, Montreal, Quebec H3A 1G5, Canada.
detemple@management.mcgill.ca

N. El Karoui, Laboratoire de Probabilités, CNRS-URA 224, Université de Paris VI, 4 Place Jussieu, 75232 Paris Cedex 05, France.
ne@ccr.jussieu.fr

E. Fournié, Caisse Autonome de Refinancement, 2 square de Luynes, 75007 Paris, France.

Claude Henin, University of Ottawa, 136 Jean-Jacques Lussier, CP450 SUCCA, Ottawa ON, K1N6N5, Canada.
henin@profs.admin.uottawa.ca

Adriaan Joubert, The London Parallel Applications Centre, Queen Mary & Westfield College, Mile End Road, London E1 4NS, United Kingdom.
A.W.Joubert@lpac.ac.uk

J.M. Lasry, Caisse Autonome de Refinancement, 2 square de Luynes, 75007 Paris, France.

P-L. Lions, Ceremade, URA CNRS 749, Université Paris Dauphine, 75775 Paris (Cedex 16) France.

Nigel J. Newton, Department of ESE, University of Essex, Wivenhoe Park, Colchester, CO4 3SQ, United Kingdom.
njn@essex.ac.uk

E. Pardoux, LATP, URA CNRS 225, Centre de Mathématiques et d'Informatique, Université de Provence, 39, rue F. Joliot Curie, F13453 Marseille Cedex 13, France.
pardoux@gyptis.univ-mrs.fr

Nathalie Pistre, Groupe CERAM, Rue Dostoievski, BP 085, 06902 Sophia Antipolis Cedex, France *and* INRIA-Sophia Antipolis, Route des Lucioles, BP 109, 06902 Sophia Antipolis Cedex, France.
Nathalie.Pistre@ceram.fr

M.C. Quenez, Equipe de Mathématiques, Université de Marne la Vallée, 2 rue de la Butte Verte, 93.166 Noisy-Le-Grand. France.
quenez@math.univ-mlv.fr

L.C.G. Rogers, School of Mathematical Sciences, University of Bath, Bath BA2 7AY, United Kingdom.
lcgr@maths.bath.ac.uk

Agnès Sulem, INRIA, Domaine de Voluceau Rocquencourt, BP105, 78153 Le Chesnay Cedex, France.
agnes.sulem@inria.fr

Agnès Tourin, CEREMADE, Université de Paris IX Dauphine, Place de Lattre-De-Tassigny, 75775 Paris Cedex 16, France.
agnes@ceremade.dauphine.fr

N. Touzi, Ceremade, URA CNRS 749, Université Paris Dauphine, 75775 Paris (Cedex 16) France, *and* CREST.

Bas Werker, Department of Economics, Tilburg University, PO Box 90153, 5000 LE Tilburg, The Netherlands.
B.J.M.Werker@kub.nl

Thaleia Zariphopoulou, Departments of Finance and Mathematics, University of Wisconsin, Madison, WI 53706, U.S.A.
zariphop@math.wisc.edu

Xiao Lan Zhang, Desk R & D, MARC, OTA, Tour Société Générale, 17 Cours Valmy, 92987 Paris-La Défense Cedex, France.
aurore@ota.societe-generale.fr

Introduction

L.C.R. Rogers and D. Talay

In April 1995, together with Alain Bensoussan and Agnès Sulem, we organized the session 'Numerical Methods in Finance' at the Isaac Newton Institute, within the framework of the 1995 Cambridge University programme on Financial Mathematics. We invited specialists in this area which is at the intersection of Probability Theory, Finance and Numerical Analysis.

Several participants worked in banks, which illustrates the needs of the practitioners for theoretical and/or numerical studies of the numerical methods they currently use (Monte Carlo procedures, approximation methods to solve PDEs appearing in option pricing, simulations, etc).

After the session, most of the lecturers agreed to write a paper on the subject of his or her talk. They also agreed not to write the paper as if it would be published in an ordinary volume of Proceedings. Each article presents the state of the art on a particular question of financial and numerical interest, with an extensive list of appropriate references, and then focuses on a new and original result (published elsewhere with complete proofs or complete numerical studies) with a pedagogical point of view: in particular, papers by mathematicians should be understandable by practitioners having a basic knowledge in the theory of stochastic processes.

To our knowledge, at the present time there does not exist any book presenting such a large variety of numerical methods in finance:

- computation of option prices, especially of American option prices, by finite difference methods;
- numerical solution of portfolio management strategies;
- statistical procedures, identification of models;
- Monte Carlo methods, simulation of the market;
- numerical implications of stochastic volatilities and of transaction costs;
- viscosity solutions of PDEs related to option pricing and portfolio management;
- backward stochastic differential equations in finance.

The book offers contributions from economists, probabilists, applied mathematicians, all of them interested in the mathematical and practical issues of numerical problems in finance, e.g. rates of convergence and numerical efficiency. Thus we hope that the book is also of interest for people working in areas other than finance such as random mechanics, physics, etc.

The book presents recent and advanced results; it also describes current practice and needs, and some of the most important open problems. Thus, it should be useful to researchers in the field and also to people developing advanced software for financial institutions.

We thank the Isaac Newton Institute for the opportunity to hold this meeting, and its staff for their cheerful hospitality.

We also thank Cambridge University Press, and particularly David Tranah, for giving us the opportunity of editing this volume and helping us to prepare it.

Bath and Sophia–Antipolis,
July 1996

Chris Rogers and Denis Talay

Convergence of Numerical Schemes for Degenerate Parabolic Equations Arising in Finance Theory

G. Barles

1 Introduction

The aim of this article is twofold: on one hand, we describe a general convergence result which applies to a wide range of numerical schemes ('monotone schemes') for nonlinear possibly degenerate elliptic (or parabolic) equation; this type of equation arises naturally in Finance Theory as we will show first. This convergence result was obtained in an article written in collaboration with P.E. Souganidis (1991).

On the other hand, we present several simple numerical schemes for computing the price of different types of 'simple' options: American options, lookback options and Asian options. These schemes are all based on 'splitting methods' and we want to emphasize the fact that this allows also easy extensions for computing the price of more complex options with complicated contracts (cap, floor, ... etc). These schemes also provide examples for which the convergence result of the first part applies. This second part reports on several works in collaboration with J. Burdeau, Ch. Daher & M. Romano (cf. references) which were done in connection with the Research and Development Department of the Caisse Autonome de Refinancement (CDC group).

The article is organized as follows: since the convergence result for numerical schemes relies strongly on the notion of 'viscosity solutions', which is a notion of weak solutions for nonlinear elliptic and parabolic equations, we are first going to present this notion of solutions. In order to introduce it, as a motivation, we examine in the first section several examples of equations arising in Finance Theory, and more particularly in options pricing, and we describe the theoretical difficulties in studying them. The second section presents the notion of viscosity solutions itself: we first introduce the notion of continuous viscosity solutions and then we give the extension to the more complicated framework of discontinuous viscosity solutions which is an unavoidable tool to obtain the general convergence result for numerical schemes which is given in the third section. Then several comments on the assumptions are given and finally, in the fourth section, we present some numerical schemes in option pricing models which are based on splitting methods.

2 Examples of Parabolic Equations Arising in Finance Theory

In the classical framework of the theory of Black and Scholes, the stock price $(S_s)_{s\geq t}$ for time $t \geq 0$ is the solution, in the risk-neutral probability, of the stochastic differential equation

$$dS_s = S_s(rds + \sigma dW_s)\,, \quad S_t = S\,, \tag{2.1}$$

where $(W_s)_{s\geq t}$ is a standard Brownian motion in $\mathbb{R}$. The constants or functions r and σ are known as being respectively the short term interest rate and the so-called volatility. In all the following examples, we will always assume we are in this framework.

We refer to Black & Scholes (1973), Cox & Rubinstein (1985), Cox, Ross & Rubinstein (1979), Duffie (1988) and Ingersoll (1987) for a complete presentation of options pricing theory and its financial background.

1. Classical European Options (Call): Black–Scholes Equation
It is well known that the price of the European call is given for $S \geq 0$ and for $0 \leq t \leq T$ by

$$u(S,t) = \mathbb{E}\left[e^{-r(T-t)}(S_T - K)^+|\mathcal{F}_t\right]$$

where T is the maturity of the option, K its strike and where $(\mathcal{F}_t)_t$ is the filtration associated to the Brownian motion. The derivation of this type of representation formula for the price of the options is described, for example, in Karatzas & Shreve (1988).

In this case, the function u is a solution of the celebrated Black–Scholes Equation

$$-\frac{\partial u}{\partial t} - \frac{1}{2}\sigma^2 S^2 \frac{\partial^2 u}{\partial S^2} - rS\frac{\partial u}{\partial S} + ru = 0 \quad \text{in } \mathbb{R}^+ \times (0,T)\,,$$

with the terminal data

$$u(S,T) = (S-K)^+ \quad \text{in } \mathbb{R}^+\,.$$

In this very simple case, there is no problem since u is given by an explicit formula. But it can be interesting to take into account more complicated models with, for example, a non-constant interest rate r and/or a non-constant volatility σ. In order to do this, one needs an adapted theoretical tool to study the equation and efficient numerical schemes to provide accurate approximations of the price u and also of $\partial u/\partial S$ which gives the hedging portfolio.

From a theoretical point of view, there is a difficulty due to the degeneracy of the equation for $S = 0$. To avoid it, a natural idea is to make the change of variable

$$v(x,t) = u(e^x,t) \qquad \text{for } x \in \mathbb{R}\,,\, t \in (0,T)\,,$$

which leads to the equation

$$-\frac{\partial v}{\partial t}-\frac{1}{2}\sigma^2\frac{\partial^2 v}{\partial x^2}-\left(r-\frac{1}{2}\sigma^2\right)\frac{\partial v}{\partial x}+rv=0 \quad \text{in } \mathbb{R}\times(0,T),$$

and then to use on this transformed equation the classical PDE theories (Sobolev spaces, ... etc). We refer the reader to the book of Lamberton & Lapeyre (1992) where this approach is described.

But using such PDE theories leads to work with weighted Sobolev spaces because of the exponential growth of the solutions at infinity and it is never pleasant to have to use these heavy techniques. Moreover, these weighted Sobolev spaces have *a priori* no clear connection with the probabilistic formula of representation for u and their use does not seem to be natural. So it would be convenient to have a theoretical tool to avoid them and to avoid also the exponential change.

2. American Options (Put): Variational Inequalities

In the pricing of American options, because of the possibility of early exercise, the price u is given by a stopping time problem. In the case of a Put, one has

$$u(S,t)=\inf_{\theta\text{ s.t.}}\mathbb{E}\left[e^{-r(\theta-t)}(K-S_\theta)^+|\mathcal{F}_t\right]$$

where 's.t.' means that θ has to be a stopping time with respect to $(\mathcal{F}_t)_t$.

It is well known that the price u of the option solves the variational inequality

$$\text{Min}\left(-\frac{\partial u}{\partial t}-\frac{1}{2}\sigma^2S^2\frac{\partial^2 u}{\partial S^2}-rS\frac{\partial u}{\partial S}+ru, u-(K-S)^+\right)=0 \quad \text{in } \mathbb{R}^+\times(0,T),$$

with the terminal data

$$u(S,T)=(K-S)^+ \quad \text{in } \mathbb{R}^+.$$

For the pricing of American options, we refer the reader to Bensoussan (1984) and Karatzas (1988); the more general theory of optimal stopping time control problems is described in Bensoussan & Lions (1978).

The problem is here obviously more complicated: indeed, there is no explicit formula for u – even for constant coefficients r and σ – and we face a nonlinear problem with the same degeneracy as in the case of the European options above.

3. Lookback Options

Lookback options are options on the running maximum of the stock price, a typical example of terminal pay-off being

$$(\max_{0\le t\le T} S_t - S_T)^+ .$$

This problem presents non-Markovian features: indeed, at time $0 < t \leq T$, in order to compute the price of the option, one has to know not only the current stock price S_t but also the value of the running maximum $\max_{0 \leq \tau \leq t} S_\tau$.

Therefore, the price of the option not only depends on t and on S_t but also on the running maximum $\max_{0 \leq \tau \leq t} S_\tau$. To take this fact into account, one has to introduce a new variable Z which carries the past information and the associated process $(Z_s)_s$ given for $s \geq t$ by

$$Z_s = \max\left(Z, \max_{t \leq \tau \leq s} S_\tau\right),$$

the idea being that, for $Z = \max_{0 \leq \tau \leq t} S_\tau$, then $Z_s = \max_{0 \leq \tau \leq s} S_\tau$.

To obtain the price of the lookback option, one has to consider the function u, depending on S, Z and t, which is a solution of the problem

$$-\frac{\partial u}{\partial t} - \frac{1}{2}\sigma^2 S^2 \frac{\partial^2 u}{\partial S^2} - rS\frac{\partial u}{\partial S} + ru = 0 \quad \text{in } \{S < Z\},$$

$$-\frac{\partial u}{\partial Z} = 0 \quad \text{in } \{S > Z\},$$

with the terminal data

$$u(S, Z, T) = (Z - S)^+ \quad \text{in } \mathbb{R}^+.$$

Indeed, the price of the lookback option at time $0 \leq t \leq T$ and for a current stock price S is given by $u(S, \max_{0 \leq \tau \leq t} S_\tau, t)$.

The main new remark here is that we have to solve a degenerate equation in the domain $\{S < Z\}$ since there are no second-order derivatives with respect to Z in the equation. In fact, it can be shown (cf. Barles, Daher & Romano (1994)) that the above problem reduces to this equation in $\{S < Z\}$ with the oblique derivative boundary condition

$$-\frac{\partial u}{\partial Z} = 0 \quad \text{on } \{S = Z\}.$$

In this simple case, there is an explicit formula for u but again as soon as we consider non-constant coefficients r and σ or if we want to consider some 'American' type features in the option, adapted analytical and numerical tools are needed to study u.

The pricing of lookback options in the case of constant coefficients is studied in Conze (1990). The optimal control problems on the running maximum of a diffusion process are considered in Heinricher & Stockbridge (1991) by probabilistic methods and in Barron (1993) and in Barles, Daher & Romano (1994). In these two last papers, the applications to lookback options are described.

4. Asian Options (Options on the Average)

In this case, the terminal pay-off of the option is typically given by

$$\left(\frac{1}{T}\int_0^T S_\tau d\tau - S_T\right)^+ .$$

The main characteristics of these options are almost the same as for the look-back options; the non-Markovian feature of the problem leads us to introduce the process

$$Z_s = Z + \frac{1}{T}\int_t^s S_\tau d\tau ,$$

for $t \leq s$.

To compute the price of the Asian option, one has to consider the solution u of the equation

$$-\frac{\partial u}{\partial t} - \frac{1}{2}\sigma^2 S^2 \frac{\partial^2 u}{\partial S^2} - rS\frac{\partial u}{\partial S} + ru - \frac{S}{T}\frac{\partial u}{\partial Z} = 0 \quad \text{in } \mathbb{R}^+ \times \mathbb{R}^+ \times (0,T) ,$$

with the terminal data

$$u(S,Z,T) = (Z-S)^+ \quad \text{in } \mathbb{R}^+ .$$

The price of the Asian option at time $0 \leq t \leq T$ and for a current stock price S is then given by $u\left(S, \frac{1}{T}\int_0^t S_\tau d\tau, t\right)$.

Again we face here a degenerate equation, but this time, even with constant coefficients, there is no explicit formula which can be used for a practical point of view.

We refer to Barles, Daher & Romano (1994), Ingersoll (1987) and to Rogers & Shi (1995) for a PDE approach of the pricing of Asian options.

5. Portfolio Management

The last example we want to give is a more complicated example taken from the work of Tourin & Zariphopoulou (1993) where a lot of difficulties are gathered. The equation satisfied by the value-function $v(x,y)$ of the optimal investment-consumption problem is the following

$$\begin{aligned}\operatorname{Min}\Big\{ \inf_{c\geq 0}\left(-\frac{1}{2}\sigma^2 y^2 v_{yy} - (rx-c)v_x - byv_y - U(c) + \beta v\right), \\ -v_y + (1+\lambda)v_x, -(1-\mu)v_x + v_y\Big\} = 0\end{aligned}$$

in the domain

$$\{x + (1-\mu)y \geq 0 \quad \text{and} \quad x + (1+\lambda)y \geq 0\} ,$$

where the subscripts in the equation mean differentiation with respect to x or y, where $\sigma, r, b, \lambda, \mu$ are constant coefficients, $\beta > 0$ and U is some given utility function.

We do not want to enter too much into the details of this equation and the underlying portfolio management problem but we want to point out the main difficulties one encounters here: we have, at the same time, a *fully nonlinear* and *degenerate* equation with *gradient constraints* and with a *state-constraint boundary condition.* All these difficulties together imply that the theoretical and numerical treatment of this equation is very delicate.

We conclude this section by summarizing the main characteristics of all the above equations: these equations are nonlinear and degenerate. This implies that they have in general no classical solutions ('smooth' solutions). Therefore, a notion of 'weak' solutions is needed in order to make sense of the equations. But as soon as one defines a notion of weak solution, several difficulties occur such as nonuniqueness problems, for example.

On the other (positive) hand, all these equations are *degenerate elliptic equations*, i.e. they can be written as

$$H(x,u,Du,D^2u) = 0 \quad \text{in } \Omega\,, \tag{2.2}$$

where Ω is a domain in $\mathbb{R}^N$ and where H is, say, a continuous, real-valued function defined on $\Omega\times\mathbb{R}\times\mathbb{R}^N\times\mathcal{S}^n$, $\mathcal{S}^n$ being the space of $N\times N$ symmetric matrices, and which satisfies the *ellipticity condition*

$$H(x,u,p,M) \le H(x,u,p,N) \quad \text{if } M \ge N\,, \tag{2.3}$$

for any $x\in\Omega$, $u\in\mathbb{R}$, $p\in\mathbb{R}^N$.

This ellipticity property is a key property for defining the notion of viscosity solutions for the equations (2.2): this fact will become clear in the next section. From now on, we will always assume it is satisfied by the equations we consider.

Remark All the above examples (except the last one) lead to parabolic equations i.e., in particular, to time-dependent equations. To rewrite these equations in the form (2.2), one has to set $x = (y,t)$ where t is the time variable and y is the space variable (typically $y = S$ in examples 1 and 2, $y = (S,Z)$ in examples 3 and 4). In these cases, D and D^2 stand respectively for the gradient and for the matrix of second derivatives with respect to $x = (y,t)$ and not only to the space variables. It is clear that parabolic equations are a particular case of degenerate elliptic equations since these equations contain no second derivatives with respect to t.

3 The Notion of Viscosity Solutions

The notion of viscosity solutions was introduced by Crandall & Lions (1983) (see also Crandall, Evans & Lions (1984)) for solving problems related to first-order Hamilton–Jacobi Equations. We refer the reader to the 'User's Guide' of Crandall, Ishii & Lions (1992) for a complete presentation of this notion of solutions and to the book of Fleming & Soner (1993) where the applications to deterministic and stochastic optimal control theory are also described.

In order to introduce the notion of viscosity solutions, we give an equivalent definition of the notion of classical solution which uses only the Maximum Principle.

Theorem: (Classical solutions and Maximum Principle) *$u \in C^2(\Omega)$ is a classical solution of*

$$H(x, u, Du, D^2u) = 0 \ \ in \ \ \Omega \ ,$$

where H is a continuous function satisfying (2.3), **if and only if**

$\forall \varphi \in C^2(\Omega)$, if $x_0 \in \Omega$ is a local maximum point of $u - \varphi$, one has

$$H(x_0, u(x_0), D\varphi(x_0), D^2\varphi(x_0)) \leq 0 \ ,$$

and

$\forall \varphi \in C^2(\Omega)$, if $x_0 \in \Omega$ is a local minimum point of $u - \varphi$, one has

$$H(x_0, u(x_0), D\varphi(x_0), D^2\varphi(x_0)) \geq 0 \ .$$

□

The proof of this result is very simple: the first part of the equivalence just comes from the classical properties

$$Du(x_0) = D\varphi(x_0) \ , \quad D^2u(x_0) \leq D^2\varphi(x_0) \ ,$$

at a maximum point x_0 of $u - \varphi$ (recall that u and φ are smooth) or

$$Du(x_0) = D\varphi(x_0) \ , \quad D^2u(x_0) \geq D^2\varphi(x_0) \ ,$$

at a minimum point x_0 of $u - \varphi$. One has just to use these properties together with the ellipticity property (2.3) of H to obtain the inequalities of the theorem.

The second part is a consequence of the fact that we can take $\varphi = u$ as test-function and therefore $H(x_0, u(x_0), Du(x_0), D^2u(x_0))$ is both positive and negative at any point x_0 of Ω since any $x_0 \in \Omega$ is both a local maximum and minimum point of $u - u$.

Now we simply remark that the equivalent definition of classical solutions which is given here in terms of test-functions φ does not require the existence of first and second derivatives of u. For example, the continuity of u is sufficient to give a meaning to this equivalent definition; so we use this formulation to define viscosity solutions.

Definition: (Continuous Viscosity Solutions) $u \in C(\Omega)$ is a viscosity solution of

$$H(x, u, Du, D^2u) = 0 \text{ in } \Omega ,$$

where H is here a continuous function satisfying (2.3), *if and only if*

$\forall \varphi \in C^2(\Omega)$, if $x_0 \in \Omega$ is a local maximum point of $u - \varphi$, one has

$$H(x_0, u(x_0), D\varphi(x_0), D^2\varphi(x_0)) \leq 0 ,$$

and

$\forall \varphi \in C^2(\Omega)$, if $x_0 \in \Omega$ is a local minimum point of $u - \varphi$, one has

$$H(x_0, u(x_0), D\varphi(x_0), D^2\varphi(x_0)) \geq 0 .$$

□

We now give a few concrete examples of equations where there is a unique viscosity solution but no smooth solutions.

The first example is

$$\frac{\partial u}{\partial t} + \left|\frac{\partial u}{\partial x}\right| = 0 \quad \text{in } \mathbb{R} \times (0, +\infty) . \tag{3.1}$$

It can be shown that the function u defined in $\mathbb{R} \times (0, +\infty)$ by

$$u(x, t) = -(|x| + t)^2 ,$$

is the unique viscosity solution of (3.1) in $C(\mathbb{R} \times (0, +\infty))$. It is worth remarking in this example that u is only continuous for $t > 0$ despite the initial data

$$u(x, 0) = -x^2 \quad \text{in } \mathbb{R} ,$$

is in $C^\infty(\mathbb{R})$. In particular, this problem has no smooth solution as it is generally the case for such nonlinear hyperbolic equations.

Moreover, if we consider (3.1) together with the initial data

$$u(x, 0) = |x| \quad \text{in } \mathbb{R} , \tag{3.2}$$

then the functions $u_1(x, t) = |x| - t$ and $u_2(x, t) = (|x| - t)^+$ are two 'generalized' solutions in the sense that they satisfy the equation almost everywhere (at each of their points of differentiability). This problem of nonuniqueness is solved by the notion of viscosity solutions since it can be shown that u_2 is the unique continuous viscosity solution of (3.1)-(3.2). In that case, the notion of viscosity solutions selects the 'good' solution which is in that example the value-function of the associated deterministic control problem (cf. Fleming & Soner (1993)).

For second-order equations, non-smooth solutions appear generally as a consequence of the degeneracy of the equation as in the following example

$$\frac{\partial u}{\partial t} - x^2 \frac{\partial^2 u}{\partial x^2} = 0 \quad \text{in } \mathbb{R} \times (0, +\infty) \tag{3.3}$$

$$u(x, 0) = |x|^\alpha \quad \text{in } \mathbb{R} , \tag{3.4}$$

where $0 < \alpha < 1$. The unique uniformly continuous viscosity solution of this problem is

$$u(x, t) = |x|^\alpha e^{\alpha(\alpha-1)t} \quad \text{in } \mathbb{R} \times (0, +\infty) ,$$

and u is only Hölder continuous in x. The singularity of u at $x = 0$ exists because the equation is degenerate at $x = 0$.

Now we turn to the problem of taking in account the *boundary conditions*: this is a well known difficulty with degenerate equations since losses of boundary data may occur.

We consider for example the Dirichlet problem

$$\begin{cases} H(x, u, Du, D^2 u) = 0 & \text{in } \Omega , \\ u = g & \text{on } \partial\Omega , \end{cases}$$

where g is a given continuous function.

In order to solve this Dirichlet problem, a classical idea consists in considering the approximate problem

$$\begin{cases} -\varepsilon \Delta u_\varepsilon + H(x, u_\varepsilon, Du_\varepsilon, D^2 u_\varepsilon) = 0 & \text{in } \Omega , \\ u_\varepsilon = g & \text{on } \partial\Omega . \end{cases}$$

Indeed, by adding a $-\varepsilon\Delta$ term, we regularize the equation in the sense that one can expect to have more regular solutions for this approximate problem – typically in $C^2(\Omega) \cap C(\overline{\Omega})$ –.

We assume that this is indeed the case, that this regularized problem has a smooth solution u_ε and, moreover, that $u_\varepsilon \to u$ in $C(\overline{\Omega})$. We forget for the moment that the uniform convergence of u_ε to u on $\overline{\Omega}$ implies that $u = g$ on $\partial\Omega$ and we look for boundary conditions for u.

It is easy to see that the continuous function u satisfies in the viscosity sense

$$\begin{cases} H(x, u, Du, D^2 u) = 0 & \text{in } \Omega, \\ \operatorname{Min}(H(x, u, Du, D^2 u), u - g) \leq 0 & \text{on } \partial\Omega, \\ \operatorname{Max}(H(x, u, Du, D^2 u), u - g) \geq 0 & \text{on } \partial\Omega , \end{cases}$$

where, for example, the 'Min' inequality on $\partial\Omega$ means

$\forall \varphi \in C^2(\overline{\Omega})$, *if* $x_0 \in \partial\Omega$ *is a maximum point of* $u - \varphi$ *on* $\overline{\Omega}$, *one has*

$$\operatorname{Min}(H(x_0, u(x_0), D\varphi(x_0), D^2\varphi(x_0)), u(x_0) - g(x_0)) \leq 0 .$$

The proof of the above claim is not difficult: it first consists in showing that *strict* local maximum of minimum points of $u - \varphi$ are limits of local maximum of minimum points of $u_\varepsilon - \varphi$ and then an easy passage to the limit concludes. It remains to remark that the definitions of viscosity solutions obtained by considering 'strict local maximum of minimum points of $u - \varphi$' or 'local maximum of minimum points of $u - \varphi$' are equivalent (cf. Crandall, Ishii & Lions (1992)).

The interpretation of this new problem can be done by setting the equation in $\overline{\Omega}$ instead of Ω. To do so, we introduce the function G defined by

$$G(x,u,p,M) = \begin{cases} H(x,u,p,M) & \text{if } x \in \Omega\ , \\ u - g & \text{if } x \in \partial\Omega\ . \end{cases}$$

The above argument shows that the function u is a viscosity solution of

$$G(x,u,Du,D^2u) = 0 \quad \text{on } \overline{\Omega}\ ,$$

iff

$$G_*(x,u,Du,D^2u) \leq 0 \quad \text{on } \overline{\Omega}$$
$$G^*(x,u,Du,D^2u) \geq 0 \quad \text{on } \overline{\Omega}$$

where G_* and G^* stand respectively for the lower semicontinuous and upper semicontinuous envelopes of G. Indeed, the 'Min' and the 'Max' above are nothing but G_* and G^* on $\partial\Omega$.

In the same way, for general boundary conditions,

$$F(x,u,Du) = 0 \quad \text{on } \partial\Omega\ ,$$

we introduce the function G defined by

$$G(x,u,p,M) = \begin{cases} H(x,u,p,M) & \text{if } x \in \Omega\ , \\ F(x,u,p) & \text{if } x \in \partial\Omega\ . \end{cases}$$

The function u is said to be a viscosity solution of

$$G(x,u,Du,D^2u) = 0 \quad \text{on } \overline{\Omega}$$

if and only if it is a viscosity solution in Ω and if

$$G_* \leq 0 \quad \text{on } \partial\Omega \iff$$
$$\text{Min}(H(x,u,Du,D^2u), F(x,u,Du)) \leq 0 \ \text{ on } \partial\Omega$$

and

$$G^* \geq 0 \quad \text{on } \partial\Omega \iff$$
$$\text{Max}(H(x,u,Du,D^2u), F(x,u,Du)) \geq 0 \ \text{ on } \partial\Omega\ .$$

Remark The above example of the Dirichlet problem shows what our convergence result should be able to do: on one hand, it should take into account in a general setting this type of passage to the limit with any kind of boundary conditions (and this is the reason for introducing the formulation of the Gs). On an other hand, it should avoid the uniform convergence property on the u_ε which does not allow boundary layers and losses of boundary data (and this is the reason for introducing discontinuous viscosity solutions now).

Now we give the general definition of discontinuous viscosity solutions.

Definition: (Discontinuous Viscosity Solutions) *A locally bounded upper semicontinuous (usc in short) function u is a viscosity subsolution of the equation*

$$G(x, u, Du, D^2u) = 0 \quad \text{on } \overline{\Omega}$$

if and only if

$\forall \varphi \in C^2(\overline{\Omega})$, *if $x_0 \in \overline{\Omega}$ is a maximum point of $u - \varphi$, one has*

$$G_*(x_0, u(x_0), D\varphi(x_0), D^2\varphi(x_0)) \leq 0 \ .$$

A locally bounded lower semicontinuous (lsc for short) function v is a viscosity supersolution of the equation

$$G(x, u, Du, D^2u) = 0 \quad \text{on } \overline{\Omega}$$

if and only if

$\forall \varphi \in C^2(\overline{\Omega})$, *if $x_0 \in \overline{\Omega}$ is a minimum point of $u - \varphi$, one has*

$$G^*(x_0, u(x_0), D\varphi(x_0), D^2\varphi(x_0)) \geq 0 \ .$$

A solution is a function whose usc and lsc envelopes are respectively viscosity sub- and supersolutions of the equation. □

The first reason for introducing such a complicated formulation is to unify the convergence result we present in the next section: we incorporate in the function G the equation together with the boundary condition and this avoids the need of having a different result for each type of boundary condition. The possibility of handling discontinuous sub- and supersolutions is a key point in the convergence proof.

4 Convergence of Numerical Schemes

A numerical scheme approximating the equation

$$G(x, u, Du, D^2u) = 0 \ \text{ on } \overline{\Omega} \ , \tag{4.1}$$

is written in the following way

$$S(\rho, x, u^\rho(x), u^\rho) = 0 \text{ on } \overline{\Omega}$$

where S is a real-valued function defined on $\mathbb{R}^+ \times \overline{\Omega} \times \mathbb{R} \times B(\overline{\Omega})$ where $B(\overline{\Omega})$ is the set of bounded functions defined pointwise on $\overline{\Omega}$. We do not denote this space by $L^\infty(\Omega)$ since there is no measure theory involved here and, moreover, we are considering functions defined pointwise and not only almost everywhere.

We assume that the scheme satisfies the following assumptions.

Stability

For any $\rho > 0$, the scheme has a solution u^ρ. Moreover, u^ρ is uniformly bounded, i.e. there exists a constant $C > 0$ s.t.

$$-C \leq u^\rho \leq C \quad \text{on } \overline{\Omega}\,,$$

for any $\rho > 0$.

Consistency

For any smooth function ϕ, one has:

$$\liminf_{\substack{\rho \to 0 \\ y \to x \\ \xi \to 0}} \frac{S(\rho, y, \phi(y) + \xi, \phi + \xi)}{\rho} \geq G_*(x, \phi(x), D\phi(x), D^2\phi(x))$$

and

$$\limsup_{\substack{\rho \to 0 \\ y \to x \\ \xi \to 0}} \frac{S(\rho, y, \phi(y) + \xi, \phi + \xi)}{\rho} \leq G^*(x, \phi(x), D\phi(x), D^2\phi(x)).$$

Monotonicity

$$S(\rho, x, t, u) \leq S(\rho, x, t, v) \quad \text{if } u \geq v$$

for any $\rho > 0$, $x \in \overline{\Omega}$, $t \in \mathbb{R}$ and $u, v \in B(\overline{\Omega})$.

Strong Comparison Result

If u is a usc subsolution of the equation (4.1) and if v is an lsc supersolution of the equation (4.1), then

$$u \leq v \qquad \text{on } \overline{\Omega}\,.$$

The result is the following.

Theorem *Under the above assumptions, the solution u^ρ of the scheme converges uniformly on each compact subset of $\overline{\Omega}$ to the unique viscosity solution of the equation.* □

Sketch of the proof We just describe the main steps.

1. We set

$$\underline{u}(x) = \liminf_{\substack{\rho \to 0 \\ y \to x}} u^\rho(y) \ ,$$

and

$$\overline{u}(x) = \limsup_{\substack{\rho \to 0 \\ y \to x}} u^\rho(y) \ .$$

The monotonicity and consistency assumptions on S imply that $\overline{u}$ and $\underline{u}$ are respectively sub and supersolutions of the limiting equation.

2. By the Strong Comparison Result for the equation (4.1), we have

$$\overline{u} \le \underline{u} \quad \text{on } \overline{\Omega} \ .$$

3. But, by definition

$$\underline{u} \le \overline{u} \quad \text{on } \overline{\Omega} \ .$$

Therefore

$$\overline{u} = \underline{u} \quad \text{on } \overline{\Omega} \ ,$$

and this equality implies the uniform convergence of u^ρ to $u := \overline{u} = \underline{u}$ as a simple variation on the proof of Dini's Theorem. □

Remark The above proof is based on the so-called 'half-relaxed limits method' which was introduced by Perthame and the author (1987). This method allows passages to the limit in fully nonlinear elliptic PDEs with only a uniform bound on the solutions. It is worth mentioning that – because one needs only this uniform bound – this method lets us treat problems where boundary layers occur.

Now we discuss the assumptions on the scheme.

Consistency

For the 'interior points' of Ω where the function G is generally continuous, the consistency requirement is equivalent to

$$\frac{S(\rho, x, \phi(x), \phi)}{\rho} \to G(x, \phi(x), D\phi(x), D^2\phi(x))$$

when $\rho \to 0$ uniformly on compact subsets of Ω, for any smooth function ϕ. We recover here a more standard formulation and the apparent complexity of the consistency assumption above just comes from the fact that we want to handle at the same time the boundary conditions in a general setting which leads to a discontinuous function G.

Strong Comparison Result

This is the key result to get the convergence. Such results exist in the following cases:

– For first-order equations: optimal Strong Comparison Results have been proved for all kinds of 'classical' equations and boundary conditions (cf. Barles (1994)).

– For second-order equations: optimal results are available for 'Neumann' type boundary conditions; for 'Dirichlet' type boundary conditions, optimal results exist when the boundary condition is assumed in the classical sense (cf. Crandall, Ishii & Lions (1992) and references therein). For the 'generalized' Dirichlet boundary conditions – where losses of boundary data may occur (which implies that the equation is degenerate) – only the semilinear case is well understood (cf. Barles & Burdeau (1994)). The case of the general Dirichlet problem (fully nonlinear degenerate equation) is still open.

Monotonicity

This assumption can be understood with the following table:

$S(\rho, x, u(x), u) = 0$	$G(x, u, Du, D^2u) = 0$
If $x \in \Omega$ is a maximum point of $u - \phi$, one has	If $x \in \Omega$ is a maximum point of $u - \phi$, one has
Discrete Maximum Principle	**Maximum Principle**
$u \leq \phi + \xi$ where $\xi = u(x) - \phi(x)$.	$Du(x) = D\phi(x)$ $D^2u(x) \leq D^2\phi(x)$
Monotonicity	**Ellipticity**
$S(\rho, x, \phi(x) + \xi, \phi + \xi) \leq \cdots$	$G(x, u, D\phi(x), D^2\phi(x)) \leq \cdots$
$S(\rho, x, u(x), u) = 0$	$G(x, u, Du(x), D^2u(x)) = 0$

It is clear enough from this table that the monotonicity assumption plays, for the numerical scheme, exactly the same role as the ellipticity assumption for the nonlinear PDEs we consider. Therefore

$$\text{Monotonicity} \Longleftrightarrow \text{Discrete Ellipticity}$$

First examples

For the sake of simplicity, we first consider classical schemes approximating the heat equation in one dimension

$$\frac{\partial u}{\partial t} - \frac{\partial^2 u}{\partial x^2} = 0 \quad \text{in } \mathbb{R} \times (0,T) ,$$

with a given initial condition

$$u(x,0) = u_0(x) \quad \text{in } \mathbb{R} ,$$

where u_0 is, say, a continuous bounded function.

We use below the standard notation in numerical analysis: u_j^n denotes an approximation of $u(n\Delta t, j\Delta x)$ for $n \in \mathbb{N}$ and $j \in \mathbb{Z}$ where Δt and Δx are respectively the mesh size in t and in x. We refer the reader to Glowinski, Lions & Tremolieres (1976) and to Raviart & Thomas (1983) for an introduction to basic methods in numerical analysis.

- **The Standard Implicit Scheme**

If $u^n := (u_j^n)_j$ is known, one can compute $u^{n+1} = (u_j^{n+1})_j$ by solving

$$u_j^{n+1} - u_j^n - \frac{\Delta t}{(\Delta x)^2}(u_{j+1}^{n+1} - 2u_j^{n+1} + u_{j-1}^{n+1}) = 0 .$$

The above equation has to be read as

$$S\left((n+1)\Delta x, j\Delta t, u_j^{n+1}, u_{j+1}^{n+1}, u_{j-1}^{n+1}, u_j^n\right) = 0 ;$$

in other words, the above equation is the equation of the scheme at the point $((n+1)\Delta x, j\Delta t)$, the role of the variable ‘ u^ρ ’ is played here by $(u_{j+1}^{n+1}, u_{j-1}^{n+1}, u_j^n)$.

It is clear since we have a ‘$-$’ in front of u_{j+1}^{n+1}, u_{j-1}^{n+1} and u_j^n that this scheme is an unconditionally monotone scheme (there is no condition either on Δt or on Δx). A consequence of this property is also that it is an unconditionally stable scheme since the monotonicity property implies that the scheme satisfies the Maximum Principle i.e.

$$\max_j |u_j^n| \le \max_j |u_j^0| ,$$

for any $n \in \mathbb{N}$. And the boundedness of the initial data implies the boundedness of each u^n.

- **The Standard Explicit Scheme**

$$u_j^{n+1} - u_j^n - \frac{\Delta t}{(\Delta x)^2}(u_{j+1}^n - 2u_j^n + u_{j-1}^n) = 0 .$$

We interpret this equality as above; in order to have the monotonicity property, in particular with respect to u_j^n, one should have the classical Courant–Friedrichs–Levy condition

$$\frac{2\Delta t}{(\Delta x)^2} \leq 1 \ .$$

If this condition holds, we have a monotone and stable scheme.

Remark Before giving examples of numerical schemes which can be used in options pricing, we want to mention that a general class of schemes for which the above convergence result applies are, in optimal control theory, those which are based on the Dynamic Programming Principle; we refer to Kushner (1977, 1984) for the description of these schemes.

5 Numerical Schemes in Options Pricing: Splitting Methods

We present in this section numerical schemes for computing the price of different types of classical options; these schemes – or more sophisticated schemes based on similar ideas – were implemented at the Caisse Autonome de Refinancement (CDC group). We do not pretend that these schemes are the most efficient in each cases (indeed they are not!). Our aim is to present simple examples to emphasize the advantages of splitting methods.

The reason for using splitting methods was the following: we wanted to build a program for computing the price of a wide variety of options and the use of splitting methods allows us to have a very modular program since the idea is to treat the equations and the constraints separately, one after the other in the right order. In that way, to add more constraints (cap, floor, partially American options, ... etc) is almost costless.

For the sake of simplicity, we are going to present these schemes on simplified equations presenting the same features: we essentially replace below the Black–Scholes equation by the heat equation. Moreover we also present them under the form of approximation schemes; to deduce numerical schemes from them being completely straightforward. It will be clear in each case that the assumptions for the convergence of these schemes are satisfied and we leave the checking to the reader.

1. American options: Variational Inequalities

$$\operatorname{Max}\left(\frac{\partial u}{\partial t} - \Delta u, u - \psi\right) = 0 \quad \text{in } \mathbb{R}^N \times (0,T) \ ,$$

with a given initial condition.

The scheme
1st Step given u^n, we solve

$$\begin{cases} \dfrac{\partial w}{\partial t} - \Delta w = 0 & \text{in } \mathbb{R}^N \times (0, \Delta t), \\ w(x,0) = u^n(x) & \text{in } \mathbb{R}^N, \end{cases}$$

and we set

$$u^{n+\frac{1}{2}}(x) = w(x, \Delta t) \quad \text{in } \mathbb{R}^N .$$

2nd Step

$$u^{n+1} = \mathrm{Inf}(u^{n+\frac{1}{2}}, \psi^{n+1}) ,$$

where $\psi^{n+1}(x) = \psi(x, (n+1)\Delta t)$ for $x \in \mathbb{R}^N$, i.e.

$$u^{n+1} = \mathrm{Inf}(S(\Delta t)u^n, \psi^{n+1}) ,$$

where S is the semigroup associated with the heat equation.

This scheme is a very classical one: the second step is, in fact, a projection on the convex set $\tilde{K} := \{u \in C(\mathbb{R}^N);\ u \le \psi^{n+1}\}$ and it is generally associated with a Conjugate Gradient Method.

In this example, the claim we made at the beginning of this section becomes more clear: we first treat the equation part by solving the heat equation and then the constraint part by imposing $u \le \psi$. Notice that if this constraint had been put only on some part of the space $\mathbb{R}^N \times (0, T)$, the same type of scheme could have been used. Finally we mention that cap and floor constraints are treated in that way.

2. Lookback Options

The simplified problem is

$$\frac{\partial u}{\partial t} - \Delta_x u = 0 \qquad \text{in } \{|x| < z\} ,$$

and

$$-\frac{\partial u}{\partial z} = 0 \qquad \text{in } \{|x| > z\} ,$$

with an initial condition given.

The scheme

1st Step for any fixed z, we solve

$$\begin{cases} \dfrac{\partial w}{\partial t} - \Delta_x w = 0 & \text{in } \mathbb{R}^N \times (0, \Delta t), \\ w(x,0) = u^n(x,z) & \text{in } \mathbb{R}^N, \end{cases}$$

and we set

$$u^{n+\frac{1}{2}}(x,z) = w(x,z,\Delta t).$$

2nd Step

$$u^{n+1}(x,z)=\begin{cases}u^{n+\frac{1}{2}}(x,z) & \text{if } |x|<z,\\ u^{n+\frac{1}{2}}(x,|x|) & \text{if } |x|\geq z.\end{cases}$$

This scheme takes into account the different roles of the x and z variables in the equation. It is worth noticing that, in step 1, we solve nondegenerate problems with, for each z, the same equation: in practice, this means that we have to factorize only once the matrix associated with the Δ_x operator. Of course, in concrete computations, step 1 is performed for a finite set of z, namely for the grid points.

Despite these advantages and the fact that we know that this scheme is convergent, it is not very accurate: it is a first-order accurate scheme and the error made for a reasonable number of grid points is not satisfactory. We refer the reader to Barles, Burdeau, Daher & Romano (1995) for an improvement of this scheme by still using splitting methods but in a slightly different way and with a better accuracy.

3. Options on the Average

$$\frac{\partial u}{\partial t}-\frac{\partial^2 u}{\partial x^2}-x\frac{\partial u}{\partial z}=0 \quad \text{in } \mathbb{R}\times\mathbb{R}\times(0,T),$$

with the initial condition given.

The scheme

1st Step for any fixed z, we solve

$$\begin{cases}\dfrac{\partial w}{\partial t}-\dfrac{\partial^2 w}{\partial x^2}=0 & \text{in } \mathbb{R}\times(0,\Delta t),\\ w(x,0)=u^n(x,z) & \text{in } \mathbb{R},\end{cases}$$

and we set

$$u^{n+\frac{1}{2}}(x,z)=w(x,z,\Delta t) \quad \text{in } \mathbb{R}\,.$$

2nd Step for any fixed x, we solve

$$\begin{cases}\dfrac{\partial w}{\partial t}-x\dfrac{\partial w}{\partial z}=0 & \text{in } \mathbb{R}\times(0,\Delta t),\\ w(x,0)=u^{n+\frac{1}{2}}(x,z) & \text{in } \mathbb{R}.\end{cases}$$

and we set

$$u^{n+\frac{1}{2}}(x,z)=w(x,z,\Delta t) \quad \text{in } \mathbb{R}\,.$$

This scheme is closer to the classical idea of splitting methods, which can be explained in the following way: in order to solve numerically

$$\frac{\partial u}{\partial t}+F_1([u])+F_2([u])=0 \quad \text{in } \mathbb{R}^N\times(0,\Delta t)\,,$$

where $[u]$ stands for (u, Du, D^2u), we apply successively the schemes

$$u^{1/2} = u^0 + \Delta t F_1([u^0])\,,$$

and

$$u^1 = u^{1/2} + \Delta t F_2([u^{1/2}])\,.$$

To justify this method, we perform the following formal computations, replacing $u^{1/2}$ in the second equality by its value taken from the first one

$$\begin{aligned} u^1 &= u^0 + \Delta t F_1([u^0]) + \Delta t F_2([u^0 + \Delta t F_1([u^0])]) \\ &= u^0 + \Delta t F_1([u^0]) + \Delta t F_2([u^0]) + O\left((\Delta t)^2\right) \\ &\simeq u^0 + \Delta t \left(F_1([u^0]) + F_2([u^0])\right)\,. \end{aligned}$$

Therefore within terms of order $O\left((\Delta t)^2\right)$, we have indeed a numerical approximation of the equation.

It is clear, in this example, that the splitting methods allows us to treat differently (using a variety of schemes, for instance) the different parts of an equation or of a complex problem and this is their main advantage.

To conclude this article, we want also to mention that another difficulty in numerically solving equations arising in finance comes from the fact that they are set in unbounded domains. The first step, which consists in approximating these equations by a problem posed in bounded domains, is not obvious since the solutions may not have a well known behavior at infinity (cf. for example the case of lookback options).

It is shown in Barles, Daher & Romano (1995) that the convergence when we let the domain tend to infinity is governed by phenomena of Large Deviations type and therefore completely artificial boundary conditions on the boundary of the domain of computations lead (theoretically) to an exponentially small error inside the domain (only a boundary layer is or should be observed). Nonetheless, this result is theoretical and a good guess on these artificial boundary conditions can really improve the accuracy.

References

Barles G. (1994) *Solutions de viscosité des équations de Hamilton–Jacobi*, Mathématiques et Applications (SMAI), **17**, Springer Verlag.

Barles G. & Burdeau J. (1995) 'The Dirichlet problem for semilinear second-order degenerate elliptic equtions and applications to stochastic exit time control problem', *Comm. in PDE* **20**, 129–178.

Barles G., Burdeau J., Daher Ch. & Romano M. (1995), in preparation.

Barles G., Daher Ch. & Romano M. (1994) 'Optimal control on the L^∞-norm of a diffusion process', *SIAM J. on Control and Optimization* **32**, 612–634.

Barles G., Daher Ch. & Romano M. (1995) 'Convergence of numerical schemes for problems arising in finance theory', *Math. Models and Meth. in Appl. Sciences* **5**, 125–143.

Barles G. & Perthame B. (1987) 'Discontinuous solutions of deterministic optimal stopping problems', *Math. Modelling Numer. Anal.* **21**, 557–579.

Barles G. & Souganidis P.E (1991) 'Convergence of approximation schemes for fully non-linear equations', *Asymptotic Analysis* **4**, 271–283.

Barron E.N. (1993) 'The Bellman equation for the running max of a diffusion and applications to lookback options', *Applicable Analysis* **48**, 205–222.

Bensoussan A. (1984) 'On the theory of options pricing', *Acta Applicandae Mathematicae,* **2**, 139–158.

Bensoussan A. & Lions J.-L. (1978) *Applications des Inéquations Variationnelles en Contrôle Stochastique*, Dunod.

Black F. & Scholes M. (1973) 'The pricing of options and corporate liabilities', *Journal of Political Economy,* **81**, 637–654.

Conze A. (1990), Thèse de Doctorat. Université Paris IX–Dauphine.

Cox J.C & Rubinstein M. (1985) *Options Markets*, Prentice Hall.

Cox J.C, Ross S. & Rubinstein M. (1979) 'Options pricing: a simplified approach', *Journal of Financial Economics,* **7**, 229–263.

Crandall M.G., Evans L.C. & Lions P.L. (1984) 'Some properties of viscosity solutions of Hamilton–Jacobi Equations', *Trans. Amer. Math. Soc.* **282**, 487–502.

Crandall M.G., Ishii H. & Lions P.L. (1992) 'User's guide to viscosity solutions of second order partial differential equations', *Bull. Amer. Soc.* **27**, 1–67.

Crandall M.G. & Lions P.L. (1983) 'Viscosity solutions of Hamilton–Jacobi equations', *Trans. Amer. Math. Soc.* **277**, 1–42.

Duffie D. (1988), *Security Markets, Stochastic Models*, Academic Press.

Fleming W.H. & Soner H.M. (1993) *Controlled Markov Processes and Viscosity Solutions*, Springer Verlag.

Glowinski R., Lions J.L. & Trémolières R. (1976) *Analyse Numérique des Inéquations Variationnelles*, Dunod.

Heinricher A. & Stockbridge R. (1991) 'Optimal control of the running max', *SIAM Journal on Control and Optimisation* **29**, 936–953.

Ingersoll J.E. (1987) *Theory of Financial Decision Making*, Rowman & Littlefield Publishers.

Karatzas I. (1988) 'The pricing of American options', *Appl. Math. Optimiz.* **17**, 37–60.

Karatzas I. & Shreve S.E. (1988) *Brownian Motion and Stochastic Calculus*, Springer Verlag.

Kushner H.J. (1977) *Probability Method for Approximations in Stochastic Control and for Elliptic Equations*, Academic Press.

Kushner H.J. (1984) *Approximations and Weak Convergence Methods for Random Processes with Applications to Stochastic Systems Theory*, MIT Press.

Lamberton D. and Lapeyre B. (1992) *Introduction au Calcul Stochastique Appliqué à la Finance*, Mathématiques et Applications (SMAI), **9**, Ellipses.

Raviart P.A. & Thomas J.M. (1983) *Introduction à l'Analyse Numérique des Équations aux Dérivées Partielles*, Masson.

Rogers L.C.G. & Shi Z. (1995) 'The value of an Asian option', *J. Appl. Prob.* **32**, 1077–1088.

Tourin A. & Zariphopoulou T. (1993) 'Numerical schemes for models with singular transactions', *Computational Economics*, in press.

Continuous-Time Monte Carlo Methods and Variance Reduction

Nigel J. Newton

1 Introduction

This article concerns Monte Carlo methods for a class of linear partial differential equations (PDEs). The methods are based on the simulation of solutions of Itô stochastic differential equations (SDEs). Numerical methods for SDEs, which can be used as the basis of such simulations, are discussed, and recent results concerning parametrically optimal variance reduction techniques are presented.

The general Monte Carlo method and variance reduction methods can be illustrated in the context of the integration of a real-valued function, f, defined on the unit interval of the real line:

$$I(f) = \int_0^1 f(x)dx. \tag{1.1}$$

One of the simplest Monte Carlo methods for this problem comprises the simulation of a large number of independent, uniform variates $(X_1, X_2, \ldots, X_M)$ by means of pseudo-random number generators (see, for example, Knuth (1981)), and the formation of the average value of f over these simulated values:

$$\hat{I}_b(f) = \frac{1}{M} \sum_{m=1}^{M} f(X_m). \tag{1.2}$$

According to Khinchine's strong law of large numbers, as long as f is integrable, $\hat{I}_b(f)$ will converge *almost surely* to $I(f)$ as M increases, i.e.

$$P\left(\hat{I}_b(f) \to I(f)\right) = 1.$$

If f is *square integrable* then the approximation error of this Monte Carlo method can be quantified by its variance

$$\mathrm{var}\left(\hat{I}_b(f)\right) = \frac{1}{M}\mathrm{var}(f(X)).$$

This falls in inverse proportion to the sample size, M, and so the standard deviation of the error, on which, for example, confidence intervals are based, falls as $1/\sqrt{M}$. This is a major disadvantage of the Monte Carlo method when compared with deterministic numerical integration techniques – in order to halve the statistical error we need to quadruple the sample size. However, various variance reduction techniques can be used that replace the basic variate in (1.2), $f(X)$, by one having the same mean, but smaller variance. Two commonly used techniques are *importance sampling* and *control variates.* These are described next, in the context of the integration (1.1).

With *importance sampling,* independent variates $(Y_1, Y_2, \ldots, Y_M)$ are drawn from a distribution on $[0, 1]$ having a density p_Y, which is non-zero wherever f is non-zero, and the following approximation for $I(f)$ is formed:

$$\hat{I}_i(f) = \frac{1}{M}\sum_{m=1}^{M}\frac{f(Y_m)}{p_Y(Y_m)}.$$

Like $\hat{I}_b(f)$, $\hat{I}_i(f)$ is an *unbiased* estimator of $I(f)$, in that:

$$\begin{aligned}\mathbf{E}\hat{I}_i(f) &= \mathbf{E}\frac{f(Y)}{p_Y(Y)}\\ &= \int_0^1 \frac{f(y)}{p_Y(y)}p_Y(y)dy\\ &= I(f).\end{aligned}$$

If f is non-negative, the variance of $\hat{I}_i$ can be made small by choosing p_Y to approximate f in shape (thus making the ratio $f(y)/p_Y(y)$ relatively insensitive to y). When such a p_Y is used, the variates simulated, Y_m, are more likely to take on values for which f is large than values for which it is small, and so the sampling process can be said to sample the *important* regions of $[0, 1]$ for this problem more than the unimportant regions. This process can be perfected by the use a sampling distribution with the following density

$$p_Y^*(y) = \frac{f(y)}{I(f)}.$$

This yields an unbiased variate with zero variance. Of course, we cannot use this distribution directly because the expression for its density involves the desired integral, $I(f)$; however, it is often possible to obtain approximations to this perfect sampling distribution which can be used to yield variates with low variance. If f takes on both positive and negative values then it would appear, at first sight, that there is no perfect sampling distribution. However, if f is bounded below (resp. above), then there *is* a perfect sampling distribution for $f + c$ (resp. $f - c$) for a suitable positive constant c, and this can be used in an obvious way as the basis of variance reduced Monte Carlo methods for integrating f. Alternatively, the positive and negative parts of f, for which perfect sampling distributions exist, can be integrated separately.

A *control variate* is an auxiliary variate, of finite variance and known mean, which is strongly positively correlated with the basic variate. By subtracting it from the basic variate one obtains a new variate with lower variance, whose mean differs from that of the basic variate by a known amount. Thus, if $f(X)$ is the basic variate and Y the control variate, the following approximation for $I(f)$ is used:

$$\hat{I}_c(f) = \frac{1}{M} \sum_{m=1}^{M} (f(X_m) - Y_m) + \mathbf{E}Y.$$

The control variate may be obtained from the same underlying variate as is the main variate—in the context of the integration of a function on the unit interval, this means that Y can be expressed in the form $c(X)$ for some function c, in which case $\mathbf{E}Y$ is the integral of c over the unit interval, and this must be known. In rather the same way as the sampling distribution of importance sampling, the control function, c, must have a similar shape to f, although it is not constrained to be non-negative and have unit integral as is p_Y. For example, if it is known that $f(x)$ on the whole increases with x, then the function $c(x) = cx$, for some $c > 0$, may provide a useful control variate. As with importance sampling, there is a *perfect* control variate

$$Y^* = f(X) - \mathbf{E}f(X),$$

which cannot be used directly for obvious reasons. However, good unbiased approximations to the perfect control variate can sometimes be obtained if the perfect variate is expressed in an appropriate way.

Of course, Monte Carlo methods are not appropriate for integrating functions on the unit interval—they are greatly out-performed by deterministic techniques such as the trapezium rule. They do, however, have a role to play in problems of high dimension, for example in the integration of functions with respect to large numbers of variables, or the solution of PDEs with large numbers of independent variables. The complexity of a Monte Carlo method is typically polynomial in the dimension, whereas that of deterministic methods is typically exponential. In the context of PDEs, another useful feature of Monte Carlo methods is that they allow the solution to be found at just one point, if required (with an associated saving in computation), whereas deterministic methods necessarily find the solution at a large number of points simultaneously. This property of Monte Carlo methods can be particularly useful in problems such as option pricing, where the value of an option is required only at the time of striking, and for the state of the market at that time. Monte Carlo methods can also out-perform deterministic approximations for PDEs that are subject to certain types of degeneracy or near degeneracy, for example second order elliptic PDEs with coefficients of second derivatives that become very small in some areas of the underlying space.

2 Stochastic Differential Equations

It is possible to base Monte Carlo methods for certain linear, second order PDEs on *weak solutions* of the following type of Itô stochastic differential (integral) equation (SDE):

$$X_t = X_0 + \int_0^t b(v, X_v)dv + \int_0^t \sigma(v, X_v)dW_v \quad \text{for } 0 \le t < \infty. \tag{2.1}$$

Here $(X_t \in \mathbb{R}^d)$ is a d-dimensional 'solution process', $(W_t \in \mathbb{R}^r)$ is an r-dimensional Brownian motion process, b is a vector-valued 'drift coefficient', and σ is a matrix-valued 'diffusion coefficient'. The second integral on the right-hand side of (2.1) is an Itô stochastic integral.

Definition 2.1 A Weak Solution *to equation (2.1), with initial law μ, is a probability space with a filtration, $(\Omega, \mathcal{F}, (\mathcal{F}_t), P)$, supporting an r-dimensional Brownian motion $(W_t, \mathcal{F}_t; 0 \le t < \infty)$, and a continuous d-dimensional semimartingale process, $(X_t, \mathcal{F}_t;\ 0 \le t < \infty)$, such that X_0 has the distribution μ, and the pair (X, W) satisfies (2.1) for all t.*

Implicit in this definition is the requirement that the integrals in (2.1) be well defined for the solution process. For definitions of basic terms, precise conditions on (X_t) and $(\mathcal{F}_t)$, conditions on the coefficients of equation (2.1) such that it has a weak solution, and further discussion see, for example, Rogers & Williams (1987) or Karatzas & Shreve (1991). The appropriate concept of uniqueness for weak solutions is *uniqueness in law*—equation (2.1) is unique in law (for initial law μ) if the distribution of X is the same for all its weak solutions (with initial law μ).

Weak solutions to unique-in-law SDEs provide pathwise representations of diffusion processes. If the coefficients of equation (2.1) are sufficiently smooth then it defines a (Markov) diffusion process having transition densities, $p(s, x; t, y)$, which obey the Kolmogorov backward equation

$$\frac{\partial p}{\partial s} + \mathcal{L}p = 0,$$

where, for a function $f : [0, \infty) \times \mathbb{R}^d \to \mathbb{R}$,

$$(\mathcal{L}f)(s, x) = \sum_{i=1}^{d} b_i(s, x)\frac{\partial f}{\partial x_i}(s, x) + \frac{1}{2}\sum_{i,j=1}^{d} (\sigma\sigma')_{i,j}(s, x)\frac{\partial^2 f}{\partial x_i \partial x_j}(s, x)$$

(σ' is the matrix transpose of σ), and the Kolmogorov forward (or Fokker-Planck) equation

$$-\frac{\partial p}{\partial t} + \mathcal{L}^* p = 0,$$

where $\mathcal{L}^*$ is the adjoint of $\mathcal{L}$,

$$(\mathcal{L}^* f)(t,y) = -\sum_{i=1}^{d} \frac{\partial}{\partial y_i}(b_i f)(t,y) + \frac{1}{2}\sum_{i,j=1}^{d} \frac{\partial^2}{\partial y_i \partial y_j}((\sigma\sigma')_{i,j} f)(t,y).$$

The solutions of various initial and boundary value problems associated with these and related PDEs can be represented as means of simple functionals of a weak solution of (2.1). Consider, for example, the following Cauchy problem: find a continuous function $u(s,x)$

$$u : (0,T) \times \mathbb{R}^d \to \mathbb{R},$$

which has continuous first derivative with respect to s and continuous first and second derivatives with respect to the components of x, and which solves the PDE

$$\frac{\partial u}{\partial s}(s,x) + (\mathcal{L}u)(s,x) + c(s,x)u(s,x) = g(s,x) \quad \text{for } (s,x) \in (0,T) \times \mathbb{R}^d \tag{2.2}$$

subject to the initial value constraint

$$\lim_{(s,x)\to(T,y)} u(s,x) = h(y) \quad \text{for } y \in \mathbb{R}^d. \tag{2.3}$$

(It is convenient, for the stochastic representation formulae discussed below, to reverse the time variable, s: the more usual form for this Cauchy problem, with an initial condition rather than a final condition, is obtained by substituting $t = T - s$.) Under appropriate conditions on b, σ, c, g and h (see, for example, Friedman (1975)), this has a unique solution with the following Feynman-Kac stochastic representation:

$$u(s,x) = \mathbf{E}\left\{ Z_{s,T} h(X_T^{s,x}) - \int_s^T Z_{s,t} g(t, X_t^{s,x}) dt \right\},$$

where

$$Z_{s,t} = \exp\left(\int_s^t c(v, X_v^{s,x}) dv \right), \tag{2.4}$$

and $X^{s,x}$ is a weak solution of (2.1) taking the value x at (initial) time s.

Similarly, the following initial-boundary value problem also admits a stochastic representation: for an open, bounded set $G \subset \mathbb{R}^d$, find a continuous function $u(s,x)$,

$$u : (0,T) \times G \to \mathbb{R},$$

which has continuous first derivative with respect to s and continuous first and second derivatives with respect to the components of x, and which solves (2.2) in the domain $(0,T) \times G$ subject to the initial-boundary constraint

$$\lim_{(s,x)\to(t,y)} u(s,x) = h(t,y) \quad \text{for all } (t,y) \in \overline{B} \cup \overline{S}. \tag{2.5}$$

(See the comments above regarding time reversal.) Here, $B = \{T\} \times G$, $S = (0,T) \times \partial G$, where ∂G is the boundary of G, and the over-bar notation indicates the closure of a set. Under appropriate conditions (again, see Friedman (1975)), this has a unique solution which admits the stochastic representation:

$$u(s,x) = \mathbf{E}\left\{Z_{s,\tau}h(\tau, X_\tau^{s,x}) - \int_s^\tau Z_{s,t}g(t, X_t^{s,x})dt\right\},$$

where $X^{s,x}$ and Z are as defined above, x is a point in G, and τ is the hitting time of the boundary, ∂G, or the final time, T, whichever is smaller:

$$\tau = \inf\{t \geq s : X_t^{s,x} \in \partial G\} \wedge T.$$

Stochastic representation formulae of this type exist for a variety of initial and boundary value problems including those with constraints on derivatives of the solution at the boundary, and such formulae can be used to construct Monte Carlo methods for the associated problems. All of these Monte Carlo methods involve the simulation of weak solutions of SDEs such as (2.1), which cannot be done exactly, and so a need arises for discretisation schemes for SDEs. Section 3 provides a brief introduction to such schemes.

A weak solution of an SDE is the appropriate concept for these Monte Carlo methods. However, there are other applications of SDEs in which a stronger definition of solution is required.

Definition 2.2 A Strong Solution *to equation (2.1) is a function*

$$F : \mathbb{R}^d \times C([0,\infty); \mathbb{R}^r) \to C([0,\infty); \mathbb{R}^d),$$

satisfying certain measurability conditions, such that, for any filtered probability space $(\Omega, \mathcal{F}, (\mathcal{F}_t), P)$ *supporting an* r*-dimensional Brownian motion,* $(W_t, \mathcal{F}_t)$*, and an* $\mathbb{R}^d$*-valued random variable,* X_0*, the pair* $(X\ (:= F(X_0, W)), W)$ *satisfies (2.1) for all* t*.*

The measurability conditions on F ensure that the solution process at time t depends only on the initial condition, X_0, and the past of W. (See Rogers & Williams (1987) or Karatzas & Shreve (1991) for precise formulations.) The appropriate concept of uniqueness for strong solutions is as follows: equation (2.1) has a *unique strong solution*, F, if F is a strong solution and, for every weak solution, $(\Omega, \mathcal{F}, (\mathcal{F}_t), P, X, W))$, (with no matter what initial law) $P(X = F(X_0, W)) = 1$.

An SDE with a unique strong solution can be thought of as a model for a causal system with initial state X_0, input W and output X. For example, in the following, classical model for the continuous-time evolution of the unit value of a stock, the value at time t, X_t, depends only on the value at time zero, X_0, and the (scalar) driving Brownian motion between times 0 and t.

$$X_t = X_0 + \int_0^t b(v)X_v dv + \int_0^t \sigma(v)X_v dW_v. \tag{2.6}$$

More formally, if the mean rate of growth, $b(t)$, and the volatility, $\sigma(t)$, are sufficiently smooth (e.g. continuously differentiable), then (2.6) has the following unique strong solution:

$$X_t = F_t(X_0, W) = \exp\left(\int_0^t \sigma(v)dW_v + \int_0^t \left(b(v) - \frac{1}{2}\sigma^2(v)\right) dv\right) X_0.$$

By contrast, the following SDE (Tanaka's SDE) has only a weak solution:

$$X_t = X_0 + \int_0^t \operatorname{sgn}(X_v)dW_v, \tag{2.7}$$

where $W_t, X_t \in \mathbb{R}$ and

$$\operatorname{sgn}(x) = \begin{cases} +1 & \text{if } x > 0, \\ -1 & \text{if } x \leq 0. \end{cases}$$

To see this, suppose that $(X_t, \mathcal{F}_t)$ is a Brownian motion on some probability space $(\Omega, \mathcal{F}, (\mathcal{F}_t), P)$. The process $(W_t, \mathcal{F}_t)$, defined by

$$W_t = \int_0^t \operatorname{sgn}(X_v)dX_v$$

is a continuous martingale with quadratic variation process

$$[W]_t = \int_0^t \operatorname{sgn}^2(X_v)dv = t,$$

and so, according to Lévy's theorem, is itself a Brownian motion. Also

$$\int_0^t \operatorname{sgn}(X_v)dW_v = \int_0^t \operatorname{sgn}^2(X_v)dX_v = X_t,$$

and so $(\Omega, \mathcal{F}, (\mathcal{F}_t), P, X, W)$ is a weak solution of (2.7) with an 'atomic' initial law at 0. However, it is not difficult to show that $(\Omega, \mathcal{F}, (\mathcal{F}_t), P, -X, W)$ is also a weak solution with the same initial law. Since X is a Brownian motion, $P(X \equiv -X) = 0$, and so (2.7) does not have a unique strong solution. However, it is easy to show that (2.7) is unique in law. Notice that in this example the process W is derived from the process X rather than the other way around; in effect X possesses more randomness than W, and so certainly cannot be represented as a function of W.

3 Approximations for SDEs

This section reviews some of the techniques for discretising SDEs of the type (2.1). The techniques fall broadly into two classes: those intended for approximating unique strong solutions, which must approximate the function F of Definition 2.2, and those intended for approximating the distribution of weak solutions of unique-in-law SDEs, which need only furnish variates from approximating distributions. Different convergence criteria are appropriate for these two classes, and the following are examples of such criteria based on the discretisation error at some fixed time $T < \infty$.

Definition 3.1 A Weak Convergence Criterion. *A sequence of approximations to X_T, $\hat{X}_N$, converges in the weak sense with order α if, for any multi-variate polynomial, $p : \mathbb{R}^d \to \mathbb{R}$, there exists a constant, $K_p < \infty$, such that*

$$|\mathbf{E}p(X_T) - \mathbf{E}p(\hat{X}_N)| < K_p N^{-\alpha} \quad \textit{for } N = 1, 2, \ldots$$

Definition 3.2 A Strong Convergence Criterion. *A sequence of approximations to X_T, $\hat{X}_N$, converges in the strong sense with order α if there exists a constant, $K < \infty$, such that*

$$\left(\mathbf{E}|X_T - \hat{X}_N|^2\right)^{1/2} < KN^{-\alpha} \quad \textit{for } N = 1, 2, \ldots$$

There are, of course, many other convergence criteria for the two types of solution. For example, for strong solutions, it may in some situations be appropriate to use an *almost sure* convergence criterion, or some criterion based on the supremum norm of the error over an interval of time. Similarly, for weak solutions, it may be appropriate to use a convergence criterion based on the means of *bounded* (but not necessarily smooth) functions of the solution rather than the polynomial moments of Definition 3.1. The literature on the subject is now large and contains results formulated in terms of a variety of such criteria.

The simplest approximation for (2.1) is obtained by a stochastic generalisation of the *Euler* method for ordinary differential equations, which was first proposed for use with SDEs (in a form including interpolation between discretisation points) by Maruyama (1955); with a regular time mesh, $(0 < h < 2h < \cdots < Nh = T)$, this takes the form:

$$\begin{aligned} Y_0^N &= X_0 \\ Y_{n+1}^N &= Y_n^N + b(nh, Y_n^N)h + \sigma(nh, Y_n^N)\Delta W_{n+1}^N \quad \text{for } 0 \le n \le N-1, \\ \hat{X}_N &= Y_N^N, \end{aligned} \tag{3.1}$$

where $\Delta W_{n+1}^N := W_{(n+1)h} - W_{nh}$.

The generic rate of convergence of this approximation is 1 in the sense of the weak criterion, but only 1/2 in the sense of the strong criterion. (See Milshtein (1978).) In order for the approximation to converge in the strong sense, the ΔW must be increments of the specific Brownian motion giving rise to X, because in this case $\hat{X}_N$ has to approximate X_T outcome by outcome. However, for the result on convergence in the weak sense, they can be any set of independent Gaussian variates with mean zero and variance h (usually defined on the probability space residing in a computer). In fact they need only approximate Gaussian variates in the sense of the first few moments (Milshtein (1978), Talay (1986)), and so, for example, binary random variables taking the values $\pm\sqrt{h}$ with equal probability can be used.

An approximation which converges in the sense of the weak criterion does not actually need to converge to a weak solution. The point, of course, is that we are interested in approximating the *law* of the solution of a unique-in-law SDE, not a specific weak solution. This is illustrated by Tanaka's SDE. For this equation, and the initial condition $X_0 = 0$, Euler's method yields the approximation

$$\hat{X}_N = \sum_{n=0}^{N-1} \operatorname{sgn}(Y_n^N)\Delta W_{n+1}^N.$$

If the ΔW are distributed exactly as increments of a Brownian motion, then $\hat{X}_N$ and X_T will have identical distributions, meaning that the Euler approximation is exact in the sense of the weak criterion for this example with any value of N. It seems, moreover, that the Euler formula should approximate a strong solution (and, therefore, a specific weak solution) as for each value of N it represents a causal system, yielding an approximation, $\hat{X}_N$, that depends only on X_0 and W. However, if we sample a fixed Brownian motion, W, to produce the increments, ΔW, for different values of the mesh parameter, N, and apply these as inputs to the Euler 'system', the output does not converge in any strong sense as N increases. In fact, every time that a reduction in the step size, h, causes the first increment, ΔW_1^N, to change in sign, the whole process (Y_n^N) changes in sign. Since W crosses the t axis infinitely often in any neighbourhood of the origin, this behaviour never ceases as N increases. The same phenomenon occurs wherever (Y_n^N) crosses the t axis. As this example and the different convergence rates of the Euler method illustrate, the design of good approximations according to the weak and strong convergence criteria are very different objectives and so different families of schemes for the two criteria have evolved.

The disappointing strong rate of convergence of the Euler scheme is a result of the fractal nature of the paths of Brownian motion. This nature is inherited by the stochastic integral in (2.1) (considered as a function of the upper limit), and so, in order to obtain higher orders of convergence, it seems reasonable to add to the Euler approximation higher order terms from an expansion of this integral. Consider the following scheme for equations driven by one-dimensional Brownian motion, which is due to Milshtein (1974):

$$\begin{aligned} Y_0^N &= X_0 \\ Y_{n+1}^N &= Y_n^N + b(nh, Y_n^N)h + \sigma(nh, Y_n^N)\Delta W_{n+1}^N \\ &\quad + \frac{1}{2}\sum_{k=1}^{d} \frac{\partial \sigma}{\partial x_k}\sigma_k(nh, Y_n^N)((\Delta W_{n+1}^N)^2 - h) \quad \text{for } 0 \le n \le N-1, \\ \hat{X}_N &= Y_N^N, \end{aligned} \tag{3.2}$$

where the ΔW_n^N are as defined above. This has strong order of convergence 1, but does not generalise for equations driven by multi-dimensional Brownian motion unless the values of the Lévy area integrals between the components of W, $(W_t^{(i)}, \ i = 1, 2, \ldots, r)$,

$$A_t^{(i,j)} := \int_0^t W_v^{(i)} dW_v^{(j)} - \int_0^t W_v^{(j)} dW_v^{(i)},$$

at the discretisation points, $0, h, 2h, \ldots$, are also available. In fact any approximation with generic strong order of convergence greater that $1/2$ must include these integrals. (See Clark & Cameron (1980).) This presents serious problems for approximations to strong solutions, because any such approximations must be derived from a particular Brownian motion, either measured or simulated, and in either case it is not easy to obtain the Lévy area integrals. (See Gaines & Lyons (1994).) However, when the columns of σ 'commute' in a certain sense, the Lévy area integrals are not needed and higher orders of convergence can be achieved. (See Clark & Cameron (1980).)

The limitation in the orders of convergence that can be achieved for strong solutions is a result of the dependence of the solutions on the highly erratic behaviour of the paths of the driving process between sample points, and the inability of approximations based on simple sampling strategies to 'see' this behaviour. The best that can be achieved by such methods is *asymptotic efficiency*. An asymptotically efficient approximation is one that achieves the maximum possible order of convergence and has, in a certain sense, the smallest possible leading coefficient in the power series expansion of its error. (See Newton (1986, 1991) for details.)

Higher rates of convergence in terms of the weak criterion are more easily obtained, however, because the statistics used to 'drive' a numerical method need only approximate the corresponding statistics of a Brownian motion in the sense of their first few moments. For example, for the purposes of weak sense second-order methods, the increments of a multi-dimensional Brownian motion and its Lévy area integrals can be jointly approximated by appropriately defined ternary and binary variates (Talay (1986)). Thus, for example, the multi-dimensional generalisation of Milshtein's scheme is easy to implement for weak solutions, although it convergences no faster, in the weak sense, than does the Euler scheme. (Milshtein's method is specifically designed for approximating strong solutions.) Several authors have suggested schemes with weak order of convergence 2. (See, for example, Kloeden & Platen (1992), Mackevicius (1993), Milshtein (1978, 1986), Platen (1984) and Talay (1986).) Among these are methods that require the evaluation of several derivatives of the coefficients, b and σ, and others that are 'derivative sparse' or 'derivative free'. It seems, though, that a less computationally intensive way of obtaining higher orders of convergence is to use extrapolation techniques based on simple methods such as the Euler method. Talay & Tubaro (1990) showed that, under appropriate smoothness assumptions on the coefficients, b and σ, the weak sense error of the Euler approximation admits a well defined expansion as a power series in the step size, h. Under these conditions, for any multivariate polynomial p,

$$\mathbf{E}p(X_T) - \mathbf{E}p(\hat{X}_N) = C(b, \sigma, p, T)h + O(h^2),$$

where the coefficient C does not depend on the discretisation step size, h. (Under a uniform non-degeneracy condition on the operator $\mathcal{L}$, the same type of expansion can be shown to exist for functions p that are just measurable and bounded (Bally & Talay (1996)) allowing a detailed study of the distribution of X_T.) Suppose that the Euler method is applied with two different step sizes, h and $h/2$, to yield approximations $\hat{X}_N$ and $\hat{X}_{2N}$. The extrapolated approximation

$$\tilde{X}_N = 2\hat{X}_{2N} - \hat{X}_N$$

will clearly yield an error of order h^2 in the sense of the weak convergence criterion. With this approach one can also estimate discretisation errors, allowing run-time error control. In fact the leading coefficient in the error expansion can be estimated as follows:

$$\hat{C}(b, \sigma, p, T) = 2\mathbf{E}(p(\hat{X}_N) - p(\hat{X}_{2N}))/h.$$

Extrapolated approximations with higher orders of convergence are also possible (see Talay & Tubaro (1990), Kloeden, Platen & Hofmann (1995)), although it is unclear whether or not they offer any practical advantages over the second order method described above. There is no upper bound on the order of convergence of approximations to (2.1) in the sense of the weak criterion. In fact, Wagner (1989) has produced a scheme, which has an infinite order of convergence in the sense of Definition 3.1. However it is somewhat complex in that it involves an auxiliary simulation to estimate the solution of an integral equation at each time step.

The approximations discussed above are all based on *uniform* discretisations of the time parameter, t. A simple variation on them can be obtained by discretising t according to some non-uniform sampling density

$$\rho : [0, \infty) \to \mathbb{R}^+;$$

i.e. by using the mesh $(0, t_1, t_2, \ldots, t_n, \ldots)$, where

$$t_{n+1} = t_n + h\rho^{-1}(t_n).$$

The sampling density might be chosen such that it is large when either $\mathbf{E}\|b(t, X_t)\|$ or $\mathbf{E}\|\sigma(t, X_t)\|$ is large, and small when both of these are small (if this is *a priori* known). A more productive development, though, is to use *random*, non-uniform meshes, in which the sampling density is outcome dependent. An especially useful mesh for weak approximations comprises the passage times of the solution process, (X_t), through the points of a deterministic *spatial* mesh. In this, the discretisation times, τ_n, are defined by

$$\tau_{n+1} = \tau_n + \inf\{t \geq 0 : X_{\tau_n + t} \in N(X_{\tau_n})\},$$

where $N(x)$ is the set of points on the mesh neighbouring the point x (also on the mesh). Of course, it is impractical to approximate strong solutions of SDEs on such a mesh as the determination of the sampling times would require exact knowledge of the solution process at all times. However, it is relatively easy to design approximations that converge in a weak sense similar to that of Definition 3.1. (See Kushner (1977) and Kushner & Dupuis (1992).) These have more controlled spatial errors than the methods described above because the discretisation times are more closely spaced when the path of the diffusion is changing 'rapidly' and more widely spaced when there is little movement in X_t. The price paid for this, though, is that errors are introduced in the time variable t—typically it is not possible to simulate the times τ_n exactly. The methods are ideally suited to the Monte Carlo solution of elliptic boundary value problems, because the time variable plays no role in such problems. They are also useful for parabolic, mixed initial-boundary value problems of the type (2.2), (2.5) when spatial errors are more costly than errors in t. Of course approximations can also be defined on hybrid discretisation meshes which in some sense try to find an appropriate balance between spatial and temporal errors. (Again, see Kushner (1977) and Kushner & Dupuis (1992).)

There is now an extensive literature on the subject of approximations for stochastic differential equations, and this section has mentioned only a few aspects. In particular, it has concentrated on SDEs driven by multi-dimensional Brownian motion of the type (2.1), as these are appropriate for Monte Carlo methods in options pricing, and market modelling. Even within this class of equations there are many aspects which have not been mentioned. For example: approximations for diffusions with reflecting boundary conditions (see Slominski (1994) and Lépingle (1995)); approximations for the invariant laws of certain ergodic diffusions (see, for example, Talay (1987)); *robust* approximations for strong solutions (which have application in nonlinear signal processing and stochastic control) (see Clark (1978) and Pardoux & Talay (1985)); the numerical stability of approximations (see, for example, Hernandez & Spigler (1992, 1993), Hofmann (1995) and Schurz (1996)); etc. For further information and an extensive list of references the reader should consult the book by Kloeden & Platen (1992).

4 Variance Reduction

This section describes some recent results concerning variance reduced Monte Carlo methods for the PDE problems discussed in Section 2. Parametrically optimal techniques of control variates and importance sampling are described, which are rooted in an integral representation of the perfect control variate. The techniques yield stochastic representation formulae of the Feynman-Kac type, but involving variates with lower variance. As is the case with the Feynman-Kac formulae, the variates cannot be simulated exactly but can be approximated by the techniques discussed in Section 3.

Consider first the Cauchy problem (2.2), (2.3). Theorem 4.1, below, formalises the discussion, in Section 2, on stochastic representation formulae for this problem. It also provides a perfect control variate for use with that representation in the form of a stochastic integral. It is based on the following hypotheses.

(H1) There exists a constant $\epsilon > 0$, such that

$$\theta'\sigma\sigma'(s,x)\theta \geq \epsilon|\theta|^2 \quad \text{for all } (s,x) \in (0,T) \times \mathbb{R}^d \quad \text{and all } \theta \in \mathbb{R}^d.$$

(H2) The coefficients b, σ and c are bounded and uniformly Hölder continuous; i.e. for some positive constants $K, q < \infty$,

$$\|b(s,x) - b(t,y)\| \leq K|(s,x) - (t,y)|^q \quad \text{for all } (s,x),(t,y) \in (0,T) \times \mathbb{R}^d,$$

etc.

(H3) $D_x b$, $D_x \sigma_k$ (for $k = 1, 2, \ldots, d$) and $D_x c$ are continuous and bounded. (D_x signifies the row vector Jacobian operator $[\partial/\partial x_1 \; \partial/\partial x_2 \; \cdots \; \partial/\partial x_d]$, and σ_k is the kth column of σ.) (In this section r, the dimension of W, is equal to d, the dimension of X.)

(H4) The functions g, h, $D_x g$ and $D_x h$ are continuous and satisfy a polynomial growth condition in x; i.e. for some constants $K, q < \infty$

$$|D_x g(s,x)| \leq K(1 + |x|^q) \quad \text{for all } (s,x) \in (0,T) \times \mathbb{R}^d,$$

etc., and g is Hölder continuous in x, uniformly in $(0,T) \times \mathbb{R}^d$.

Theorem 4.1 *Suppose that (H1)–(H4) are satisfied; then the Cauchy problem (2.2), (2.3) has a unique solution, which admits the stochastic representation*

$$u(s,x) = \mathbf{E}\Lambda_{s,x} \quad \textit{for all } (s,x) \in (0,T) \times \mathbb{R}^d,$$

where

$$\Lambda_{s,x} = Z_{s,T} h(X_T^{s,x}) - \int_s^T Z_{s,t} g(t, X_t^{s,x}) dt, \tag{4.1}$$

and $Z_{s,t}$ is as given in (2.4). Furthermore,

$$u(s,x) = \Lambda_{s,x} - \int_s^T Z_{s,t}\pi_t \sigma(t, X_t^{s,x}) dW_t, \tag{4.2}$$

where

$$\begin{aligned} \pi_t &:= \mathbf{E}(\rho_t | X_t^{s,x}), \\ \rho_t &:= Z_{t,T} D_x h(X_T^{s,x}) \Phi_{t,T} - \int_t^T Z_{t,v} D_x g(v, X_v^{s,x}) \Phi_{t,v} dv \\ &\quad + \int_t^T \left(Z_{t,T} h(X_T^{s,x}) - \int_v^T Z_{t,w} g(w, X_w^{s,x}) dw \right) D_x c(v, X_v^{s,x}) \Phi_{t,v} dv, \end{aligned} \tag{4.3}$$

Φ is the unique strong solution of the equation of first-order variation associated with (2.1): for $0 \le s \le t \le T$,

$$\Phi_{s,t} = I_d + \int_s^t D_x b(v, X_v^{s,x}) \Phi_{s,v} dv + \sum_{i=1}^d \int_s^t D_x \sigma_i(v, X_v^{s,x}) \Phi_{s,v} dW_v^{(i)}, \tag{4.4}$$

I_d is the identity matrix of order d, and $W^{(i)}$ is the ith component of W.

The first part of Theorem 4.1 is proved, for example, in Friedman (1975); it formalises the disccussion of stochastic representation formulae of Section 2. The second part is a corollary to the Funke-Shevlyakov-Haussmann integral representation theorem for functionals of Itô processes. (See Ocone (1984) and Newton (1994).) It states that the perfect control variate for use with $\Lambda_{s,x}$ is the stochastic integral in (4.2)—if we subtract the integral from $\Lambda_{s,x}$ we get a variate that is (almost surely) equal to $u(s,x)$. Of course, the term π in the theorem is as difficult to evaluate as $u(s,x)$ itself, if not more so, and so the perfect control variate cannot generally be computed exactly. However, since it is expressed in the form of a stochastic inegral it can be approximated in a way which yields an unbiased statistic.

Suppose we use the form of the perfect control variate described in Theorem 4.1, but with the term π replaced by an approximation, $\hat{\pi}$. If $\hat{\pi}_t$ is $X_t^{s,x}$-measurable for each t and $\mathbf{E}\int_s^T Z_{s,t}^2 |\hat{\pi}_t \sigma(t, X_t^{s,x})|^2 dt < \infty$ then the resulting controlled variate

$$\Lambda_{c,s,x} := \Lambda_{s,x} - \int_s^T Z_{s,t} \hat{\pi}_t \sigma(t, X_t^{s,x}) dW_t$$

will be unbiased and will have variance

$$\begin{aligned} \operatorname{var}(\Lambda_{c,s,x}) &= \mathbf{E} \int_s^T Z_{s,t}^2 |(\pi_t - \hat{\pi}_t) \sigma(t, X_t^{s,x})|^2 dt \\ &= \|\pi - \hat{\pi}\|_H^2. \end{aligned}$$

The norm here is that of the Hilbert space, H, of (row) d-vector stochastic processes ($\alpha_t \in \mathbb{R}^d,\ s \le t \le T$) with inner product

$$\langle \alpha, \beta \rangle_H := \mathbf{E} \int_s^T Z_{s,t}^2 \alpha_t \sigma\sigma'(t, X_t^{s,x}) \beta_t' dt.$$

A good approximation to π is one for which the error, $\pi - \hat{\pi}$, is small in the sense of this Hilbert space. Now, both π and $\hat{\pi}$ are adapted to the filtration generated by X, $(\mathcal{F}_t^X)$. In fact π is the projection, in H, of ρ onto the subspace of $(\mathcal{F}_t^X)$-adapted processes ($\rho \in H$ by virtue of hypotheses (H1)–(H4)), and so $\rho - \pi$ is orthogonal to $\pi - \hat{\pi}$, and

$$\mathrm{var}(\Lambda_{c,s,x}) = \|\rho - \hat{\pi}\|_H^2 - \|\rho - \pi\|_H^2.$$

We must therefore choose $\hat{\pi}$ so that $\|\rho - \hat{\pi}\|_H$ is small, and one way of doing this is to use projection.

Consider the following class of approximations, parametrised by the l coefficients $c_1, c_2, \ldots, c_l$:

$$\hat{\pi}_t = \sum_{i=1}^{l} c_i \gamma_i(t, X_t^{s,x}), \tag{4.5}$$

where the (row vector) functions $\gamma_i : [s, T] \times \mathbb{R}^d \to \mathbb{R}^d$ are such that the processes $(\gamma_i(\cdot, X_\cdot);\ i = 1, 2, \ldots, l)$ form a linearly independent set of elements of H. This class spans an l-dimensional subspace of H, and so the optimal approximations in this class can be found by projecting ρ onto this subspace. The optimal coefficients are given by the following projection equations:

$$\begin{aligned} C^* &= \mathbf{E} \int_s^T Z_{s,t}^2 \rho(t, X_t^{s,x}) \sigma\sigma'(t, X_t^{s,x}) \Gamma'(t, X_t^{s,x}) dt \\ &\quad \times \left(\mathbf{E} \int_s^T Z_{s,t}^2 \Gamma(t, X_t^{s,x}) \sigma\sigma'(t, X_t^{s,x}) \Gamma'(t, X_t^{s,x}) dt \right)^{-1} \end{aligned} \tag{4.6}$$

where Γ is the $l \times d$ matrix whose rows are $\gamma_1, \gamma_2, \ldots, \gamma_l$, and $C^* = [c_1^*\ c_2^* \ldots c_l^*]$. The terms on the right-hand side of this equation can be evaluated by means of an auxiliary simulation involving the discretisation of (2.1) and (4.4). (See Newton (1994) for details.) This represents an increase in computational burden, of course, but the increase need not be great; the auxilliary simulation to solve the system (4.6) can be carried out with a much coarser time discretisation and far fewer outcomes than the main simulation, and still yield useful variance reduction. (Again, see Newton (1994) for details.) It is worth approximating the (c_i^*) accurately only when the use of their exact values would yield a large reduction in variance, in which case the rewards for the extra computation are great. Of course, if the approximation is too inaccurate we may get little variance reduction, or even increase the variance. However, no matter what values for the coefficients are used in (4.5), the resulting variate, $\Lambda_{c,s,x}$, will be unbiased. In addition, the general technique described above simplifies considerably for many problems. If, for example, the terms $c(s,x)$ and $g(s,x)$ of the Cauchy problem (2.2), (2.3) are zero then ρ_t has the simplified form

$$\rho_t = D_x h(X_T^{s,x})\Phi_{t,T}.$$

On the other hand, if the coefficients in the differential operator $\mathcal{L}$ of (2.2) are constant then the associated diffusion, defined by (2.1), is Gaussian and the equation of first order variation, (4.4), becomes trivial; in fact in this case

$$\Phi_{s,t} = I_d \quad \text{for all} \quad s, t.$$

The functional of this Gaussian process to be integrated, $\Lambda_{s,T}$, may not be easy to integrate by deterministic means if the dimension d is large.

In order for this projective method of control variates to be successful a good choice of functions (γ_i) must be made; they clearly need to represent significant components of π. Various ways of choosing them are discussed in Newton (1994). One method of choosing the first function, γ_1, is to make it the mean of ρ_t according to some crude approximation of its $X_t^{s,x}$-conditional distribution, based on the particular nature of the problem in question. If, for a particular problem, the $X_t^{s,x}$-conditional distribution of ρ_t can be approximated with reasonable accuracy then the following direct approximation for π_t could be used:

$$\hat{\pi}_t = \hat{\mathbf{E}}(\rho_t | X_t^{s,x}).$$

This avoids the need to carry out the projection and thus saves on computational cost. Other direct (non-projective) approximations for π_t are possible for certain problems. For example, if σ is small then the value of π_t corresponding to a linearised version of equation (2.1) can be used. This is described in more detail in Newton (1994).

It is likely, with the projective method, that a certain amount of experimentation with different functions (γ_i) and different discretisation parameters for the auxilliary and main simulations will be necessary in order to find good values. The simulations in Newton (1994) suggest that the method can offer significant savings in computational effort.

A similar approach can be followed for the initial-boundary value problem (2.1), (2.5). (See Newton (1996).)

Theorem 4.1 also provides a representation for the perfect system of importance sampling for these problems. Suppose that $(\Omega, \mathcal{F}, (\mathcal{F}_t), P, X, W)$ is a weak solution of equation (2.1), and that $(Y_t,\ 0 \le t \le T)$ is an $\mathbb{R}^d$-valued, $(\mathcal{F}_t)$-adapted process with

$$\mathbf{E} \exp\left(\frac{1}{2}\int_0^T |Y_t|^2 dt\right) < \infty, \tag{4.7}$$

and set

$$\mu_{s,t} = \exp\left(\int_s^t Y_v' dW_v + \frac{1}{2}\int_s^t |Y_v|^2 dv\right)$$
$$= \exp\left(\int_s^t Y_v' d\tilde{W}_v - \frac{1}{2}\int_s^t |Y_v|^2 dv\right), \qquad (4.8)$$

where

$$\tilde{W}_t = W_t + \int_0^t Y_v dv.$$

Because of the bound (4.7)

$$P(\inf_{0\le s\le t\le T} \mu_{s,t} > 0) = P(\sup_{0\le s\le t\le T} \mu_{s,t} < \infty) = 1,$$

and (Novikov (1972))

$$\mathbf{E}\mu_{s,t}^{-1} = 1, \quad \text{for all } 0 \le s \le t \le T.$$

Girsanov's theorem (see, for example, Karatzas & Shreve (1991) or Rogers & Williams (1987)) shows that $(\tilde{W}_t, \mathcal{F}_t;\ 0 \le t \le T)$ is a Brownian motion under the probability measure $\tilde{P}$ defined on $\mathcal{F}_T$ by

$$\tilde{P}(A) = \int_A \mu_{0,T}^{-1} dP \quad \text{for all } A \in \mathcal{F}_T. \qquad (4.9)$$

We are now in a position to sample the process X according to its distribution under $\tilde{P}$ (the importance sampling distribution)—in fact this can be achieved by a discretisation of the following SDE, in which $\tilde{W}$ is regarded as a Brownian motion:

$$X_t = X_0 + \int_0^t (b(v, X_v) - \sigma(v, X_v)Y_v)\, dv + \int_0^t \sigma(v, X_v) d\tilde{W}_v. \qquad (4.10)$$

The 'density' of the importance sampling distribution with respect to that induced by P is, according to (4.9), $\mu_{0,T}^{-1}$. Thus the importance sampled stochastic representation formula is as follows

$$u(s,x) = \tilde{\mathbf{E}}\mu_{s,T}\Lambda_{s,T},$$

where $(\Lambda_{s,t})$ is as defined in (4.1) with $(X_t^{s,x})$ a weak solution of (4.10) taking the value x at initial time s. We clearly want to choose the process (Y_t) such that the following variance is small:

$$\tilde{\text{var}}(\mu_{s,T}\Lambda_{s,T}) = \tilde{\mathbf{E}}(\mu_{s,T}\Lambda_{s,T})^2 - (\tilde{\mathbf{E}}\mu_{s,T}\Lambda_{s,T})^2$$
$$= \tilde{\mathbf{E}}(\mu_{s,T}\Lambda_{s,T})^2 - u(s,x)^2.$$

The optimal process, which yields the perfect system of importance sampling, can be found from Theorem 4.1 under the following additional hypothesis.

(H5) There exists a constant $\epsilon > 0$ such that either $P(\Lambda_{s,T} \geq \epsilon) = 1$ or $P(\Lambda_{s,T} \leq -\epsilon) = 1$.

Corollary 4.2 *Suppose that (H1)–(H5) are satisfied, then*

$$P\left(u(s,x) = \mu^*_{s,T}\Lambda_{s,T}\right) = 1,$$

where $\mu^*_{s,T}$ is defined by (4.8) with

$$Y_t = Y^*_t = -\frac{\sigma'(t, X^{s,x}_t)\pi'_t}{\mathbf{E}(\Lambda_{s,T}|\mathcal{F}^X_t)} \quad \textit{for all} \quad s \leq t \leq T,$$

and (π_t) is as defined in (4.3).

Thus, as was the case for the perfect control variate, the perfect method of importance sampling involves the process (π_t), which is not explicitly known; it also involves the process $\mathbf{E}(\Lambda_{s,T}|\mathcal{F}^X_t)$. The corollary is proved in Newton (1994), where optimal parametric approximations to Y^* are also developed. These are less satisfactory than the optimal parametric approximations for the control variate described above in that they involve a stochastic approximation procedure to solve a set of nonlinear equations.

There is, apparently, an additional problem with importance sampling in the requirement of (H5) that $\Lambda_{s,T}$ be strictly bounded away from 0. However, if $\Lambda_{s,T}$ is bounded above or below (almost surely) then the corollary can be applied to $\Lambda_{s,T} + \alpha$ for an appropriate constant α. In fact

$$P\left(\mu^*_{s,T,\alpha}(\Lambda_{s,T} + \alpha) - \alpha = u(s,x)\right) = 1,$$

where $\mu^*_{s,T,\alpha}$ is defined by (4.8) with

$$Y_t = Y^*_{t,\alpha} = -\frac{\sigma'(t, X^{s,x}_t)\pi'_t}{\mathbf{E}(\Lambda_{s,T}|\mathcal{F}^X_t) + \alpha} \quad \text{for all} \quad s \leq t \leq T.$$

((π_t) does not depend on α.) For more details, see the discussion in Newton (1994). Alternatively $\Lambda_{s,T}$ can be split into its positive and negative parts

$$\Lambda^+_{s,T} = \max\{\Lambda_{s,T}, 0\},$$

and

$$\Lambda^-_{s,T} = -\min\{\Lambda_{s,T}, 0\},$$

and these could be integrated separately by means of appropriate importance sampling strategies, as discussed in Section 1.

For the reasons given above, unless for a specific problem there is an easy way of approximating the process (Y^*_t) well, and no equally good method of approximating (π_t) alone, then it would almost certainly be better to use the method of control variates described above than that of importance sampling.

References

Bally, V. & Talay, D. (1996) 'The law of the Euler scheme for stochastic differential equations (I): convergence rate of the distribution function', *Probability Theory and Related Fields* **104**, 43–60.

Clark, J.M.C. (1978) 'The design of robust approximations to the stochastic differential equations of nonlinear filtering', in: J. Skwirzynskii (ed.): *Communication Systems and Random Process Theory*, NATO Advanced Study Institute Series, Sijthoff and Noorshoff.

Clark, J.M.C. & Cameron, R.J. (1980) 'The maximum rate of convergence of discrete approximations for stochastic differential equations', in: B. Grigelionis (ed.): *Stochastic Differential Systems—Filtering and Control*; Lecture Notes in Control and Information Sciences **25**, Springer-Verlag.

Friedman, A., (1975) *Stochastic Differential Equations and Applications—Volume 1*, Academic Press.

Gaines, J.G. & Lyons, T.J. (1994) 'Random generation of stochastic area integrals', *SIAM J. Appl. Math.* **54**, No.4, 1132–1146.

Hernandez, D.B. & Spigler, R. (1992) 'A-stability of implicit Runge-Kutta methods for systems with additive noise', *BIT* **32**, 620–633.

Hernandez, D.B. & Spigler, R. (1993) 'Convergence and stability of implicit Runge-Kutta methods for systems with multiplicative noise', *BIT* **33**, 654–669.

Hofmann, N. (1995) 'Stability of weak numerical schemes for stochastic differential equations', *Mathematics and Computers in Simulation* **38**, 63–68.

Karatzas, I. & Shreve, S.E. (1991) *Brownian Motion and Stochastic Calculus*, Springer-Verlag.

Kloeden, P.E. & Platen, E. (1992) *Numerical Solution of Stochastic Differential Equations*, Springer-Verlag.

Kloeden, P.E., Platen, E. & Hofmann, N. (1995) 'Extrapolation methods for the weak approximation of Itô diffusions', *SIAM J. Num. Anal.* **32**, 1519–1534.

Knuth, D.E. (1981) *The Art of Computer Programming, Volume 2: Semi-Numerical Algorithms*, Addison-Wesley.

Kushner, H.J. (1977) *Probability Methods for Approximations in Stochastic Control and for Elliptic Equations*, Academic Press.

Kushner, H.J. & Dupuis, P.G. (1992) *Numerical Methods for Stochastic Control Problems in Continuous Time*, Springer-Verlag.

Lépingle, D. (1995) 'Euler scheme for reflected stochastic differential equations', *Mathematics and Computers in Simulation* **38**, 119–126.

Mackevicius, V. (1993) 'Second order weak approximations for Stratonovich stochastic differential equations', Preprint: Department of Mathematics, Vilnius University, Lithuania.

Maruyama, G., (1955) 'Continuous Markov processes and stochastic equations', *Rend. Circ. Mat. Palermo* **4**, 48–90.

Milshtein, G.N. (1974) 'Approximate integration of stochastic differential equations', *Theory Probab. Appl.* **19**, 557–562.

Milshtein, G.N. (1978) 'A method of second-order accuracy integration of stochastic differential equations', *Theory Probab. Appl.* **23**, 396–401.

Milshtein, G.N. (1986) 'Weak approximations of the solution of systems of stochastic differential equations', *Theory Probab. Appl.* **30**, 750–766.

Newton, N.J. (1986) 'An asymptotically efficient difference formula for solving stochastic differential equations', *Stochastics* **19**, 175–206.

Newton, N.J. (1991) 'Asymptotically efficient Runge-Kutta methods for a class of Itô and Stratonovich equations', *SIAM J. Appl. Math.* **51**, No. 2, 542–567.

Newton, N.J. (1994) 'Variance reduction for simulated diffusions', *SIAM J. Appl. Math.* **54**, 1780–1805.

Newton, N.J. (1996) 'Variance reduced Monte Carlo methods for PDEs', *Zeitschrift fur Angewandte Mathematik und Mechanik*, to appear.

Novikov, A.A. (1972) 'On an identity for stochastic integrals', *Theory Probab. Appl.* **17**, 717–720.

Ocone, D. (1984) 'Malliavin's calculus and stochastic integral representations of functionals of diffusion processes', *Stochastics* **12**, 161–185.

Pardoux, E. & Talay, D. (1985) 'Discretisation and simulation of stochastic differential equations', *Acta Appl. Math.* **3**, 23–47.

Platen, E. (1984) 'Zur zeitdiskreten approximation von Itoprozessen', *Diss. B., IMath, Akad. der Wiss. der DDR, Berlin.*

Rogers, L.C.G. & Williams, D.W. (1987) *Diffusions, Markov Processes and Martingales—Volume 2: Itô Calculus*, Wiley.

Schurz, H. (1996) 'Asymptotical mean square stability of an equilibrium point of some linear numerical solutions with multiplicative noise', *Stoch. Anal. Appl.* **14**, to appear.

Slominski, L., (1994) 'On approximation of solutions of multidimensional SDEs with reflecting boundary conditions', *Stoch. Proc. Appl.* **50**, 197–219.

Talay, D. (1986) 'Discrétisation d'une équation différentielle stochastique et calcul approché d'espérances de fonctionelles de la solution', *Math. Model. Numer. Anal.* **20**, 141–179.

Talay, D. (1987) 'Classification of discretisation schemes of diffusions according to an ergodic criterium', *Lecture Notes in Control and Information Sciences* **91**, 207–218.

Talay, D. & Tubaro, L. (1990) 'Expansion of the global error for numerical schemes solving stochastic differential equations', *Stoch. Anal. Appl.* **8**, 483–509.

Wagner, W. (1989) 'Unbiased Monte Carlo estimators for functionals of weak solutions of stochastic differential equations', *Stochastics and Stochastics Reports* **28**, 1–20.

Recent Advances in Numerical Methods for Pricing Derivative Securities

M. Broadie and J. Detemple

1 Introduction

In the past two decades there has been an explosion in the use of derivative securities by investors, corporations, mutual funds, and financial institutions. Exchange traded derivatives have experienced unprecedented growth in volume while 'exotic' securities (i.e., securities with nonstandard payoff patterns) have become more common in the over-the-counter market. Using the most widely accepted financial models, there are many types of securities which cannot be priced in closed-form. This void has created a great need for efficient numerical procedures for security pricing.

Closed-form prices are available in a few special cases. One example is a European option (i.e., an option which can only be exercised at the maturity date of the contract) written on a single underlying asset. The European option valuation formula was derived in the seminal papers of Black & Scholes (1973) and Merton (1973). In the case of American options (i.e., options which can be exercised at any time at or before the maturity date) analytical expressions for the price have been derived, but there are no easily computable, explicit formulas currently available. Researchers and practitioners must then resort to numerical approximation techniques to compute the prices of these instruments. Further complications occur when the payoff of the derivative security depends on multiple assets or multiple sources of uncertainty. Analytical solutions are often not available for options with path-dependent payoffs and other exotic options.

In this paper we provide a survey of recent numerical methods for pricing derivative securities. Section 2 focuses on standard American options on a single underlying asset. Section 3 briefly treats barrier and lookback options. Options on multiple assets are covered in Section 4. New computational results are also presented.

2 American Options on a Single Underlying Asset

In the standard model for pricing options, the price of the underlying security is assumed to follow a lognormal process. To fix notation, suppose that the price of the underlying asset is S_t at time t. Then S_t satisfies

$$dS_t = S_t[(\mu - \delta)dt + \sigma dW_t], \tag{2.1}$$

where W_t is a standard Brownian motion process. The parameter μ is the expected return of the asset, δ is the dividend rate, and σ is the volatility of the asset price, which are all taken to be constant. In the standard model money can be invested in a risk-free asset which has a constant interest rate r. For an overview of this model in particular, and derivatives in general, see the textbooks by Cox & Rubinstein (1985), Hull (1993), Stoll & Whaley (1993), and Jarrow & Turnbull (1996).

We first consider a *European* call option with maturity T and strike price K. This means that its payoff at expiration is $(S_T - K)^+$.[1] The value of the European call option at time 0 can be written as

$$C^E(S_0) = \mathbf{E}^*[e^{-rT}(S_T - K)^+] \tag{2.2}$$

where $\mathbf{E}^*$ denotes the expectation relative to the risk-neutral process for S_t, i.e., where r replaces μ in (2.1). This risk-neutral valuation approach was pioneered by Cox & Ross (1976); its theoretical foundations are identified and characterized in the seminal papers of Harrison & Kreps (1979) and Harrison & Pliska (1981). The solution to (2.2) was first derived in Black & Scholes (1973) and Merton (1973) and is given by

$$C^E(S_0) = S_0 N(d_1) - e^{-rT} K N(d_2) \tag{2.3}$$

with

$$d_1 = \frac{\ln(S_0/K) + (r + \frac{1}{2}\sigma^2)T}{\sigma\sqrt{T}} \tag{2.4}$$

$$d_2 = \frac{\ln(S_0/K) + (r - \frac{1}{2}\sigma^2)T}{\sigma\sqrt{T}} = d_1 - \sigma\sqrt{T} \tag{2.5}$$

where $N(\cdot)$ denotes the standard normal cumulative distribution. This solution is considered closed-form because the cumulative normal distribution is easily computed. See Abramowitz & Stegun (1972) or Moro (1995) for methods to approximate the cumulative normal distribution.[2]

---

[1]The operator x^+ denotes $\max(x, 0)$.

[2]Moro (1995) proposes the approximation

$$N(x) \approx \begin{cases} 0.5 + x(\sum_{i=0}^{2} a_i x^{2i}/1 + \sum_{i=1}^{3} b_i x^{2i}) & \text{when } 0 \le x \le 1.87 \\ 1 - \left(\sum_{i=0}^{2} c_i x^i / \sum_{i=0}^{3} d_i x^i\right)^{16} & \text{when } 1.87 < x < 6 \\ 1 & \text{when } x \ge 6 \end{cases}$$

where $a_0 = 0.398942270991$, $a_1 = 0.020133760596$, $a_2 = 0.002946756074$, $b_1 = 0.217134277847$, $b_2 = 0.018576112465$, $b_3 = 0.000643163695$, $c_0 = 1.398247031184$, $c_1 =$

An *American* call option with maturity T and strike price K can be exercised at any time at or prior to maturity. Its payoff is $(S_\tau - K)^+$ if it is exercised at time $\tau \leq T$. The value of the American call option at time 0 can be written as

$$C(S_0) = \max_{\tau} \mathbf{E}^*[e^{-r\tau}(S_\tau - K)^+] \tag{2.6}$$

where the max is over all stopping times $\tau \leq T$. For a rigorous justification of (2.6) as the appropriate pricing formula, see Bensoussan (1984) and Karatzas (1988). Finding the optimal stopping policy is equivalent to determining the points (t, S_t) for which early exercise is optimal. The boundary which separates the early exercise region from the continuation region is the *optimal exercise boundary.* Analytical solutions in the case of call options with discrete dividends were derived in Roll (1977), Geske (1979), and Whaley (1981). Early work in the non-dividend American put case is given in Johnson (1983) and Blomeyer (1986).[3]

The literature concerning the numerical solution of equation (2.6) is vast. Major approaches include binomial (or lattice) methods, techniques based on solving partial differential equations, integral equations, variational inequalities, Monte Carlo simulation, and others.[4]

The binomial method for the valuation of American options was introduced by Cox, Ross, & Rubinstein (1979). Trinomial methods have been proposed in Parkinson (1977) and Boyle (1988) and further analyzed in Omberg (1988). Other generalizations and variations of the binomial approach are given in Rendleman & Bartter (1979), Jarrow & Rudd (1983), Hull & White (1988), Amin (1991), Trigeorgis (1991), Tian (1993), and Leisen & Reimer (1995).[5] Implementation improvements are given in Kim & Byun (1994) and Curran (1995). Applications to computing price derivatives appear in Pelsser

-0.360040248231, $c_2 = 0.022719786588$, $d_0 = 1.460954518699$, $d_1 = -0.305459640162$, $d_2 = 0.038611796258$, and $d_3 = -0.003787400686$. Moro (1995) shows that this approximation, properly implemented, is faster and more accurate than previous methods. Proper implementation includes using multiplication rather than exponentiation wherever possible. For example, rather than computing $z = ax^4 + bx^2 + c$ using the power function, it is more efficient to compute $y = x * x$ and then $z = (ay + b)y + c$.

[3] The payoff of a put option is $(K - S_\tau)^+$ if it is exercised at time $\tau \leq T$. McDonald & Schroder (1990) and Chesney & Gibson (1995) derive an interesting put-call symmetry result. They show that in the standard model (geometric Brownian motion setting), the value of an American call option with parameters S, K, r, δ, T is related to the value of an American put option by

$$C(S, K, r, \delta, T) = P(K, S, \delta, r, T). \tag{2.7}$$

Thus, the American put price equals the American call price with the identification of parameters: $S \to K$, $K \to S$, $r \to \delta$, and $\delta \to r$.

[4] A comparison of some early methods is given in Geske & Shastri (1985).

[5] There is also a large literature on lattice methods with alternative specifications of the stochastic process and for pricing interest rate sensitive securities. See, e.g., Nelson & Ramaswamy (1990), Hull & White (1994a, 1994b), Tian (1992, 1994), Amin (1995), Amin & Bodurtha (1995), and Li *et al.* (1995).

& Vorst (1994). A convergence proof for these types of lattice methods for pricing American options is provided in Amin & Khanna (1994). The convergence of the optimal exercise boundary is proved in Lamberton (1993). Convergence rates for European option pricing are given in Leisen & Reimer (1995). Empirical convergence rate evidence for American option pricing is provided in Broadie & Detemple (1996). Analytical convergence rate bounds for American option pricing using binomial methods are derived in Lamberton (1995).

Black & Scholes (1973) and Merton (1973) showed that the price of any contingent claim, in particular a call option, must satisfy what is now called the Black–Scholes fundamental partial differential equation (PDE):

$$\frac{\partial C(S,t)}{\partial t} + (r-\delta)S\frac{\partial C(S,t)}{\partial S} + \frac{1}{2}\sigma^2 S^2 \frac{\partial^2 C(S,t)}{\partial S^2} - rC(S,t) = 0 \tag{2.8}$$

subject to the appropriate boundary conditions. For an American option some of the boundary conditions are related to the early exercise event. Finite difference methods for the numerical solution of this PDE and its associated boundary conditions in the American option case were introduced in Schwartz (1977) and Brennan & Schwartz (1977, 1978). Convergence of the Brennan–Schwartz method is proved in Jaillet, Lamberton, & Lapeyre (1990) and Zhang (1997). Related numerical approaches include Courtadon (1982) and Hull & White (1990). The quadratic method of MacMillan (1986) and Barone–Adesi & Whaley (1987) and the method of lines of Carr & Faguet (1995) are based on exact solutions to approximations of the Black–Scholes PDE.

Geske & Johnson (1984) present an exact analytical solution for the American option pricing problem. They write the continuous option price as the sum of prices of simpler options which can be exercised only at discrete points in time. However, their formula is an infinite series involving multidimensional cumulative normals (that can only be evaluated approximately by numerical methods) and an unknown exercise boundary (which must also be determined numerically). In the same paper, Geske & Johnson (1984) introduced the method of Richardson extrapolation to the option pricing problem. Richardson extrapolation has also been used in Breen (1991), Bunch & Johnson (1992), Ho, Stapleton, & Subrahmanyam (1994), Huang, Subrahmanyam, & Yu (1995), and Carr & Faguet (1995). For an extensive treatment of Richardson extrapolation see Marchuk & Shaidurov (1983). Other extrapolation techniques (see, e.g., Press *et al.* 1992) have not been extensively tested in this context.

Jaillet, Lamberton, & Lapeyre (1990) introduced the variational inequality approach to American option pricing. A discretization of this formulation leads to a linear complementarity problem (LCP) which can be solved by linear programming-type methods (see Cottle, Pang, & Stone (1992) for a complete treatment of LCPs). Numerical results with this approach are given in Dempster (1994). For an overview of differential equations and variational inequality approaches to option pricing, see the textbook by Wilmott, Dewynne, & Howison (1993).

McKean (1965) first derived an integral representation of the option price. Kim (1990), Jacka (1991), and Carr, Jarrow, & Myneni (1992) derive an alternate integral representation which expresses the value of the American option as the value of the corresponding European option plus an integral which represents the present value of the gains from early exercise:

$$C(S_0) = C^E(S_0) + \int_{s=0}^{T} [\delta S_0 e^{-\delta s} N(d_3(S_0, B_s, s)) - rKe^{-rs} N(d_4(S_0, B_s, s))] ds, \tag{2.9}$$

where $C^E(S_0)$ is the corresponding European call option value, B_s is the optimal exercise boundary, and

$$\begin{aligned} d_3(S_0, B_s, s) &= \frac{1}{\sigma\sqrt{s}}[\log(S_0/B_s) + (r - \delta + \tfrac{1}{2}\sigma^2)s] \\ d_4(S_0, B_s, s) &= d_3(S_0, B_s, s) - \sigma\sqrt{s}. \end{aligned}$$

This representation can be used to solve for the optimal exercise boundary (see, e.g., Kim 1990). Numerical results using equation (2.9) are given in Kim (1994) and Huang, Subrahmanyam, & Yu (1995).

2.1 Evaluation criteria for numerical methods

Numerical solution procedures can be compared on many dimensions. Important factors to consider when evaluating and choosing a solution algorithm include:

- Numerical accuracy
- Computation speed
- Error bounds or error estimates
- Algorithm complexity
- Flexibility
- Availability of price derivatives (the 'Greeks')
- Memory/storage requirements

Accuracy and speed are often the most important of these factors. The accuracy of a method can be measured in many ways, including average or worst-case error measures. Speed requirements vary depending on the intended application. Are answers required in real-time? How many securities need to be priced? Do implied parameters (e.g., implied volatility) need to be computed? For example, algorithms used to generate daily risk reports may have less stringent speed requirements than those used in a real-time trading support system.

Many other factors are important in the design and implementation of numerical algorithms for security pricing. Since numerical methods generate only approximate answers, error estimates or exact error bounds are highly desirable. Although algorithm implementation seems like a one-time cost, in many real applications the solution procedures are continually modified and updated, e.g., to incorporate algorithm enhancements or to extend the algorithm to price new securities. For this reason, simple and straightforward algorithms are highly preferred to more complicated, difficult to implement methods. Similarly, flexible algorithms, i.e., those which are easily adapted to new securities, are desirable. In the options context, the 'Greeks' are often as important to compute as the prices themselves. Hence, those algorithms which generate price derivatives as a by-product of the pricing calculation are desirable. Finally, computer memory and disk storage requirements can be important considerations in choosing an algorithm. (One reasonable, though not very elegant, approach to American option pricing is to precompute a large table of suitably parameterized option prices. Then the pricing procedure involves only table lookup and interpolation.)

We begin our analysis by giving a brief description of lattice methods and the approximation procedures proposed in Broadie & Detemple (1996). We then present performance results for several methods which quantify the speed–accuracy tradeoff.

2.2 Lattice methods

The idea of binomial (and other lattice) methods is to discretize the risk-neutral process specified in equation (2.1) and then to use dynamic programming to solve for the option price. A three-step tree is illustrated in Figure 1.

In the Cox, Ross, & Rubinstein (1979) binomial method, the stock price parameters are set to $u = e^{\sigma\sqrt{\Delta t}}$, $d = 1/u$, where $\Delta t = T/n$, and n is the number of time steps between time 0 and T. The probability of an upmove is set to $p = (e^{r\Delta t} - d)/(u - d)$. With these choices, the binomial process converges to the geometric Brownian motion model as $n \to \infty$. The choice of $ud = 1$ is not only convenient, but it reduces the number of numerical computations required. Other binomial variants use slightly different values for these parameters.

The dynamic programming routine is initialized by setting the call option price to $C_T(S_T) = (S_T - K)^+$ at each of the terminal nodes. For example, at the top-right node in Figure 1, $C_T(u^3S)$ is set to $(u^3S - K)^+$. At the previous

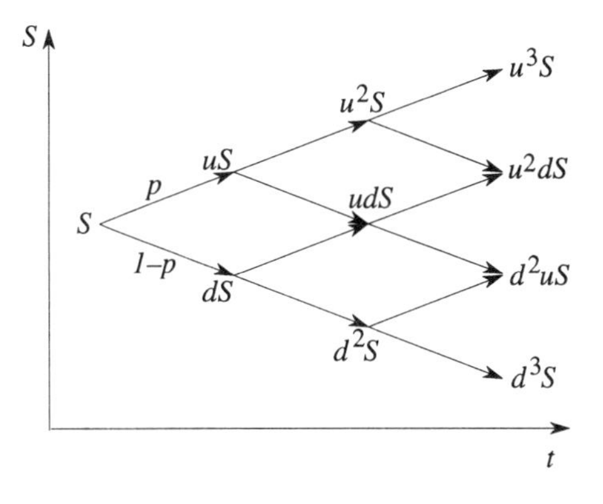

Figure 1: Binomial tree illustration for $n = 3$

node corresponding to stock price u^2S at time $T - \Delta t$, the call option value $C_{T-\Delta t}(u^2S)$ is set to

$$\max\{(u^2S - K)^+, e^{-r\Delta t}(pC_T(u^3S) + (1-p)C_T(u^2dS))\}. \tag{2.10}$$

That is, the American call value is the maximum of the immediate exercise value and the present value of continuing. The call values at the remaining nodes are determined in a similar recursive fashion.

Figure 2 shows the binomial price as a function of the number of time steps.[6] The well-known 'oscillatory convergence' of the binomial method is evident in the figure. This has led many practitioners to use a variation of the binomial method where the n- and $(n+1)$-step binomial prices are averaged. We term this the 'binomial average' method.

Broadie & Detemple (1996) suggest two modifications to the binomial method. In the first modification, the Black–Scholes formula replaces the usual 'continuation value' at the time step just before option maturity. This method is termed BBS (for *b*inomial with a *B*lack-*S*choles modification). Figure 3 shows the BBS price as a function of the number of time steps. Notice that the error is substantially reduced for the same number of time steps and the convergence to the true value is smoother. The smoother convergence suggests that Richardson extrapolation may be useful. The second modification adds Richardson extrapolation to the BBS method, and we refer to it as the BBSR method. In particular, the BBSR method with n steps computes the BBS prices corresponding to $m = n/2$ steps (say C_m) and n steps (say C_n) and then sets the BBSR price to $C = 2C_n - C_m$.

[6]The parameters for this American call option are $S = 105$, $K = 100$, $r = 0.05$, $\delta = 0.02$, $\sigma = 0.30$, and $T = 0.2$. The true value of this option is 8.679.

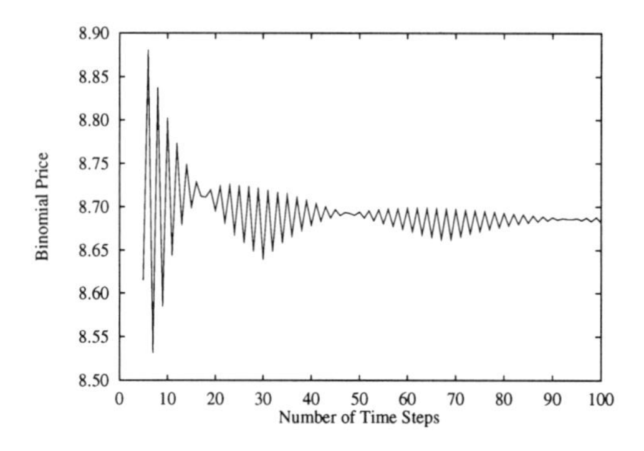

Figure 2: Binomial price versus number of time steps

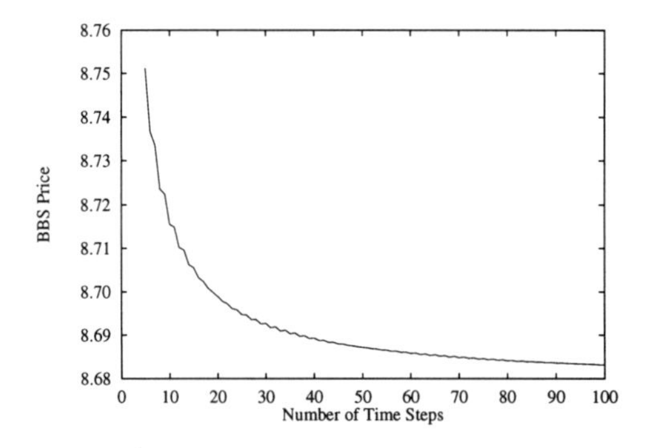

Figure 3: BBS price versus number of time steps

2.3 LBA and LUBA methods

Broadie & Detemple (1996) propose two approximation methods based on lower and upper bounds for the American option price. The lower bound is based on easily computable 'capped call' option values.[7] Then capped call option values are used in a different way to generate an approximation to the optimal exercise boundary. Unlike other pricing procedures, this approximate boundary (which is shown to lie uniformly below the optimal boundary) can be computed without recursion. An upper bound is then derived by substituting this approximate boundary in the integral equation (2.9).

The payoff of a capped call option with cap L is

$$(\min(S_t, L) - K)^+$$

if it is exercised at time $t \le T$. Under the policy 'exercise at the cap', the current value of the capped option, denoted $C_0(S_0, L)$, can be written explicitly (in terms of univariate cumulative normals). Since the 'exercise at the cap' policy is a feasible but suboptimal strategy for the American option, $C_0(S_0, L)$ provides a simple lower bound on the American option price $C(S_0)$.[8] A good lower bound is given by solving the univariate optimization problem:

$$\max_{L \ge S_0} C_0(S_0, L).$$

The lower bound approximation, LBA, is given by multiplying the lower bound by a weight $\lambda \ge 1$.

The optimal exercise boundary can be approximated by the following procedure. Define the derivative of the capped call option value with respect to the constant cap L, evaluated as S_t approaches L from below:

$$D(L, t) \equiv \lim_{S_t \uparrow L} \frac{\partial C_t(S_t, L)}{\partial L}.$$

An explicit formula for $D(L, t)$ is available. Define L_t^* to be the solution to

$$D(L, t) = 0.$$

Note that this equation does *not have to be solved recursively* and it can be solved *very fast* for any given t. The function L_t^* lies below the optimal exercise boundary B_t for all $t \in [0, T]$. Using L_t^* in place of B in equation (2.9) leads to an upper bound for the American option value. LUBA, the lower and upper bound approximation, is a convex combination of these lower and upper bounds. Details are given in Broadie & Detemple (1996).

[7] See Broadie & Detemple (1995a) for a discussion of capped call options.

[8] Similar ideas were independently proposed in Omberg (1987) and Bjerksund & Stensland (1992). We thank D. Lamberton for pointing out the latter reference to us.

2.4 Performance results

To compare the performance of different methods, we follow the procedure in Broadie & Detemple (1996). We first choose a large test set of options by randomly selecting parameters from a pre-determined distribution which is of practical interest.[9] Then for each method we price the test set of options and compute speed and error measures. Speed is measured by the number of option prices computed per second.[10] Two error measures are computed. First, root-mean-squared (RMS) relative error is defined by:

$$\text{RMS} = \sqrt{\frac{1}{m}\sum_{i=1}^{m} e_i^2}$$

where $e_i = |\hat{C}_i - C_i|/C_i$ is the absolute relative error, C_i is the 'true' American option value (estimated by a 15,000-step binomial tree), $\hat{C}_i$ is the estimated option value, and the index i refers to the ith option in the test set. To make relative error meaningful, the summation is taken over options in the dataset satisfying $C_i \geq 0.50$. Out of a sample of 5,000 options, $m = 4{,}592$ satisfied this criterion. Second, the 'maximum' relative error is defined to be observation e_i such that 99.5% of the sample observations are below e_i. We do not take the largest observation, because estimating the maximum of a distribution is very difficult.[11]

We test the binomial method with the original Cox, Ross, & Rubinstein (1979) parameters (Binom CRR) and with the parameters suggested in Hull & White (1988, footnote 4) modified to account for dividends (Binom HW). We also test the 'binomial average' method, the BBSR method, and the LBA and LUBA methods. The speed versus RMS-error results are shown in Figure 4.[12] The binomial CRR and HW methods perform almost identically. For 200 time steps, their RMS-error is about 0.1%, or about one cent on a \$10 option. This confirms the result in the folklore that using 200 binomial time steps produces 'penny accuracy.' The binomial average method performs insignificantly better than the standard binomial method. Apparently, the gain in accuracy is just about offset by the doubling of the work to compute prices at n and $n+1$ time steps. The BBSR method performs significantly

[9] The distribution of parameters for the test is: σ is distributed uniformly between 0.1 and 0.6; T is, with probability 0.75, uniform between 0.1 and 1.0 years and, with probability 0.25, uniform between 1.0 and 5.0 years; $K = 100$, S_0 is uniform between 70 and 130; δ is uniform between 0.0 and 0.10; r is, with probability 0.8, uniform between 0.0 and 0.10 and, with probability 0.2, equal to 0.0. Finally, each parameter is selected independently of the others. Note that relative errors do not change if S_0 and K are scaled by the same factor, i.e., only the ratio S_0/K is of interest.

[10] The computations were done on a PC with a 133-MHz Pentium processor.

[11] We found that the sample maximum varies so widely within subsamples as to be an unreliable tool for comparing various methods. Results using the 99.5 percentile of the observations seem to be much less sensitive to the random test set used.

[12] Numbers next to each method indicate the number of time steps.

better than the other binomial methods in this speed-error tradeoff. Better still are the LBA and LUBA methods. The LUBA method has an RMS-error of about 0.02% (less than a 1000-step binomial tree) and a speed of over 1000 options per second (faster than a 50-step binomial tree).

The computational effort (work) with the standard binomial method increases as $O(n^2)$. Figure 4 shows the interesting result that RMS-error decreases approximately *linearly* with the number of time steps. Hence, the binomial error decreases as $O(1/\sqrt{\text{work}})$.[13] Leisen & Reimer (1995) show analytically that the binomial method has order one (i.e., linear) convergence for European options.[14] They also suggest an interesting modification of the binomial method which appears to have order two convergence for European options and order one convergence (with a smaller constant) for American options.

Figure 5 shows the tradeoff between computation speed and the maximum error (recall that the 'maximum' error is defined as the 99.5 percentile of the ordered absolute relative errors). The ranking of the methods is the same, however, the maximum error is approximately five times larger than the RMS-error for each method.

Comparative results of several other methods are given in Broadie & Detemple (1996). Of the other methods tested, only the method of lines of Carr & Faguet (1995) has an RMS-error of 0.1% or less. AitSahlia (1995) and AitSahlia & Lai (1996) describe a pricing method for American options which uses a continuity correction technique for estimating the optimal exercise boundary. Their method also appears to be very promising.[15] Recent methods represent orders of magnitude improvement over earlier approaches in terms of speed and/or accuracy. The BBSR method is a simple modification of the binomial method which is simple to program and performs very well. The LUBA method is the only method tested which also provides upper and lower bounds. The binomial method is very easy to program and the algorithm can easily be adapted to many alternative contract specifications. All of the methods tested can generate prices as well as price derivatives. Finally, the storage requirements of the tested methods are minimal.

The determination of a closed-form solution for the optimal stopping boundary and the corresponding American option price remains an open question. However, we conclude from these recent results that from a numerical viewpoint, the single asset American option pricing problem in the standard model is essentially solved. Many challenges remain for the pricing of path-dependent options, multi-asset options, interest-rate sensitive securities, and

[13] This is also the convergence rate typically associated with simulation methods!

[14] They also show that the same order of convergence holds for the parameters used in the Cox, Ross, & Rubinstein (1979), Jarrow & Rudd (1983), and Tian (1993) binomial variants.

[15] It was not tested because it has not yet been extended to handle dividends.

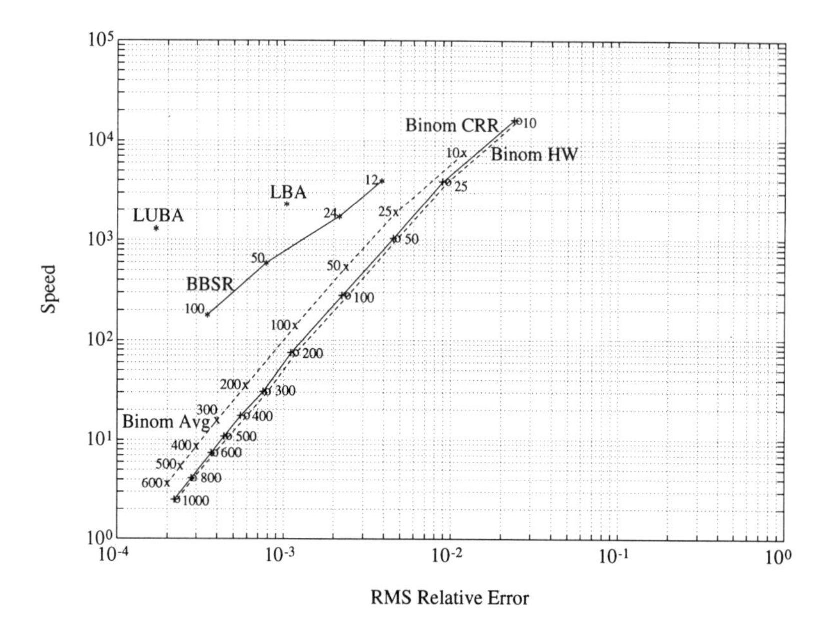

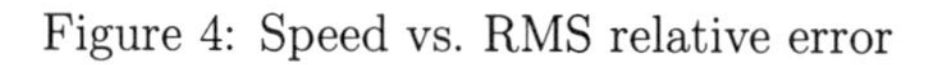
Figure 4: Speed vs. RMS relative error

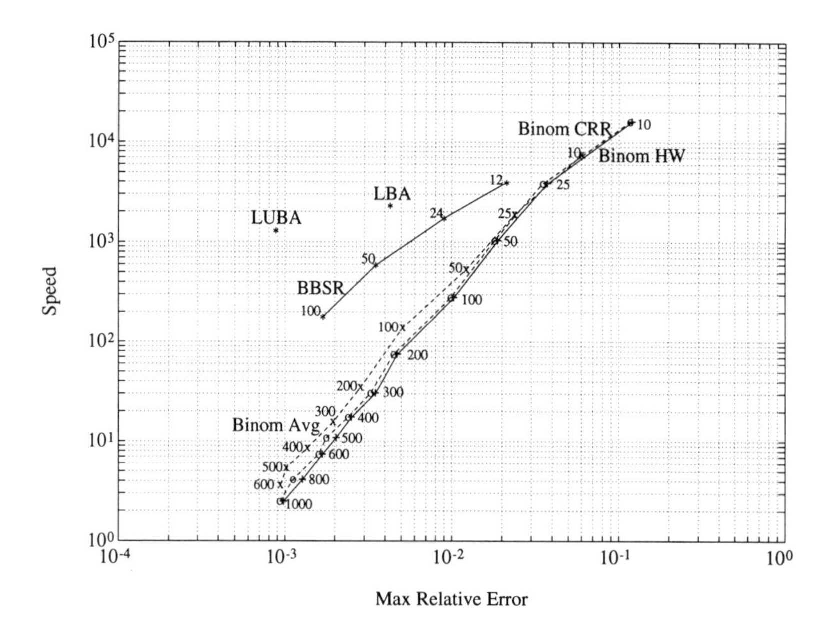

Figure 5: Speed vs. 'Maximum' relative error

options in more general models (e.g., non-constant volatility). Some of these issues are explored briefly in the next two sections.

3 Barrier and Lookback Options

Capped call options are one example of barrier options – options whose payoff depends on the value of the underlying asset relative to a barrier level. Knock-in options are another example. These options have a zero payoff, unless the underlying asset price crosses a pre-determined barrier which makes the option come 'alive.' Barrier options are treated in Rubinstein & Reiner (1991) and Rich (1994). For an overview of these and other types of exotic options, see Jarrow (1995) and Nelken (1995).

Cox & Rubinstein (1985) describe a straightforward modification of the binomial method for pricing certain barrier options. Broadie & Detemple (1993) and Boyle & Lau (1994) first pointed out the slow convergence of the binomial method for pricing barrier options. For a comparable number of time steps, the binomial pricing error for barrier options can be two orders of magnitude larger than for standard options.

Boyle & Lau (1994) identify the cause of the problem and suggest a remarkably simple and effective solution. As the number of time steps in the binomial method changes, the placement of the barrier relative to the layers of nodes of the tree changes. They recommend choosing the number of time steps n so that there is a layer of nodes at or just beyond the barrier. These 'good values' of n can easily be determined in advance of the pricing computation. Their results show that these choices for n restore the original error properties of the binomial method.

Numerical pricing of barrier options is also studied in Derman, Kani, Ergener & Bardhan (1995), and in Ritchken (1995). Derman *et al.* (1995) suggest an interpolation scheme for improving the pricing error of lattice methods applied to barrier options. This approach is especially useful when the volatility of the underlying asset is not constant. Ritchken (1995) suggests using a trinomial procedure, where the trinomial 'stretch' parameter is chosen so that the barrier coincides with a layer of nodes.

The payoff of a lookback call option is $(S_T - \min_{0\leq t\leq T} S_t)$ and a lookback put is $(\max_{0\leq t\leq T} S_t - S_T)$. Analytical solutions have been given for European versions of these options in the standard model (see, e.g., Goldman, Sosin, & Gatto (1979) and Conze & Viswanathan (1991)). Numerical techniques must be used for American lookbacks, to handle discrete dividends, when volatility is not constant, or for other variations of the standard model. The standard binomial approach does not apply to the case of lookbacks because of the path-dependent payoff.

Babbs (1992) and Cheuk & Vorst (1994) suggest a clever change of numeraire so that a version of the binomial method is again applicable. Hull

& White (1993) resolve the path dependency by the standard technique of adding an additional state variable. This adds an extra dimension to the binomial method, which considerably increases its computation time. The resulting method, however, is very flexible. Kat (1995) offers a summary and comparison of these approaches.

For many path-dependent option contracts, the payoff does not depend on the continuous price path, but rather it depends on the price of the underlying asset at discrete points in time. For barrier options, it is often the case that the barrier-crossing event can only be triggered at specific dates or times. For lookback options, the maximum or minimum price might be determined at daily closings, for example. The implications of ignoring the difference between continuous and discrete monitoring is discussed in Flesaker (1992), Chance (1994), and Kat & Verdonk (1995). Numerical methods and analytical approximations for discrete path-dependent options are given in Broadie, Glasserman, & Kou (1995, 1996) and Levy & Mantion (1995).

4 Methods for Multiple State Variables

Options on multiple assets ('rainbow options') are being traded with increasing frequency. For example, in 1994 the New York Mercantile Exchange began trading options on *crack spreads* (e.g., the difference between unleaded gasoline and crude oil futures prices, or the difference between heating oil and crude oil futures prices). Other examples include options on the maximum of two or more asset prices, dual-strike options, and portfolio or basket options.[16]

In the multi-asset context, the standard model is a straightforward generalization of (2.1):

$$dS_t^i = S_t^i[(\mu_i - \delta_i)dt + \sigma_i dW_t^i], \tag{4.1}$$

where S_t^i is the price of asset i at time t and where the W^i are standard Brownian motion processes ($i = 1, \ldots, n$) and the correlation between W^i and W^j is ρ_{ij}. With a constant rate of interest r, the risk-neutral form of (4.1) is given by replacing each μ_i by r.

Multinomial approaches to pricing options with two or more state variables are given in Boyle (1988), Boyle, Evnine, & Gibbs (1989), Madan, Milne, & Shefrin (1989), Cheyette (1990), He (1990), Kamrad & Ritchken (1991), and Rubinstein (1994). The basic idea of the multinomial approaches is the same as in the single asset case, namely, to discretize the risk-neutral process specified in equation (4.1) and then to use dynamic programming to solve for the option price. A tree with four branches per node in the two-asset case is illustrated in Figure 6.

[16]Closed-form solutions for some European multi-asset options are given in Boyle (1993). Properties of American option prices and optimal exercise boundaries are investigated in

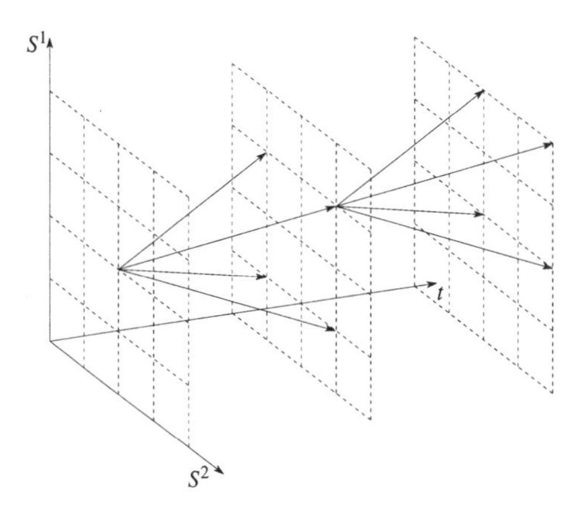

Figure 6: Evolution of a two-dimensional binomial tree (4-branch method)

Boyle, Evnine, & Gibbs (1989), hereafter BEG, proposed a general lattice method to price contingent claims on k assets. The BEG method has four branches per node in the two-asset case, and 2^k branches per node in the k-asset case. For the two-asset case, the node (S^1, S^2) is connected to $(u_1 S^1, u_2 S^2)$ with probability p_{uu}, to $(u_1 S^1, d_2 S^2)$ with probability p_{ud}, to $(d_1 S^1, u_2 S^2)$ with probability p_{du}, and to $(d_1 S^1, d_2 S^2)$ with probability p_{dd}. As in the single-asset case, $u_i = e^{\sigma_i \sqrt{\Delta t}}$ and $d_i = 1/u_i$ for $i = 1, 2$. The transition probabilities are defined by

$$
\begin{aligned}
p_{uu} &= \frac{1}{4}\left(1 + \rho + \sqrt{\Delta t}\left(\frac{\nu_1}{\sigma_1} + \frac{\nu_2}{\sigma_2}\right)\right) \\
p_{ud} &= \frac{1}{4}\left(1 - \rho + \sqrt{\Delta t}\left(\frac{\nu_1}{\sigma_1} - \frac{\nu_2}{\sigma_2}\right)\right) \\
p_{du} &= \frac{1}{4}\left(1 - \rho + \sqrt{\Delta t}\left(-\frac{\nu_1}{\sigma_1} + \frac{\nu_2}{\sigma_2}\right)\right) \\
p_{dd} &= \frac{1}{4}\left(1 + \rho - \sqrt{\Delta t}\left(\frac{\nu_1}{\sigma_1} + \frac{\nu_2}{\sigma_2}\right)\right),
\end{aligned}
$$

where $\nu_i = r - \delta_i - \frac{1}{2}\sigma_i^2$, for $i = 1, 2$, and $\rho = \rho_{12}$. For the test results which follow, we refer to this BEG approach as the '4-Branch' method.

Boyle (1988) proposed a lattice method in the two-asset case which has five branches per node, where the additional branch represents a horizontal move, i.e., a transition from (S^1, S^2) to the same node (S^1, S^2) one time-period later. Kamrad & Ritchken (1991) proposed a general lattice method for k assets. In the case of two assets, their method has five branches per node. Like the trinomial method in the single asset case, their method has

Broadie & Detemple (1997) in the multi-asset context.

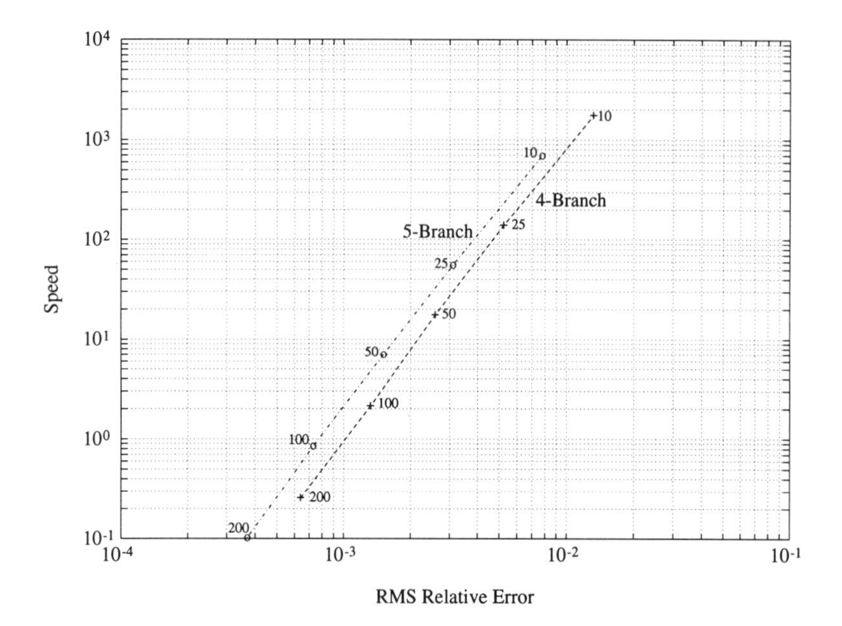

Figure 7: Speed vs. RMS relative error for European call options on the maximum of two assets

an additional 'stretch' parameter, denoted λ. When $\lambda = 1$, the Kamrad and Ritchken method reduces to the BEG method. In the two asset case, Kamrad and Ritchken recommend using $\lambda = 1.11803$, and for the test results which follow, we refer to this as the '5-Branch' method.

To compare the performance of the 4-Branch and the 5-Branch methods, we price European max-options on two assets. The payoff of the max-option is $(\max(S_T^1, S_T^2) - K)^+$. We test the methods in the European case because the true price can be determined by the analytical formula given in Johnson (1981, 1987) and Stulz (1982). We chose a test set of 5,000 options by randomly selecting parameters from a pre-determined distribution.[17] Then for each method we price the test set of options and compute the usual speed and RMS-error measures. The results are shown in Figure 7.[18]

For both methods, Figure 7 shows that the RMS-error decreases approx-

[17]The distribution of parameters for the test is: σ_i is distributed uniformly between 0.1 and 0.6; T is, with probability 0.75, uniform between 0.1 and 1.0 years and, with probability 0.25, uniform between 1.0 and 5.0 years; $K = 100$, S_0^i is uniform between 70 and 130; δ_i is uniform between 0.0 and 0.10; r is, with probability 0.8, uniform between 0.0 and 0.10 and, with probability 0.2, equal to 0.0, ρ follows a triangular distribution between -1 and 1 ($i = 1, 2$, where applicable). Finally, each parameter is selected independently of the others.

[18]Numbers next to each method indicate the number of time steps.

imately linearly with the number of time steps. The RMS-error in the two-asset case is comparable to the single-asset case with the same number of time steps.[19] However, the computational effort with both two-asset methods increases as $O(n^3)$. With current computing technology, these lattice methods are practical for problems of at most three or four dimensions. For higher dimensions, the computation time and the memory/storage requirements become prohibitive.

4.1 Simulation methods

To overcome the 'curse of dimensionality' of current lattice methods, recent work has focused on simulation-based approaches. The convergence rate of Monte Carlo simulation methods is typically independent of the number of state variables, and so this approach should be increasingly attractive as the dimension of the problem grows. The simulation approach was introduced to finance in Boyle (1977). For a recent survey see Boyle, Broadie, & Glasserman (1995).

While the simulation approach has been used extensively to price European-style contingent claims, only recently have there been attempts to extend the method to price American-style claims. The first attempt to price American options using simulation is given in Tilley (1993). This effort created considerable interest by demonstrating the potential practicality of using simulation in this context. More recent developments are given in Barraquand & Martineau (1995) and Broadie & Glasserman (1995).

References

Abramowitz, W. & Stegun, I. (1972) *Handbook of Mathematical Functions*, Dover Publications.

AitSahlia, F. (1995) *Optimal Stopping and Weak Convergence Methods for Some Problems in Financial Economics*, Ph.D. dissertation, Stanford University.

AitSahlia, F. & Lai, T. (1996) 'Approximations for American Options', working paper, Cornell University.

Amin, K. (1991) 'On the Computation of Continuous Time Option Prices using Discrete Approximations', *Journal of Financial and Quantitative Analysis* **26**, 477–496.

Amin, K. & Khanna, A. (1994) 'Convergence of American Option Values from Discrete- to Continuous-Time Financial Models', *Mathematical Finance* **4**, 289–304.

[19]This two-asset test if for the easier European option case, while the single-asset test was for American options.

Amin, K. (1995) 'Option Pricing Trees', *Journal of Derivatives* **2**, 34–46.

Amin, K. & Bodurtha, J. (1995) 'Discrete-Time American Option Valuation with Stochastic Interest Rates', *Review of Financial Studies* **8**, 193–234.

Babbs, S. (1992) 'Binomial Valuation of Lookback Options', working paper, Midland Global Markets, London, England.

Barone-Adesi, G. & Whaley, R. (1987) 'Efficient Analytical Approximation of American Option Values', *Journal of Finance* **42**, 301–320.

Barraquand, J. & Martineau, D. (1995) 'Numerical Valuation of High Dimensional Multivariate American Securities', *Journal of Financial and Quantitative Analysis* **30**, 383–405.

Bensoussan, A. (1984) 'On the Theory of Option Pricing', *Acta Applicandae Mathematicae* **2**, 139–158.

Bjerksund, P. & Stensland, G. (1992) 'Closed Form Approximation of American Option Prices', working paper, Norwegian School of Economics and Business Administration, Bergen, Norway, to appear in the *Scandinavian Journal of Management.*

Black, F. & Scholes, M. (1973) 'The Pricing of Options and Corporate Liabilities', *Journal of Political Economy* **81**, 637–654.

Blomeyer, E. (1986) 'An Analytic Approximation for the American Put Price for Options on Stocks with Dividends', *Journal of Financial and Quantitative Analysis* **21**, 229–233.

Boyle, P. (1977) 'Options: A Monte Carlo Approach', *Journal of Financial Economics* **4**, 323–338.

Boyle, P. (1988) 'A Lattice Framework for Option Pricing with Two State Variables', *Journal of Financial and Quantitative Analysis* **23**, 1–12.

Boyle, P. (1993) 'New Life Forms on the Options Landscape', *Journal of Financial Engineering* **2**, 217–252.

Boyle, P., Evnine, J., & Gibbs, S. (1989) 'Numerical Evaluation of Multivariate Contingent Claims', *Review of Financial Studies* **2**, 241–250.

Boyle, P. & Lau, S. (1994) 'Bumping Up Against the Barrier with the Binomial Method', *Journal of Derivatives* **1**, 6–14.

Boyle, P., Broadie, M., & Glasserman, P. (1995) 'Monte Carlo Methods for Security Pricing', working paper, Columbia University, to appear in the *Journal of Economic Dynamics and Control.*

Breen, R. (1991) 'The Accelerated Binomial Option Pricing Model', *Journal of Financial and Quantitative Analysis* **26**, 153–164.

Brennan, M. & Schwartz, E. (1977) 'The Valuation of American Put Options', *Journal of Finance* **32**, 449–462.

Brennan, M. & Schwartz, E. (1978) 'Finite Difference Methods and Jump Processes Arising in the Pricing of Contingent Claims: A Synthesis', *Journal of Financial and Quantitative Analysis* **13**, 461–474.

Broadie, M. & Detemple, J. (1993) 'The Valuation of American Capped Call Options', working paper, Columbia University.

Broadie, M. & Detemple, J. (1995a) 'American Capped Call Options on Dividend Paying Assets', *Review of Financial Studies* **8**, 161–191.

Broadie, M. & Detemple, J. (1996) 'American Option Valuation: New Bounds, Approximations, and a Comparison of Existing Methods', working paper, Columbia University, to appear in the *Review of Financial Studies.*

Broadie, M. & Detemple, J. (1997) 'The Valuation of American Options on Multiple Assets', working paper, Columbia University, to appear in *Mathematical Finance.*

Broadie, M. & Glasserman, P. (1995) 'Pricing American-Style Securities Using Simulation', working paper, Columbia University, to appear in *Journal of Economic Dynamics and Control.*

Broadie, M., Glasserman, P., & Kou, S. (1995) 'A Continuity Correction for Discrete Barrier Options', working paper, Columbia University, to appear in *Mathematical Finance.*

Broadie, M., Glasserman, P., & Kou, S. (1996) 'Connecting Discrete and Continuous Path-Dependent Options', working paper, Columbia University.

Bunch, D. & Johnson, H. (1992) 'A Simple and Numerically Efficient Valuation Method for American Puts Using a Modified Geske-Johnson Approach', *Journal of Finance* **47**, 809–816.

Carr, P. & Faguet, D. (1995) 'Fast Accurate Valuation of American Options', working paper, Cornell University.

Carr, P., Jarrow, R., & Myneni, R. (1992) 'Alternative Characterizations of American Put Options', *Mathematical Finance* **2**, 87–106.

Chance, D. (1994) 'The Pricing and Hedging of Limited Exercise Caps and Spreads', *Journal of Financial Research* **17**, 561–584.

Chesney, M. & Gibson, R. (1995) 'State Space Symmetry and Two-Factor Option Pricing Models', *Advances in Futures and Options Research* **8**, 85–112.

Cheuk, T. & Vorst, T. (1994) 'Lookback Options and the Observation Frequency', working paper, Erasmus Univerity, Rotterdam, The Netherlands.

Cheyette, O. (1990) 'Pricing Options on Multiple Assets', *Advances in Futures and Options Research* **4**, 69–81.

Conze, A. & Viswanathan, R. (1991) 'Path Dependent Options: The Case of Lookback Options', *Journal of Finance* **46**, 1893–1907.

Cottle, R., Pang, J.-S.,& Stone, R. (1992) *The Linear Complementarity Problem*, Academic Press.

Courtadon, G. (1982) 'A More Accurate Finite Difference Approximation for the Valuation of Options', *Journal of Financial and Quantitative Analysis* **17**, 697–705.

Cox, J. & Ross, S. (1976) 'The Valuation of Options for Alternative Stochastic Processes', *Journal of Financial Economics* **3**, 145–166.

Cox, J., Ross, S., & Rubinstein, M. (1979) 'Option Pricing: A Simplified Approach', *Journal of Financial Economics* **7**, 229–263.

Cox, J. & Rubinstein, M. (1985) *Options Markets*, Prentice-Hall.

Curran, M., (1995) 'Accelerating American Option Pricing in Lattices', *Journal of Derivatives* **3**, 8–17.

Dempster, M. (1994) 'Fast Numerical Valuation of American, Exotic and Complex Options', Department of Mathematics research report, University of Essex, Colchester, England.

Derman, E., Kani, I., Ergener, D., & Bardhan, I. (1995) 'Enhanced Numerical Methods for Options with Barriers', *Financial Analysts Journal* **51**, 65–74.

El Karoui, N. & Karatzas, I. (1991) 'A New Approach to the Skorohod Problem and its Applications', *Stochastics and Stochastics Reports* **34**, 57–82.

Flesaker, B. (1992) 'The Design and Valuation of Capped Stock Index Options', working paper, Department of Finance, University of Illinois, Champaign, IL.

Geske, R. (1979) 'A Note on an Analytical Valuation Formula for Unprotected American Options on Stocks with Known Dividends', *Journal of Financial Economics* **7**, 375–380.

Geske, R. & Johnson, H. (1984) 'The American Put Options Valued Analytically', *Journal of Finance* **39**, 1511–1524.

Geske, R. & Shastri, K. (1985) 'Valuation by Approximation: A Comparison of Alternative Option Valuation Techniques', *Journal of Financial and Quantitative Analysis* **20**, 45–71.

Goldman, M., Sosin, H., & Gatto, M. (1979) 'Path Dependent Options: Buy at the Low, Sell at the High', *Journal of Finance* **34**, 1111–1127.

Harrison, M. & Kreps, D. (1979) 'Martingales and Arbitrage in Multiperiod Security Markets', *Journal of Economic Theory* **20**, 381–408.

Harrison, M. & Pliska, S. (1981) 'Martingales and Stochastic Integrals in the Theory of Continuous Trading' *Stochastic Processes and their Applications* **11**, 215–260.

He, H. (1990) 'Convergence from Discrete- to Continuous-Time Contingent Claim Prices', *Review of Financial Studies* **3**, 523–546.

Ho, T.S., Stapleton, R., & Subrahmanyam, M. (1994) 'A Simple Technique for the Valuation and Hedging of American Options', *Journal of Derivatives* **2**, 52–66.

Huang, J., Subrahmanyam, M., & Yu, G. (1995) 'Pricing and Hedging American Options: A Recursive Integration Method', working paper, New York University, *Review of Financial Studies* **9**, 277-300.

Hull, J. (1993) *Options, Futures, and Other Derivative Securities*, 2nd edition, Prentice-Hall.

Hull, J. & White, A. (1988) 'The Use of the Control Variate Technique in Option Pricing', *Journal of Financial and Quantitative Analysis* **23**, 237–251.

Hull, J. & White, A. (1990) 'Valuing Derivative Securities Using the Explicit Finite Difference Method', *Journal of Financial and Quantitative Analysis* **25**, 87–100.

Hull, J. & White, A. (1993) 'Efficient Procedures for Valuing European and American Path-Dependent Options', *Journal of Derivatives* **1**, 21–32.

Hull, J. & White, A. (1994a) 'Numerical Procedures for Implementing Term Structure Models I: Single-Factor Models', *Journal of Derivatives* **2**, 7–16.

Hull, J. & White, A. (1994b) 'Numerical Procedures for Implementing Term Structure Models II: Two-Factor Models', *Journal of Derivatives* **2**, 37–48.

Jacka, S. (1991) 'Optimal Stopping and the American Put', *Mathematical Finance* **1**, 1–14.

Jaillet, P., Lamberton, D., & Lapeyre, B. (1990) 'Variational Inequalities and the Pricing of American Options', *Acta Applicandae Mathematicae* **21**, 263–289.

Jarrow, R., ed. (1995) *Over the Rainbow*, Risk Publications.

Jarrow, R. & Rudd, A. (1983) *Option Pricing*, Dow Jones-Irwin.

Jarrow, R. & Turnbull, S. (1996) *Derivative Securities*, South-Western Publishing.

Johnson, H. (1981) 'The Pricing of Complex Options', working paper, Louisiana State University.

Johnson, H. (1983) 'An Analytic Approximation for the American Put Price', *Journal of Financial and Quantitative Analysis* **18**, 141–148.

Johnson, H. (1987) 'Options on the Maximum or the Minimum of Several Assets', *Journal of Financial and Quantitative Analysis* **22**, 227–283.

Kamrad, B. & Ritchken, P. (1991) 'Multinomial Approximating Models for Options with k State Variables', *Management Science* **37**, 1640–1652.

Karatzas, I. (1988) 'On the Pricing of American Options', *Applied Mathematics and Optimization* **17**, 37–60.

Kat, H. (1995) 'Pricing Lookback Options Using Binomial Trees: An Evaluation', *Journal of Financial Engineering* **4**, 375–398.

Kat, H. & Verdonk, L. (1995) 'Tree Surgery', *RISK* **8**, 53–56.

Kim, I. (1990) 'The Analytic Valuation of American Options', *Review of Financial Studies* **3**, 547–572.

Kim, I. (1994) 'Analytical Approximation of the Optimal Exercise Boundaries for American Futures Option', *Journal of Futures Markets* **14**, 1–24.

Kim, I. & Byun, S. (1994) 'Optimal Exercise Boundary in a Binomial Option Pricing Model', *Journal of Financial Engineering* **3**, 137–158.

Lamberton, D. (1993) 'Convergence of the Critical Price in the Approximation of American Options', *Mathematical Finance* **3**, 179–190.

Lamberton, D. (1995) 'Error Estimates for the Binomial Approximation of American Put Options', working paper, Université de Marne-la-Vallée, Noisy-le-Grand, France.

Leisen, D. & Reimer, M. (1995) 'Binomial Models for Option Valuation — Examining and Improving Convergence', Working Paper B–309, University of Bonn.

Levy, E. & Mantion, F. (1995) 'Approximate Valuation of Discrete Lookback Options', HSBC Midland, London, England.

Li, A., Ritchken, P., & Sankarasubramanian, L. (1995) 'Lattice Models for Pricing American Interest Rate Claims', *Journal of Finance* **50**, 719–737.

MacMillan, L.W. (1986) 'An Analytic Approximation for the American Put Price', *Advances in Futures and Options Research* **1**, 119–139.

Madan, D., Milne, F., & Shefrin, H. (1989) 'The Multinomial Option Pricing Model and its Brownian and Poisson Limits', *Review of Financial Studies* **4**, 251–266.

Marchuk, G. & Shaidurov, V. (1983) *Difference Methods and Their Extrapolations*, Springer Verlag.

McDonald, R. & Schroder, M. (1990) 'A Parity Result for American Options', working paper, Northwestern University.

McKean, Jr., H. (1965) 'Appendix: A Free Boundary Problem for the Heat Equation Arising from a Problem in Mathematical Economics', *Industrial Management Review* **6**, 32–39.

Merton, R. (1973) 'Theory of Rational Option Pricing', *Bell Journal of Economics and Management Science* **4**, 141–183.

Moro, B. (1995) 'Fast Computation of Cumulative Normal Distribution Function', working paper, TMG Financial Products, Greenwich, Connecticut.

Nelken, I., ed. (1995) *The Handbook of Exotic Options*, Dow Jones-Irwin.

Nelson, D. & Ramaswamy, K. (1990) 'Simple Binomial Processes as Diffusion Approximations in Financial Models', *Review of Financial Studies* **3**, 393–430.

Omberg, E. (1987) 'The Valuation of American Puts with Exponential Exercise Policies', *Advances in Futures and Options Research* **2**, 117–142.

Omberg, E. (1988) 'Efficient Discrete Time Jump Process Models in Option Pricing', *Journal of Financial and Quantitative Analysis* **23**, 161–174.

Parkinson, M. (1977) 'Option Pricing: The American Put', *Journal of Business* **50**, 21–36.

Pelsser, A. & Vorst, T. (1994) 'The Binomial Model and the Greeks', *Journal of Derivatives* **1**, 45–49.

Press, W., Teukolsky, S., Vetterling, W., & Flannery, B. (1992) *Numerical Recipes in C: The Art of Scientific Computing*, 2nd edition, Cambridge University Press.

Rendleman, R. & Bartter, B. (1979) 'Two-State Option Pricing', *Journal of Finance* **34**, 1093–1110.

Rich, D. (1994) 'The Mathematical Foundations of Barrier Option Pricing Theory', *Advances in Futures and Options Research* **7**, 267–312.

Ritchken, P. (1995) 'On Pricing Barrier Options', *Journal of Derivatives* **3**, 19–28.

Roll, R. (1977) 'An Analytic Valuation Formula for Unprotected American Call Options on Stocks with Known Dividends', *Journal of Financial Economics* **5**, 251–258.

Rubinstein, M. (1994) 'Return to Oz', *RISK* **7**, 67–71.

Rubinstein, M. & Reiner, E. (1991) 'Breaking Down the Barriers', *RISK* **4**, 28–35.

Schwartz, E. (1977) 'The Valuation of Warrants: Implementing a New Approach', *Journal of Financial Economics* **4**, 79–93.

Stoll, H. & Whaley, R. (1993) *Futures and Options: Theory and Applications*, South-Western Publishing.

Stulz, R. (1982) 'Options on the Minimum or the Maximum of Two Risky Assets', *Journal of Financial Economics* **10**, 161–185.

Tian, Y. (1992) 'A Simplified Binomial Approach to the Pricing of Interest-Rate Contingent Claims', *Journal of Financial Engineering* **1**, 14–37.

Tian, Y. (1993) 'A Modified Lattice Approach to Option Pricing', *Journal of Futures Markets* **13**, 563–577.

Tian, Y. (1994) 'A Reexamination of Lattice Procedures for Interest Rate-Contingent Claims', *Advances in Futures and Options Research* **7**, 87–111.

Tilley, J. (1993) 'Valuing American Options in a Path Simulation Model', *Transactions of the Society of Actuaries* **45**, 83–104.

Trigeorgis, L. (1991) 'A Log-Transformed Binomial Numerical Analysis Method for Valuing Complex Multi-Option Investments', *Journal of Financial and Quantitative Analysis* **26**, 309–326.

Whaley, R.E. (1981) 'On the Valuation of American Call Options on Stocks with Known Dividends', *Journal of Financial Economics* **9**, 207–211.

Wilmott, P., Dewynne, J., & Howison, S. (1993) *Option Pricing: Mathematical Models and Computation*, Oxford Financial Press.

Zhang, X. (1997) 'Valuation of American Options in a Jump-Diffusion Model', this volume.

American Options: A Comparison of Numerical Methods

F. AitSahlia and P. Carr

1 Introduction

The overwhelming majority of traded options are of American type. Yet their valuation, even in the standard case of a lognormal process for the underlying asset, remains a topic of active research. This situation stems from the nature of the solution which requires the determination of the optimal exercise strategy as well as the value of the option. In contrast the European option, which can only be exercised at its expiration date, has been valued by the celebrated Black–Scholes formula (Black & Scholes 1973) for the standard financial model.

Due to a lack of closed–form solutions to American option valuation problems, a vast array of approximation schemes has been advanced. The Broadie & Detemple article in this volume provides a summary of some experimental results. The present article is a detailed account of comparative experiments conducted with numerical schemes including the recent method of Carr & Faguet 1996. It is organized as follows: Section 2 reviews the basic Black–Scholes model, Section 3 presents the approximation approaches and Section 4 concludes with some benchmark comparisons.

2 The Standard Model

The prototypical definition of an American option is that of a contract giving its holder the right to buy (call option) or sell (put option) one unit of an underlying security (e.g. stock) at a pre–arranged price K. This right can be exercised at any time before an expiration date T. In contrast, a European option can be exercised at the expiration date only.

In the standard model, also called the Black–Scholes/Merton environment (Black & Scholes 1973, Merton 1973), the market consists of the option, its underlying security, labelled the stock, and a riskless security, labelled the bond. This market is populated by equally informed traders who do not incur transaction cost, among other simplifying assumptions. At any time t an amount β_t in the bond will evolve according to the differential equation

$$d\beta_t = r\beta_t dt,$$

where r is the riskless rate of lending and borrowing. The randomness of the continuous stock price process $\{P_t\}$ is modelled as the geometric Brownian motion:

$$P_t = P_0 e^{(\alpha-\delta-\frac{\sigma^2}{2})t+\sigma\tilde{W}_t} \tag{2.1}$$

where P_0 is the initial stock price, α is the mean rate of the stock return over an infinitesimal interval, σ is the associated standard deviation (volatility) and δ is the dividend rate paid by the stock. Here $\{\tilde{W}\}$ is a standard Brownian motion defined on a filtered probability space $(\Omega, \mathcal{F}, \{\mathcal{F}_t\}, P)$.

For this model there exists a probability measure Q_T equivalent to P such that P_t is a martingale in $(\Omega, \mathcal{F}, \{\mathcal{F}_t\}, Q_T)$. By Girsanov's theorem

$$W_t = \frac{\alpha - r}{\sigma} t + \tilde{W}_t$$

is a standard Brownian motion in $(\Omega, \mathcal{F}, \{\mathcal{F}_t\}, Q_T)$. Thus

$$P_t = P_0 e^{(r-\delta-\frac{\sigma^2}{2})t+\sigma W_t}$$

and

$$d\{e^{-rt}P_t\} = e^{-rt}P_t\left(\sigma dW_t - \delta dt\right),$$

indicating that, in the absence of dividend, the discounted stock price $\{e^{-rt}P_t\}$ is a martingale on $(\Omega, \mathcal{F}, \{\mathcal{F}_t\}, Q_T)$. For this property Q_T is sometimes called an 'equivalent martingale measure'. Note that if the dividend rate is non-negative then the discounted price process is a supermartingale under Q_T. For further details we refer the reader to Bensoussan (1984), Duffie (1988), Karatzas (1988) and Myneni (1992).

The characterization of American option valuation as an optimal stopping problem goes back at least to McKean (1965) who based his work on Samuelson's (1965) pricing model. However, it was not until Bensoussan (1984) and Karatzas (1988) that an arbitrage argument was provided. They show that the option price $U_t \equiv U(t, P_t)$ at time $t \in [0, T]$ is

$$U_t = ess\ sup_{\tau \in \mathcal{T}_{t,T}} E_{Q_T}(e^{-r(\tau-t)} f(P_\tau) | \mathcal{F}_t)$$

where $\mathcal{T}_{t,T}$ is the set of all stopping times in $[t, T]$, and $f(x) = (K - x)^+$ ($f(x) = (x - K)^+$) for a put (respectively a call) with exercise price K.

Under this optimal stopping problem formulation there exists, for the put, a function $t \longmapsto \overline{P}(t)$ in $C[0, T)$, non-decreasing, and independent of the initial stock price, such that

$$\text{if } P_t \leq \overline{P}(t) \text{ then } U(t, P_t) = (K - P_t)^+,$$
$$\text{if } P_t > \overline{P}(t) \text{ then } U(t, P_t) > (K - P_t)^+.$$

Symmetrically, for a call there corresponds a function $t \longmapsto \underline{P}(t)$ in $\mathcal{C}[0,T)$, non-increasing and independent of the initial stock price, such that

$$\text{if } P_t \leq \underline{P}(t) \text{ then } U(t,P_t) > (P_t - K)^+,$$
$$\text{if } P_t > \underline{P}(t) \text{ then } U(t,P_t) = (P_t - K)^+.$$

We refer to Friedman (1975), Van Moerbeke (1976), Jacka (1991) for the supporting arguments for the above statements.

Given that the stochastic dynamics are governed by a Brownian motion the determination of the critical (or exercise) price $\overline{P}(t)$ or $\underline{P}(t)$ rests on solving a free–boundary problem for the heat equation.

3 Approximation Methods

There are no known closed–form solutions to the above optimal stopping problems except in the infinite horizon case (McKean 1965). Many authors have therefore followed the approximate solution path. Recent efforts have indicated directions in which some closed–form expressions can be obtained for the optimal exercise boundary (AitSahlia 1995, AitSahlia & Lai 1996) or the value function at discrete dates (Carr & Faguet 1996).

Carr & Faguet (1996) present an approach that is comparable to the extant methods. They investigate the analytic method of lines as a way to evaluate American options. In common with the numerical method of lines (Meyer 1979, Meyer & Van der Hoek 1994), this approach discretizes the time derivative in the Black–Scholes partial differential equation. In contrast with the numerical method of lines, the authors then solve the resulting series of ordinary differential equations analytically. This analytic solution is protected against any truncation and spatial discretization error implicit in numerical schemes. A notable feature of this approach is that the critical stock price at each time step may be solved for explicitly, so long as the underlying stock either pays no dividends or has constant continuous payout. In common with several other papers (e.g. Geske & Johnson 1984, Subrahmanyam & Yu 1993, Breen 1991), the authors use Richardson extrapolation to speed up the computation of the solution.

In the next section this approach is compared against the following numerical schemes:

1. the standard binomial model (Cox, Ross & Rubinstein 1979, Rendleman & Bartter 1979) based on the archetypical random walk approximation to the Brownian motion,

2. the 'accelerated' binomial model (Breen 1991) which uses Richardson's extrapolation technique (Marchuk & Shaidurov 1983) to reduce the number of steps,

3. the BBS binomial method of Broadie & Detemple (1996) which is based on using the Black–Scholes formula 1 step before expiration,

4. the capped option formulae of Broadie and Detemple (1996) which result in a lower bound labelled LBA and a weighted average of lower and upper bounds labelled LUBA,

5. the trinomial method (Parkinson 1977, Kamrad & Ritchken 1991),

6. the quadratic formula (MacMillan 1986, Barone-Adesi & Whaley 1987) which seeks an approximate solution to the Black–Scholes PDE by neglecting a quadratic term for the exercise premium,

7. the regression formula (Johnson 1983) where, as in the LUBA approximation above, the American option value is regressed against lower and upper bounds,

8. the exponential formulae (Omberg 1987 and Chesney 1989) where the value function is approximated on the assumption that the optimal stopping boundary is of exponential form,

9. the implicit finite difference scheme (Brennan & Schwartz 1977) which is classically applied to the variational inequality formulation of the free-boundary problem,

10. the two point GJ formula with Richardson extrapolation (Geske & Johnson 1984) and with exponential extrapolation (Ho, Stapleton & Subrahmanyam 1994) where an American option is viewed as a compound option (option on an option), exerciseable only at a series of discrete dates,

11. the modified GJ approach (Bunch & Johnson 1992),

12. the numerical integral representation of the early exercise premium, with and without Richardson extrapolation (Subrahmanyam & Yu 1993).

4 Some Numerical Comparisons

Because of its simplicity and monotonic convergence the binomial method has attracted the most attention and has been modified into a number of variants which resulted in improved accuracy, sometimes at the expense of speed. The results of Table 1b support the empirical reliance of practitioners on the penny accuracy of 200–step binomial methods. Therefore for all practical purposes we have found it reasonable to use the average of the 1,000 and 1,001 time binomial values as 'exact'. Against this benchmark we use four accuracy criteria, the most prominent of which we deem is the mean squared relative error as defined in Broadie & Detemple (in this volume).

The speed of each of these methods is measured in calculation time (in seconds) per put.

The results of preliminary experiments appear to be in line with those of Broadie and Detemple. Our results are displayed in Tables 2 and 3 and are further summarized in Figures 1 and 2. With no dividend, the five dominant approaches ranked in terms of increasing computation time and accuracy are:

1. Johnson's regression formula,
2. Omberg's exponential formula,
3. Method of lines, especially the modified 3–point extrapolation,
4. Broadie & Detemple's lower bound and their approximation based on both bounds,
5. Modified binomial method.

We also considered options on dividend–paying stocks with an average of three years to maturity and obtained broadly similar results. However, of the five dominant approaches above, the first two are not defined for positive dividends, so only the latter three dominate in this case.

In short, the method of lines increases in accuracy as maturity increases, but decreases in speed as dividends are added. The accurracy increase is due to better approximation of the exercise boundary, while the speed decrease is due to the required numerical solution for the critical stock price.

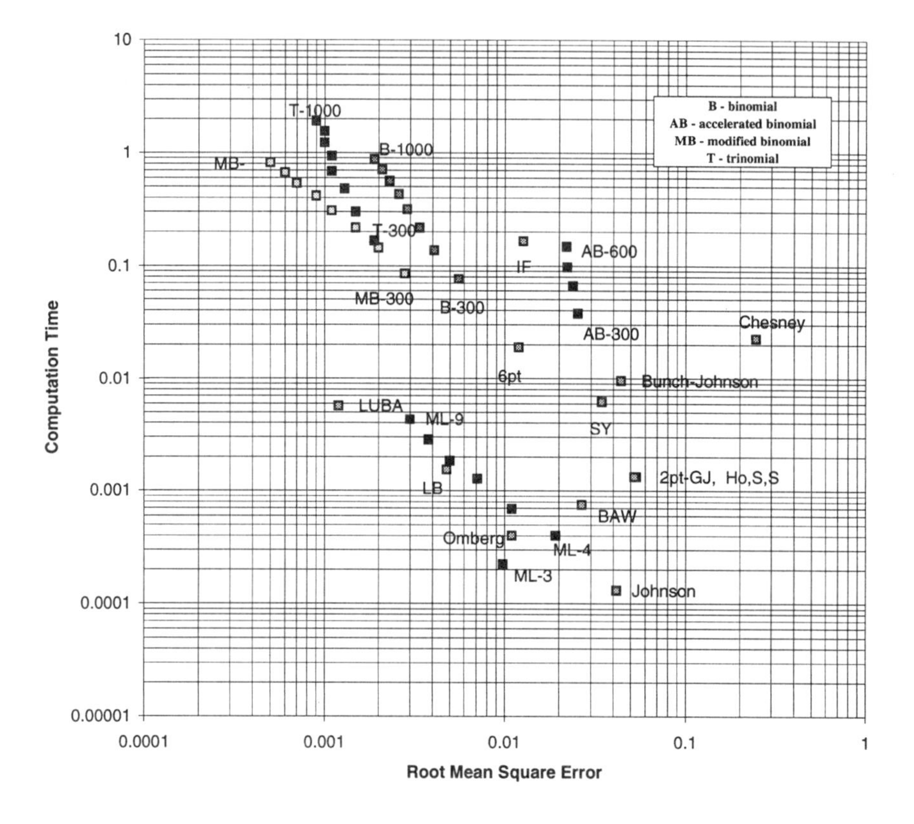

Figure 1: Short Term Options with No Dividend

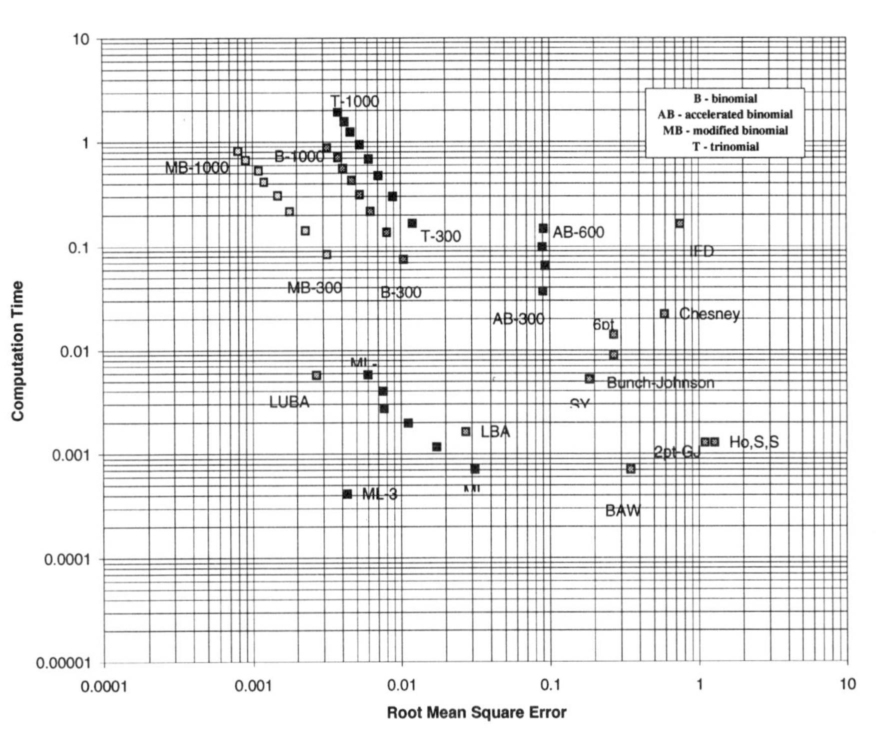

Figure 2: Long Term Options with Dividend

Table 1a: Convergence of Method of Lines with Richardson Extrapolation S = 100, K = 100, T = 1, r = 0.1, $\delta = 0$, $\sigma = 0.3$

Number of Steps n or Points N	Unextrapolated Put Value $P^{(n)}$	Extrapolated Put Value $P^{1:N}$
1	7.0405	7.0405
2	7.6175	8.1946
3	7.8353	8.3089
4	7.9505	8.3257
5	8.0220	8.3311
6	8.0709	8.3333
7	8.1065	8.3345
8	8.1335	8.3353
9	8.1548	8.3358
10	8.1720	8.3362
11	8.1862	8.3365
12	8.1981	8.3367
13	8.2082	8.3369
14	8.2169	8.3370
15	8.2246	8.3371

Table 1b: Convergence of the Binomial Model S = 100, K = 100, T = 1, r = 0.1, $\delta = 0$, $\sigma = 0.3$

N	Even Value	N	Odd Value	Average
50	8.315025	51	8.367897	8.341461
100	8.325495	101	8.352275	8.338885
150	8.329729	151	8.347580	8.338655
200	8.331951	201	8.345306	8.338628
250	8.333084	251	8.343817	8.338450
300	8.333922	301	8.342853	8.338388
350	8.334424	351	8.342078	8.338251
400	8.334808	401	8.341514	8.338161
450	8.335136	451	8.341099	8.338118
500	8.335404	501	8.340771	8.338087
550	8.335620	551	8.340500	8.338060
600	8.335805	601	8.340278	8.338041
650	8.335954	651	8.340083	8.338018
700	8.336085	701	8.339919	8.338002
750	8.336199	751	8.339779	8.337989
800	8.336293	801	8.339650	8.337971
850	8.336372	851	8.339532	8.337952
900	8.336446	901	8.339429	8.337938
950	8.336520	951	8.339347	8.337933
1000	8.336577	1001	8.339263	8.337920
1050	8.336627	1051	8.339185	8.337906
1100	8.336678	1101	8.339120	8.337899
1150	8.336721	1151	8.339057	8.337889
1200	8.336759	1201	8.338998	8.337879
1250	8.336799	1251	8.338948	8.337873
1300	8.336834	1301	8.338900	8.337867
1350	8.336865	1351	8.338854	8.337860
1400	8.336896	1401	8.338815	8.337855
1450	8.336920	1451	8.338773	8.337847
1500	8.336947	1501	8.338738	8.337843
1550	8.336970	1551	8.338704	8.337837
1600	8.336993	1601	8.338672	8.337832
1650	8.337014	1651	8.338643	8.337829
1700	8.337033	1701	8.338613	8.337823
1750	8.337053	1751	8.338589	8.337821
1800	8.337070	1801	8.338563	8.337816
1850	8.337088	1851	8.338540	8.337814
1900	8.337103	1901	8.338517	8.337810
1950	8.337119	1951	8.338497	8.337808
2000	8.337133	2001	8.338476	8.337804

Table 2a: Short Term Put Values, Strike Price = $100, No Dividends

Stock Price in $	Time to Exp'n	Vol-ati-lity	Risk-free Rate	Average Bin'l,N= 1,000	Method of Lines 3 pts	Quad-ratic Formula	Regres-sion Formula
80	0.500	40	6	21.6059	21.6257	21.5077	21.4678
85	0.500	40	6	18.0374	18.0402	17.9606	17.9802
90	0.500	40	6	14.9187	14.9190	14.8673	14.9102
95	0.500	40	6	12.2314	12.2318	12.2033	12.2498
100	0.500	40	6	9.9458	9.9417	9.9376	9.9780
105	0.500	40	6	8.0281	8.0228	8.0334	8.0638
110	0.500	40	6	6.4352	6.4233	6.4508	6.4704
115	0.500	40	6	5.1265	5.1047	5.1488	5.1586
120	0.500	40	6	4.0611	4.0328	4.0875	4.0890
100	0.500	40	2	10.7742	10.7899	10.7613	10.7757
100	0.500	40	4	10.3450	10.3478	10.3317	10.3649
100	0.500	40	6	9.9458	9.9417	9.9376	9.9780
100	0.500	40	8	9.5716	9.5642	9.5711	9.6086
100	0.500	40	10	9.2195	9.2109	9.2275	9.2548
100	0.500	30	6	7.2117	7.2060	7.2052	7.2286
100	0.500	35	6	8.5782	8.5731	8.5706	8.6036
100	0.500	40	6	9.9458	9.9417	9.9376	9.9780
100	0.500	45	6	11.3127	11.3100	11.3043	11.3505
100	0.500	50	6	12.6778	12.6770	12.6697	12.7202
100	0.083	40	6	4.3860	4.3926	4.3785	4.3762
100	0.167	40	6	6.0688	6.0737	6.0576	6.0639
100	0.250	40	6	7.3090	7.3117	7.2964	7.3131
100	0.333	40	6	8.3201	8.3205	8.3077	8.3340
100	0.417	40	6	9.1849	9.1830	9.1740	9.2083
100	0.500	40	6	9.9458	9.9417	9.9376	9.9780
100	0.583	40	6	10.6280	10.6218	10.6235	10.6681
100	0.667	40	6	11.2482	11.2400	11.2480	11.2951
100	0.750	40	6	11.8176	11.8075	11.8225	11.8702
100	0.833	40	6	12.3447	12.3327	12.3553	12.4020
100	0.917	40	6	12.8357	12.8220	12.8525	12.8967
Sec's	per	Put		1.74871	0.00022	0.00075	0.00013
Mean	Rel.	Err.		0.0000	-0.0006	-0.0005	0.0024
Root	Mean	Sqr.	Err.	0.0000	0.0098	0.0269	0.0417
Mean	Abs.	Rel.	Err.	0.0000	0.0009	0.0016	0.0033
Max.	Rel.	Err.		0.0000	-0.0070	0.0065	0.0069

Table 2b: Short Term Put Values, Strike Price = $100, No Dividends

Stock Price in $	Time to Exp'n	Vol-ati-lity	Risk-free Rate	Omberg Exp'l Formula	Broadie Detemple LBA	Broadie Detemple LUBA
80	0.500	40	6	21.5832	21.6174	21.6 034
85	0.500	40	6	18.0186	18.0400	18.0 346
90	0.500	40	6	14.9033	14.9167	14.9 174
95	0.500	40	6	12.2189	12.2274	12.2 311
100	0.500	40	6	9.9365	9.9421	9.9 466
105	0.500	40	6	8.0198	8.0238	8.0 279
110	0.500	40	6	6.4286	6.4317	6.4 349
115	0.500	40	6	5.1212	5.1237	5.1 261
120	0.500	40	6	4.0569	4.0592	4.0 607
100	0.500	40	2	10.7699	10.7800	10.7 743
100	0.500	40	4	10.3391	10.3435	10.3 457
100	0.500	40	6	9.9365	9.9421	9.9 466
100	0.500	40	8	9.5585	9.5695	9.5 721
100	0.500	40	10	9.2030	9.2213	9.2 190
100	0.500	30	6	7.2002	7.2018	7.2 114
100	0.500	35	6	8.5680	8.5711	8.5 785
100	0.500	40	6	9.9365	9.9421	9.9 466
100	0.500	45	6	11.3040	11.3130	11.3 141
100	0.500	50	6	12.6694	12.6826	12.6 797
100	0.083	40	6	4.3843	4.3886	4.3 855
100	0.167	40	6	6.0655	6.0702	6.0 683
100	0.250	40	6	7.3041	7.3091	7.3 088
100	0.333	40	6	8.3137	8.3188	8.3 202
100	0.417	40	6	9.1770	9.1824	9.1 853
100	0.500	40	6	9.9365	9.9421	9.9 466
100	0.583	40	6	10.6174	10.6234	10.6 293
100	0.667	40	6	11.2363	11.2426	11.2 498
100	0.750	40	6	11.8046	11.8111	11.8 195
100	0.833	40	6	12.3305	12.3374	12.3 469
100	0.917	40	6	12.8204	12.8276	12.8 382
Sec's	per	Put		0.00040	0.00154	0.00 567
Mean	Rel.	Err.		-0.0010	-0.0002	0.0 000
Root	Mean	Sqr.	Err.	0.0110	0.0048	0.0 012
Mean	Abs.	Rel.	Err.	0.0010	0.0004	0.0 001
Max.	Rel.	Err.		-0.0018	-0.0014	0.0 002

Table 2c: Short Term Put Values, Strike Price = $100, No Dividends

Stock Price in $	Time to Exp'n	Vol-ati-lity	Risk-free Rate	IFD N=200 M=200	Method of Lines 6 points	Ho et al. 2pt.GJ	Geske Johnson 2 points	Chesney Exp'l Formula
80	0.500	40	6	21.5989	21.6022	21.7875	21.7733	22.7094
85	0.500	40	6	18.0274	18.0317	18.1473	18.1384	17.8090
90	0.500	40	6	14.9066	14.9116	14.9606	14.9554	14.7108
95	0.500	40	6	12.2179	12.2235	12.2220	12.2192	12.0530
100	0.500	40	6	9.9327	9.9384	9.9051	9.9037	9.8000
105	0.500	40	6	8.0146	8.0198	7.9708	7.9701	7.9123
110	0.500	40	6	6.4230	6.4273	6.3740	6.3736	6.3501
115	0.500	40	6	5.1160	5.1193	5.0684	5.0682	5.0672
120	0.500	40	6	4.0526	4.0549	4.0099	4.0098	4.0202
100	0.500	40	2	10.7626	10.7684	10.7399	10.7399	10.7379
100	0.500	40	4	10.3324	10.3379	10.2997	10.2993	10.2486
100	0.500	40	6	9.9327	9.9384	9.9051	9.9037	9.8000
100	0.500	40	8	9.5585	9.5645	9.5457	9.5420	9.3913
100	0.500	40	10	9.2066	9.2127	9.2145	9.2072	9.0134
100	0.500	30	6	7.2007	7.2063	7.1887	7.1863	7.0801
100	0.500	35	6	8.5662	8.5718	8.5457	8.5439	8.4389
100	0.500	40	6	9.9327	9.9384	9.9051	9.9037	9.8000
100	0.500	45	6	11.2985	11.3045	11.2651	11.2638	11.1618
100	0.500	50	6	12.6622	12.6688	12.6241	12.6230	12.5223
100	0.083	40	6	4.3773	4.3834	4.3705	4.3705	4.4103
100	0.167	40	6	6.0594	6.0647	6.0428	6.0427	6.0457
100	0.250	40	6	7.2987	7.3039	7.2759	7.2756	7.2467
100	0.333	40	6	8.3088	8.3141	8.2825	8.2819	8.2233
100	0.417	40	6	9.1727	9.1781	9.1448	9.1439	9.0632
100	0.500	40	6	9.9327	9.9384	9.9051	9.9037	9.8000
100	0.583	40	6	10.6142	10.6202	10.5884	10.5863	10.4590
100	0.667	40	6	11.2336	11.2399	11.2109	11.2080	11.0565
100	0.750	40	6	11.8021	11.8089	11.7840	11.7801	11.6039
100	0.833	40	6	12.3281	12.3357	12.3158	12.3108	12.1098
100	0.917	40	6	12.8179	12.8263	12.8126	12.8063	12.5804
Sec's	per	Put		0.16513	0.00127	0.00133	0.00133	0.02244
Mean	Rel.	Err.		-0.0013	-0.0007	-0.0034	-0.0036	-0.0112
Root	Mean	Sqr.	Err.	0.0127	0.0071	0.0537	0.0524	0.2480
Mean	Abs.	Rel.	Err.	0.0013	0.0007	0.0045	0.0046	0.0149
Max.	Rel.	Err.		-0.0021	-0.0015	-0.0126	-0.0126	0.0511

Table 2d: Short Term Put Values, Strike Price = $100, No Dividends

Stock Price in $	Time to Exp'n	Vol-ati-lity	Risk-free Rate	Method of Lines 9 points	Bunch Johnson Mod. GJ	Integral Method N=6	Sub'n Yu 3 points
80	0.500	40	6	21.6044	21.8225	21.6516	21.7099
85	0.500	40	6	18.0349	18.1384	18.0661	18.0502
90	0.500	40	6	14.9156	14.9554	14.9333	14.8870
95	0.500	40	6	12.2278	12.2451	12.2362	12.1948
100	0.500	40	6	9.9428	9.9430	9.9449	9.9259
105	0.500	40	6	8.0241	8.0170	8.0222	8.0276
110	0.500	40	6	6.4315	6.4249	6.4271	6.4504
115	0.500	40	6	5.1230	5.1160	5.1174	5.1498
120	0.500	40	6	4.0581	4.0505	4.0521	4.0863
100	0.500	40	2	10.7719	10.7703	10.7667	10.7862
100	0.500	40	4	10.3421	10.3386	10.3384	10.3466
100	0.500	40	6	9.9428	9.9430	9.9449	9.9259
100	0.500	40	8	9.5686	9.5774	9.5797	9.5279
100	0.500	40	10	9.2166	9.2320	9.2389	9.1533
100	0.500	30	6	7.2094	7.2140	7.2170	7.1807
100	0.500	35	6	8.5756	8.5780	8.5802	8.5530
100	0.500	40	6	9.9428	9.9430	9.9449	9.9259
100	0.500	45	6	11.3093	11.3075	11.3094	11.2977
100	0.500	50	6	12.6742	12.6705	12.6724	12.6673
100	0.083	40	6	4.3850	4.3841	4.3828	4.3912
100	0.167	40	6	6.0672	6.0655	6.0644	6.0735
100	0.250	40	6	7.3069	7.3040	7.3045	7.3100
100	0.333	40	6	8.3177	8.3129	8.3162	8.3154
100	0.417	40	6	9.1821	9.1796	9.1822	9.1730
100	0.500	40	6	9.9428	9.9430	9.9449	9.9259
100	0.583	40	6	10.6248	10.6282	10.6294	10.5996
100	0.667	40	6	11.2448	11.2516	11.2521	11.2107
100	0.750	40	6	11.8140	11.8245	11.8246	11.7710
100	0.833	40	6	12.3409	12.3551	12.3550	12.2887
100	0.917	40	6	12.8318	12.8498	12.8496	12.7704
Sec's	per	Put		0.00429	0.00956	0.01886	0.00619
Mean	Rel.	Err.		-0.0003	0.0003	0.0000	-0.0012
Root	Mean	Sqr.	Err.	0.0030	0.0440	0.0120	0.0346
Mean	Abs.	Rel.	Err.	0.0003	0.0012	0.0008	0.0026
Max.	Rel.	Err.		-0.0007	0.0100	-0.0022	-0.0072

Table 2e: Short Term Put Values, Strike Price = $100, No Dividends

Stock Price in $	Time to Exp'n	Vol-ati-lity	Risk-free Rate	Std. Bin'l N=300	Acc'd Bin'l N=300	BBS Bin'l N=300	Tri-nomial N=300
80	0.500	40	6	21.6062	21.6204	21.6071	21.6056
85	0.500	40	6	18.0333	18.0088	18.0390	18.0369
90	0.500	40	6	14.9219	14.8888	14.9206	14.9199
95	0.500	40	6	12.2383	12.2232	12.2336	12.2337
100	0.500	40	6	9.9400	9.9178	9.9489	9.9439
105	0.500	40	6	8.0357	8.0333	8.0300	8.0305
110	0.500	40	6	6.4326	6.4283	6.4370	6.4363
115	0.500	40	6	5.1298	5.1296	5.1279	5.1237
120	0.500	40	6	4.0669	4.0685	4.0621	4.0599
100	0.500	40	2	10.7667	10.7665	10.7776	10.7723
100	0.500	40	4	10.3385	10.3328	10.3483	10.3432
100	0.500	40	6	9.9400	9.9178	9.9489	9.9439
100	0.500	40	8	9.5664	9.5361	9.5745	9.5696
100	0.500	40	10	9.2147	9.1834	9.2220	9.2173
100	0.500	30	6	7.2078	7.1900	7.2139	7.2104
100	0.500	35	6	8.5734	8.5569	8.5809	8.5766
100	0.500	40	6	9.9400	9.9178	9.9489	9.9439
100	0.500	45	6	11.3059	11.2864	11.3162	11.3105
100	0.500	50	6	12.6701	12.6618	12.6817	12.6753
100	0.083	40	6	4.3830	4.3836	4.3874	4.3853
100	0.167	40	6	6.0648	6.0623	6.0708	6.0678
100	0.250	40	6	7.3044	7.3000	7.3114	7.3078
100	0.333	40	6	8.3150	8.3083	8.3228	8.3187
100	0.417	40	6	9.1794	9.1635	9.1878	9.1832
100	0.500	40	6	9.9400	9.9178	9.9489	9.9439
100	0.583	40	6	10.6220	10.5996	10.6313	10.6259
100	0.667	40	6	11.2418	11.2235	11.2516	11.2458
100	0.750	40	6	11.8110	11.7890	11.8211	11.8150
100	0.833	40	6	12.3379	12.2981	12.3483	12.3418
100	0.917	40	6	12.8287	12.7695	12.8393	12.8325
Sec's	per	Put		0.07613	0.03742	0.08484	0.16677
Mean	Rel.	Err.		-0.0003	-0.0018	0.0003	-0.0001
Root	Mean	Sqr.	Err.	0.0056	0.0255	0.0028	0.0019
Mean	Abs.	Rel.	Err.	0.0006	0.0021	0.0003	0.0002
Max.	Rel.	Err.		0.0014	-0.0052	0.0003	-0.0005

Table 3a: Long Term Put Values, Strike Price = $100

Stock Price in $	Time to Exp	Vol-ati-lity	Risk-free Rate	Div. Yld. %	Average Bin'l N=1,000	Method of Lines 3 points	Quad-ratic Formula	Broadie Detemple LBA
80	3.0	40	6	2	29.2601	29.2323	29.4377	29.2105
85	3.0	40	6	2	26.9216	26.8923	27.1423	26.8793
90	3.0	40	6	2	24.8023	24.7692	25.0614	24.7669
95	3.0	40	6	2	22.8797	22.8388	23.1704	22.8488
100	3.0	40	6	2	21.1294	21.0835	21.4484	21.1039
105	3.0	40	6	2	19.5376	19.4899	19.8774	19.5142
110	3.0	40	6	2	18.0849	18.0369	18.4418	18.0635
115	3.0	40	6	2	16.7574	16.7070	17.1280	16.7380
120	3.0	40	6	2	15.5428	15.4873	15.9239	15.5252
100	3.0	40	2	2	25.8881	25.8789	26.0215	25.9096
100	3.0	40	4	2	23.2991	23.2504	23.5325	23.2812
100	3.0	40	6	2	21.1294	21.0835	21.4484	21.1039
100	3.0	40	8	2	19.2707	19.2403	19.6423	19.2564
100	3.0	40	10	2	17.6589	17.6467	18.0490	17.6637
100	3.0	40	6	0	19.8545	19.8145	20.1088	19.8362
100	3.0	40	6	2	21.1294	21.0835	21.4484	21.1039
100	3.0	40	6	4	22.4873	22.4419	22.8834	22.4635
100	3.0	30	6	2	15.1687	15.1375	15.3873	15.1444
100	3.0	35	6	2	18.1587	18.1197	18.4274	18.1332
100	3.0	40	6	2	21.1294	21.0835	21.4484	21.1039
100	3.0	45	6	2	24.0708	24.0189	24.4408	24.0464
100	3.0	50	6	2	26.9750	26.9180	27.3966	26.9525
100	0.5	40	6	2	10.2741	10.2759	10.2728	10.2697
100	1.0	40	6	2	13.8774	13.8670	13.9142	13.8679
100	1.5	40	6	2	16.3682	16.3469	16.4627	16.3545
100	2.0	40	6	2	18.2840	18.2533	18.4476	18.2664
100	2.5	40	6	2	19.8349	19.7960	20.0743	19.8134
100	3.0	40	6	2	21.1294	21.0835	21.4484	21.1039
100	3.5	40	6	2	22.2326	22.1804	22.6327	22.2029
100	4.0	40	6	2	23.1868	23.1289	23.6681	23.1526
100	4.5	40	6	2	24.0216	23.9585	24.5829	23.9826
100	5.0	40	6	2	24.7584	24.6906	25.3979	24.7145
100	5.5	40	6	2	25.4136	25.3506	26.1288	25.3645
Sec's	per	Put			1.74467	0.00041	0.00070	0.00161
Mean	Rel.	Err.			0.0000	-0.0019	0.0148	-0.0011
Root	Mean	Sqr.	Err.		0.0000	0.0435	0.3485	0.0273
Mean	Abs.	Rel.	Err.		0.0000	0.0019	0.0148	0.0012
Max.	Rel.	Err.			0.0000	-0.0036	0.0281	-0.0019

Table 3b: Long Term Put Values, Strike Price = $100

Stock Price in $	Time to Exp'n	Vol-ati-lity	Risk-free Rate	Div. Yld. %	Broadie Detemple LUBA	IFD N=200 M=200	Method of Lines 6 points	Ho et. al. 2pt.GJ
80	3.0	40	6	2	29.2540	29.0584	29.2511	31.2905
85	3.0	40	6	2	26.9157	26.6613	26.9110	28.6945
90	3.0	40	6	2	24.7989	24.4744	24.7914	26.3315
95	3.0	40	6	2	22.8778	22.4723	22.8675	24.1829
100	3.0	40	6	2	21.1306	20.6330	21.1180	22.2304
105	3.0	40	6	2	19.5387	18.9372	19.5245	20.4563
110	3.0	40	6	2	18.0860	17.3683	18.0709	18.8438
115	3.0	40	6	2	16.7585	15.9113	16.7429	17.3777
120	3.0	40	6	2	15.5437	14.5535	15.5281	16.0436
100	3.0	40	2	2	25.8864	25.3079	25.8783	26.1582
100	3.0	40	4	2	23.3016	22.7597	23.2855	23.9900
100	3.0	40	6	2	21.1306	20.6330	21.1180	22.2304
100	3.0	40	8	2	19.2670	18.8169	19.2621	20.7281
100	3.0	40	10	2	17.6510	17.2466	17.6525	19.4074
100	3.0	40	6	0	19.8570	19.3754	19.8440	20.1638
100	3.0	40	6	2	21.1306	20.6330	21.1180	22.2304
100	3.0	40	6	4	22.4884	21.9760	22.4760	24.1792
100	3.0	30	6	2	15.1693	15.0992	15.1606	16.4188
100	3.0	35	6	2	18.1600	17.9575	18.1486	19.3269
100	3.0	40	6	2	21.1306	20.6330	21.1180	22.2304
100	3.0	45	6	2	24.0713	23.1398	24.0579	25.1172
100	3.0	50	6	2	26.9741	25.4077	26.9609	27.9775
100	0.5	40	6	2	10.2750	10.2614	10.2671	10.3211
100	1.0	40	6	2	13.8796	13.8578	13.8678	14.0636
100	1.5	40	6	2	16.3712	16.3158	16.3574	16.7424
100	2.0	40	6	2	18.2869	18.1500	18.2724	18.8788
100	2.5	40	6	2	19.8371	19.5442	19.8234	20.6739
100	3.0	40	6	2	21.1306	20.6330	21.1180	22.2304
100	3.5	40	6	2	22.2326	21.4880	22.2215	23.6089
100	4.0	40	6	2	23.1857	22.1943	23.1761	24.8487
100	4.5	40	6	2	24.0197	22.7027	24.0114	25.9769
100	5.0	40	6	2	24.7564	23.1330	24.7489	27.0131
100	5.5	40	6	2	25.4121	23.4786	25.4048	27.9721
Sec's	per	Put			0.00572	0.16261	0.00196	0.00127
Mean	Rel.	Err.			-0.0000	-0.0273	-0.0005	0.0524
Root	Mean	Sqr.	Err.		0.0027	0.7551	0.0112	1.2849
Mean	Abs.	Rel.	Err.		0.0001	0.0273	0.0005	0.0524
Max.	Rel.	Err.			-0.0004	-0.0761	-0.0009	0.1007

Table 3c: Long Term Put Values, Strike Price = $100

Stock Price in $	Time to Exp'n	Vol-ati-lity	Risk-free Rate	Div. Yld %	Geske Johnson 2 points	Chesney Exp'l Formula	Method of Lines 9 points
80	3.0	40	6	2	31.0305	28.5900	29.2649
85	3.0	40	6	2	28.4794	26.2560	26.9247
90	3.0	40	6	2	26.1543	24.1515	24.8048
95	3.0	40	6	2	24.0375	22.2477	22.8807
100	3.0	40	6	2	22.1114	20.5272	21.1310
105	3.0	40	6	2	20.3591	18.9643	19.5373
110	3.0	40	6	2	18.7646	17.5435	18.0835
115	3.0	40	6	2	17.3131	16.2432	16.7554
120	3.0	40	6	2	15.9911	15.0632	15.5405
100	3.0	40	2	2	26.1543	25.5821	25.8989
100	3.0	40	4	2	23.9521	22.8125	23.2937
100	3.0	40	6	2	22.1114	20.5272	21.1310
100	3.0	40	8	2	20.4765	18.6322	19.2814
100	3.0	40	10	2	18.9743	17.0258	17.6564
100	3.0	40	6	0	20.0641	19.2356	19.8500
100	3.0	40	6	2	22.1114	20.5272	21.1310
100	3.0	40	6	4	24.0547	21.9160	22.4828
100	3.0	30	6	2	16.2614	14.6968	15.1652
100	3.0	35	6	2	19.1926	17.6211	18.1770
100	3.0	40	6	2	22.1114	20.5272	21.1310
100	3.0	45	6	2	25.0085	23.4065	24.0781
100	3.0	50	6	2	27.8760	26.2564	26.9693
100	0.5	40	6	2	10.3195	10.1600	10.2694
100	1.0	40	6	2	14.0553	13.6417	13.8750
100	1.5	40	6	2	16.7200	16.0236	16.3583
100	2.0	40	6	2	18.8338	17.8412	18.2796
100	2.5	40	6	2	20.5970	19.3062	19.8278
100	3.0	40	6	2	22.1114	20.5272	21.1310
100	3.5	40	6	2	23.4373	21.5657	22.2201
100	4.0	40	6	2	24.6135	22.4633	23.1858
100	4.5	40	6	2	25.6670	23.2484	24.0174
100	5.0	40	6	2	26.6172	23.9418	24.7546
100	5.5	40	6	2	27.4786	24.5589	25.4102
Sec's	per	Put			0.00127	0.02202	0.00577
Mean	Rel.	Err.			0.0455	-0.0272	-0.0000
Root	Mean	Sqr.	Err.		1.1049	0.5926	0.0060
Mean	Abs.	Rel.	Err.		0.0455	0.0272	0.0002
Max.	Rel.	Err.			0.0813	-0.0358	0.0010

Table 3d: Long Term Put Values, Strike Price = $100

Stock Price in $	Time to Exp'n	Vol-ati-lity	Risk-free Rate	Div. Yld. %	Bunch Johnson modGJ	Integral Method N=6	Sub'n & Yu 3 points
80	3.0	40	6	2	29.9382	29.5987	29.7147
85	3.0	40	6	2	27.3733	27.2300	27.2386
90	3.0	40	6	2	25.1566	25.0805	25.0136
95	3.0	40	6	2	23.1413	23.1280	23.0132
100	3.0	40	6	2	21.3092	21.3522	21.2121
105	3.0	40	6	2	19.6431	19.7349	19.5870
110	3.0	40	6	2	18.1558	18.2598	18.1173
115	3.0	40	6	2	16.8100	16.9126	16.7846
120	3.0	40	6	2	15.5775	15.6807	15.5729
100	3.0	40	2	2	25.9012	25.9213	25.9831
100	3.0	40	4	2	23.3798	23.4149	23.3949
100	3.0	40	6	2	21.3092	21.3522	21.2121
100	3.0	40	8	2	19.5722	19.6108	19.3745
100	3.0	40	10	2	18.0374	18.1180	17.8117
100	3.0	40	6	0	20.0641	20.0238	19.6581
100	3.0	40	6	2	21.3092	21.3522	21.2121
100	3.0	40	6	4	22.6305	22.7249	22.7568
100	3.0	30	6	2	15.3314	15.3862	15.2220
100	3.0	35	6	2	18.3298	18.3787	18.2270
100	3.0	40	6	2	21.3092	21.3522	21.2121
100	3.0	45	6	2	24.2608	24.2967	24.1674
100	3.0	50	6	2	27.1775	27.2044	27.0850
100	0.5	40	6	2	10.2679	10.2747	10.2813
100	1.0	40	6	2	13.8904	13.9006	13.8756
100	1.5	40	6	2	16.4070	16.4277	16.3657
100	2.0	40	6	2	18.3487	18.3901	18.2948
100	2.5	40	6	2	19.9434	19.9959	19.8742
100	3.0	40	6	2	21.3092	21.3522	21.2121
100	3.5	40	6	2	22.4882	22.5226	22.3725
100	4.0	40	6	2	23.5202	23.5485	23.3967
100	4.5	40	6	2	24.4327	24.4585	24.3129
100	5.0	40	6	2	25.2456	25.2735	25.1416
100	5.5	40	6	2	25.9739	26.0091	25.8980
Sec's	per	Put			0.00895	0.01404	0.00523
Mean	Rel.	Err.			0.0095	0.0109	0.0053
Root	Mean	Sqr.	Err.		0.2692	0.2698	0.1853
Mean	Abs.	Rel.	Err.		0.0095	0.0109	0.0059
Max.	Rel.	Err.			0.0232	0.0260	0.0191

Table 3e: Long Term Put Values, Strike Price = $100

Stock Price in $	Time to Exp'n	Vol-ati-lity	Risk-free Rate	Div. Yld. %	Std. Bin'l N=300	Acc'd Bin'l N=300	BBS Bin'l N=300	Tri-nomial N=300
80	3.0	40	6	2	29.2609	29.3264	29.2613	29.2545
85	3.0	40	6	2	26.9133	26.8994	26.9226	26.9158
90	3.0	40	6	2	24.8105	24.7751	24.8042	24.7947
95	3.0	40	6	2	22.8876	22.8051	22.8816	22.8688
100	3.0	40	6	2	21.1185	21.0049	21.1329	21.1183
105	3.0	40	6	2	19.5493	19.4631	19.5401	19.5256
110	3.0	40	6	2	18.0851	17.9793	18.0871	18.0744
115	3.0	40	6	2	16.7645	16.6677	16.7594	16.7499
120	3.0	40	6	2	15.5485	15.4619	15.5452	15.5386
100	3.0	40	2	2	25.8700	25.8840	25.8948	25.8799
100	3.0	40	4	2	23.2858	23.1857	23.3044	23.2900
100	3.0	40	6	2	21.1185	21.0049	21.1329	21.1183
100	3.0	40	8	2	19.2610	19.2096	19.2724	19.2571
100	3.0	40	10	2	17.6498	17.7031	17.6587	17.6421
100	3.0	40	6	0	19.8445	19.7690	19.8571	19.8407
100	3.0	40	6	2	21.1185	21.0049	21.1329	21.1183
100	3.0	40	6	4	22.4748	22.4056	22.4914	22.4781
100	3.0	30	6	2	15.1612	15.0947	15.1710	15.1615
100	3.0	35	6	2	18.1495	18.0553	18.1616	18.1497
100	3.0	40	6	2	21.1185	21.0049	21.1329	21.1183
100	3.0	45	6	2	24.0581	23.9384	24.0747	24.0568
100	3.0	50	6	2	26.9606	26.8312	26.9793	26.9575
100	0.5	40	6	2	10.2677	10.2588	10.2773	10.2723
100	1.0	40	6	2	13.8694	13.8443	13.8814	13.8742
100	1.5	40	6	2	16.3592	16.2799	16.3725	16.3635
100	2.0	40	6	2	18.2743	18.2214	18.2882	18.2773
100	2.5	40	6	2	19.8245	19.7229	19.8388	19.8260
100	3.0	40	6	2	21.1185	21.0049	21.1329	21.1183
100	3.5	40	6	2	22.2211	22.1163	22.2354	22.2187
100	4.0	40	6	2	23.1749	23.0871	23.1891	23.1705
100	4.5	40	6	2	24.0090	23.9453	24.0231	24.0019
100	5.0	40	6	2	24.7455	24.6961	24.7594	24.7367
100	5.5	40	6	2	25.3999	25.3557	25.4134	25.3879
Sec's	per	Put			0.07485	0.03667	0.08333	0.16667
Mean	Rel.	Err.			-0.0004	-0.0037	0.0001	-0.0005
Root	Mean	Sqr.	Err.		0.0105	0.0903	0.0032	0.0121
Mean	Abs.	Rel.	Err.		0.0005	0.0040	0.0001	0.0005
Max.	Rel.	Err.			-0.0007	-0.0059	0.0003	-0.0010

References

AitSahlia, F. (1996) 'Optimal Stopping and Weak Convergence Methods for Some Problems in Financial Economics', *PhD Dissertation*, Dept. of Operations Research, Stanford University.

AitSahlia, F. & Lai, T. L. (1996) 'Approximations for American Options', *working paper*, Cornell University.

Barone-Adesi G. & Whaley, R. (1987) 'Efficient Analytic Approximation of American Option Values', *Journal of Finance* **42**, 301–320.

Bensoussan, A. (1984) 'On the Theory of Option Pricing' *Acta Applicandæ Mathematicæ***2**, 139–158.

Black, F. & Scholes, M. (1973) 'The Pricing of Options and Corporate Liabilities', *Journal of Political Economy* **81**, 637–654.

Breen, R. (1991) 'The Accelerated Binomial Option Pricing Model', *Journal of Financial and Quantitative Analysis* **26**, 153–164.

Brennan, M. & Schwartz, E. (1977) 'The Valuation of American Put Options', *Journal of Finance* **32**, 449–462.

Broadie, M & Detemple, J. (1996) 'American Option Valuation: New Bounds, Approximations, and a Comparison of Existing Methods', *Review of Financial Studies*, to appear.

Bunch, D. & Johnson, H. (1992) 'A Simple and Numerically Efficient Valuation Method for American Puts Using a Modified Geske-Johnson Approach', *Journal of Finance* **47**, 2, 809–816.

Carr, P. & Faguet, D. (1996) 'Fast Accurate Valuation of American Options', *working paper*, Cornell University.

Chesney, M. (1989) 'Pricing American Currency Options: An Analytical Approach', *working paper*, HEC.

Cox, J., Ross, S. & Rubinstein, M. (1979) 'Option Pricing: A Simplified Approach', *Journal of Financial Economics* **7**, 229–263.

Duffie, D. (1988) *Security Markets*, Academic Press.

Friedman, A. (1975) 'Parabolic Variational Inequalities in One Space Dimension and Smoothness of the Free Boundary', *Journal of Functional Analyis,* **18**, 151–176.

Geske, R. & Johnson, H. (1984) 'The American Put Options Valued Analytically', *Journal of Finance* **39**, 1511–1524.

Ho T., Stapleton, R. & Subrahmanyam, M. (1994) 'A Simple Technique for the Valuation and Hedging of American Options', *Journal of Derivatives*, Fall 1994, 52–66.

Jacka, S.D. (1991). Optimal Stopping and the American Put. *Journal of Mathematical Financial* **1**, 1–14.

Johnson, H. (1983) 'An Analytic Approximation for the American Put Price', *Journal of Financial and Quantitative Analysis* **18**, 141–148.

Kamrad, B. & Ritchken, P. (1991) 'Multinomial Approximating Models for Options with k State Variables', *working paper*, Case Western Reserve University.

Karatzas, I. (1988) 'On the Pricing of the American Option', *Appl. Math. and Optim.* **17**, 37–60.

McKean, Jr, H.P. (1965) 'Appendix:A Free Boundary problem for the Heat Equation Arising from a Problem in Mathematical Economics', *Industrial Management Review* **6**, 32–39.

MacMillan, L. (1986) 'Analytic Approximation for the American Put Option', *Advances in Futures and Options Research* **1**, 119-139.

Marchuk, G. & Shaidurov, V. (1983) *Difference Methods and Their Extrapolations*, Springer Verlag.

Merton, R. C. (1973) 'Theory of Rational Option Pricing', *Bell J. Econ. Management Sci.* **4**, 141–183.

Meyer, G. H. (1979) 'One–Dimensional Free–Boundary Problems', *SIAM Review* **19**, 17–34.

Meyer, G. H. & van der Hoek, J. (1994) 'The Evaluation of American Options with the Method of Lines', *working paper*, Georgia Tech.

Myneni, R. (1992) 'The Pricing of the American Option', *Annals of App. Prob.* **2**, 1–23.

Omberg, E. (1987) 'The Valuation of American Puts with Exponential Exercise Policies', *Advances in Futures and Options Research* **2**, 117–142.

Parkinson, M. (1977) 'Option Pricing: The American Put', *Journal of Business* **50**, 21–36.

Rendleman R. & Bartter, B. (1979) 'Two-State Option Pricing', *Journal of Finance* **34**, 1093–1110.

Samuelson, P. (1965) ' Rational Theory of Warrant Pricing', *Industrial Management Review* **6**, 13–31.

Subrahmanyam, M. & Yu, G. (1993) 'Pricing and Hedging American Options: A Unified Method and its Efficient Implementation', *working paper*, New York University.

van Moerbeke, P. (1976) 'On Optimal Stopping and Free Boundary Problems', *Arch. Ration. Mech. Anal.* **60**, 101–148.

Fast, Accurate and Inelegant Valuation of American Options

Adriaan Joubert and L.C.G. Rogers

In this short article, we describe the pricing of an American option (call or put) on a share which may pay continuous dividends, using a method described by Broadie & Detemple (1997) as 'not very elegant'. The method is simply to build a look-up table of option prices, which thus splits the problem of pricing American option prices into three sub-problems:

(A) accurate calculation of the values in the table;
(B) storage of the table, and access to it;
(C) rapid calculation of prices for given parameter values.

The most important problems are clearly B and C; in principle, we may take as long as necessary to fill up the table, since this calculation is done off-line, once only. Any of the methods discussed elsewhere in this volume by Broadie & Detemple (1997) and by AitSahalia & Carr (1997) could be used to compute the values in the table. We used the binomial method with 5000 time steps using Black-Scholes in the last step, as recommended by Broadie & Detemple.

It is worth remarking that by computing and storing the table of values, we are able to calculate greeks, and the exercise boundary with relatively little extra cost; this is a valuable advantage of this inelegant approach.

To describe the storage problem, let us first state the parametrisation which we used. The price of an American put option written on a share paying continuous dividends at rate δ is

$$P(S_0, K, r, \sigma, T, \delta) \equiv \sup_{0 \le \tau \le T} E[(Ke^{-r\tau} - S_0 e^{\sigma W(\tau) - (\delta + \sigma^2/2)\tau})^+],$$

where T is the expiry of the option, τ is a stopping time with values in $[0, T]$, K is the strike price, S_0 is the price of the share at time 0, and σ is the volatility of the share returns. Though the price ostensibly depends on six parameters, we can reduce the problem somewhat by rewriting

$$\begin{aligned} P(S_0, K, r, \sigma, T, \delta) &= \sup_{0 \le \tau \le \sigma^2 T} E[(Ke^{-r\sigma^{-2}\tau} - S_0 e^{W(\tau) - (\delta\sigma^{-2} + (1/2))\tau})^+] \\ &\equiv P(S_0, K, r\sigma^{-2}, 1, \sigma^2 T, \delta\sigma^{-2}) \\ &= KP(S_0/K, 1, r\sigma^{-2}, 1, \sigma^2 T, \delta\sigma^{-2}) \\ &\equiv Kp(S_0/K, rT, \sigma^2 T, \delta T), \end{aligned}$$

say; thus there are effectively only four parameters of the problem, S_0/K, $\sigma^2 T$, rT, and δT. We therefore only have to store a four-dimensional table,

S (\$)	T (Years)	σ (%)	r (%)	Binomial (\$)	Look-Up (\$)	Rel. Err.	Abs. Err.
80	0.500	40	6	21.6057	21.6057	0.0000	0.0000
85	0.500	40	6	18.0370	18.0368	0.0000	0.0002
90	0.500	40	6	14.9178	14.9178	0.0000	0.0000
95	0.500	40	6	12.2306	12.2303	0.0000	0.0003
100	0.500	40	6	9.9448	9.9454	-0.0001	-0.0006
105	0.500	40	6	8.0265	8.0267	0.0000	-0.0003
110	0.500	40	6	6.4337	6.4339	0.0000	-0.0002
115	0.500	40	6	5.1257	5.1253	0.0001	0.0003
120	0.500	40	6	4.0602	4.0601	0.0000	0.0001
100	0.500	40	2	10.7734	10.7739	0.0000	-0.0005
100	0.500	40	4	10.3441	10.3447	-0.0001	-0.0006
100	0.500	40	6	9.9448	9.9454	-0.0001	-0.0006
100	0.500	40	8	9.5707	9.5712	-0.0001	-0.0005
100	0.500	40	10	9.2186	9.2191	0.0000	-0.0005
100	0.500	30	6	7.2110	7.2114	-0.0001	-0.0004
100	0.500	35	6	8.5774	8.5779	-0.0001	-0.0005
100	0.500	40	6	9.9448	9.9454	-0.0001	-0.0006
100	0.500	45	6	11.3116	11.3123	-0.0001	-0.0006
100	0.500	50	6	12.6767	12.6774	-0.0001	-0.0007
100	0.083	40	6	4.3732	4.3743	-0.0003	-0.0012
100	0.167	40	6	6.0716	6.0718	0.0000	-0.0002
100	0.250	40	6	7.3070	7.3074	-0.0001	-0.0004
100	0.333	40	6	8.3149	8.3154	-0.0001	-0.0005
100	0.417	40	6	9.1869	9.1874	-0.0001	-0.0005
100	0.500	40	6	9.9448	9.9454	-0.0001	-0.0006
100	0.583	40	6	10.6247	10.6253	-0.0001	-0.0006
100	0.667	40	6	11.2500	11.2506	-0.0001	-0.0006
100	0.750	40	6	11.8172	11.8178	-0.0001	-0.0006
100	0.833	40	6	12.3424	12.3430	-0.0001	-0.0006
100	0.917	40	6	12.8374	12.8380	-0.0001	-0.0007

Seconds Per Put : 0.00027
Mean Relative Error $\left(\frac{1}{n}\sum\epsilon_i\right)$: -0.000045
Root mean square error $\left(\sqrt{\frac{1}{n}\sum\epsilon_i^2}\right)$: 0.000068
Mean Absolute Relative Error$\left(\frac{1}{n}\sum|\epsilon_i|\right)$: 0.000053
Maximum Relative Error $(\max|\epsilon_i|)$: -0.000263
Maximum Absolute Error $(\max|p_i - \hat{p}_i|)$: 0.001150
(Here $n = 30$, p_i=Binomial price, $\hat{p}_i$=Look-up Table price, $\epsilon_i = \frac{p_i - \hat{p}_i}{p_i}$)

Table 1: Results for short term puts with a strike price of \$100 and no dividends. Here S is the stock price, T the expiry of the option, σ the volatility and r the risk-free interest rate. The results in the 'Binomial' column were calculated with a 5000-step binomial tree, using Black-Scholes in the last step. The relative and absolute errors are given in the last two columns.

which is feasible. In fact, the ranges we took were:

S/K	[0.7 , 1.3]	T	[0.003, 1.0]
r	[0.01,0.10]	σ	[0.03 ,0.60]
δ	[0.00,0.10]		

There are many practical difficulties in the construction of such a look-up table. Close to the exercise boundary the option pricing function p changes very rapidly, and at the exercise boundary its second derivative is large. Interpolating in intervals that straddle the exercise boundary causes larger errors. For this reason the exercise boundary is pinpointed accurately for the S/K grid lines while the table is constructed, and its position stored in a separate table. Furthermore S/K is confined to values where options have a significant value, i.e. are bigger than 10^{-5}. Consequently the grid spacing decreases when the option has a short running time — this is essential to obtain reasonable estimates for short maturities.

The grid spacing for δT is fixed, but for every grid point in the δT dimension bounds for rT are determined, so that only the range of parameters that are of interest is covered. The reason is that choosing δT, and having a fixed range for δ imposes restrictions on the values for T that have to be considered. The same is true for the grid for $\sigma\sqrt{T}$. This simple optimisation reduces the size of the table roughly by half.

We also experimented with non-equidistant grids in order to have more points in regions where T is assumed to be small, but the slight improvement in precision did not seem to justify the additional complication in accessing the table and interpolating the option prices.

For the results presented here, a table with 21 points in the T and S/K dimensions and 16 points in the δ and σ dimensions is used. The total amount of space required to store this was 0.94 MB, which is quite acceptable (it easily fits on a single floppy disc).

It is possible to use so little storage only because we use a judicious choice of grids and a polynomial interpolation method. There are various interpolation schemes well known in numerical analysis; we used a modified Neville interpolation. For an account of the method, see, for example, the book by Stoer (1983).

For the interpolation 4 points are used in every grid dimension, giving a total of 256 points. Heuristics are use to exclude some points if they are outside the exercise boundary. If the option price is not close to the exercise boundary and has a long maturity, quite acceptable results can be achieved using only quadratic or even linear interpolation. In the results reported here the interpolation is not optimised to choose the number of interpolation points dynamically.

Even in a first implementation, this method can price around 3000 options per second with high accuracy. We used a similar test to Broadie & Detemple for 5000 options, with parameters chosen independently from a uniform distribution, such that:

K	100	S	[70 , 130]
T	[0.1, 1.0]	r	[0.01,0.10]
σ	[0.1,0.60]	δ	[0.00,0.10]

We follow them in rejecting options with a price less then 0.005, which left 4687 option prices; we similarly define the relative error as $\epsilon = (p - \hat{p})/p$, where p is the put price calculated with a 5000-step binomial algorithm, using Black-Scholes in the last step, and $\hat{p}$ is the price obtained from the look-up table. The relative root-mean-square error (RMS) is then defined as

$$\sqrt{\frac{1}{n}\sum\left(\frac{p-\hat{p}}{p}\right)^2}.$$

With the look-up table defined above, the relative root-mean-square error was 0.001001, and the maximum relative error (chosen such that 99.5% of the observations have smaller errors) was −0.006011. Warnings are issued for options for which the enclosing cube in the $(\delta T, rT, \sigma\sqrt{T})$ space (8 points) contains points which are outside the exercise boundary. Ignoring options for which warnings are issued halved the relative root-mean-square error (RMS=0.000553) and reduces the relative maximum error to 0.003394.

Table 1 shows results similar to those presented by Ait-Sahalia & Carr. For this parameter set the look-up method compares favourably with any of the direct calculation methods. The fact that no dividends are present is not taken into account explicitly, so that 256 instead of only 64 points are used in this calculation. Taking this into account would reduce the computation time significantly.

The current implementation is still far from optimal, and could easily be speeded up. Currently 89 calls to a Neville interpolation routine are made during every look-up, and this routine could be in-lined and tailored to the problem. A more important optimisation would be to dynamically change the number of points used for the interpolation.

The accuracy can be increased by choosing a larger grid, which comes at the cost of higher memory usage, or by splitting the domain and using a finer grid for short maturities. For particularly difficult cases, such as when some of the interpolation points lie outside the exercise boundary, or the life-span is very short, it is always possible to fall back on a slower direct calculation.

Conclusions

The results in this article demonstrate the feasibility of using look-up tables for pricing American options with continuous dividends. It turns out that organising the look-up table to (i) restrict it to an acceptable size, and (ii) to get high accuracy, is harder than might have been thought. A key factor for the accuracy is to take explicitly into account that the price function changes very rapidly at the exercise boundary. Accessing the table has to be thought out carefully, and cubic polynomial interpolation is used to maintain accuracy. The American put is fundamentally quite a simple option, and we find that the look-up table is broadly comparable in speed and accuracy with the best of the methods discussed elsewhere in this volume. If we were now to envisage the pricing of an American put where the assumption of a constant interest rate were to be replaced by a Vasicek model for the interest rate, the various methods used elsewhere are going to suffer badly, whereas the look-up approach is still as good; filling the look-up table will of course take much longer, but once it is filled, the interpolation method will give the same sort of speed as we have found here (thousands of options per second). In addition to its speed, a key advantage of look-up tables is that the greeks can be estimated very cheaply.

References

AitSahalia, F. & Carr, P. (1997) 'American options: a comparison of numerical methods'. This volume.

Broadie, M. & Detemple. J. (1997) 'Recent advances in numerical methods for pricing derivative securities'. This volume.

Stoer, J. (1983) *Einführung in die Numerische Mathematik I*, 4th ed., Springer.

Valuation of American Options in a Jump-diffusion Model

Xiao Lan Zhang

1 Introduction

The Black–Scholes (1973) model has been widely used in option pricing problems. In this model, the underlying stock price is assumed to be a *continuous* function of time, and can be controlled by the following stochastic differential equation (SDE):

$$\frac{dS_t}{S_t} = \mu dt + \sigma dB_t$$

where μ, σ are respectively drift and volatility terms, $(B_t)_{t \geq 0}$ is a standard real-valued Brownian motion. However, many empirical studies (see Jarrow & Rosenfeld 1984, Ball & Torous 1985, Jorion 1988) exhibited some biases in this kind of model. Cox & Ross (1976) introduced a pure jump model in which the underlying stock is modeled by a known size jump process without diffusion part. The market is then complete. Merton (1976) developed a jump-diffusion model in which the 'normal' vibration in price is modeled by a standard geometric Brownian motion and has a continuous path (diffusion part), the 'abnormal' vibration in the price being modeled by a 'jump' process. Contrary to the Black–Scholes (1973) case, the stock price, in these models, is not a continuous function of time. This allows us to take into account large changes in market prices due to rare events. Another interesting feature of the jump-diffusion model is that it yields implied volatility curves similar to the so-called 'smile curves' observed on some markets. Note that stochastic volatility models also generate smile curves (cf. Hull & White 1987).

Merton (1976) derived some tractable formulae for the pricing of *European* options. Valuing American options has been given much attention in recent economic and finance literature. Nevertheless, there is no explicit formulae for the price of an American option. The objective of this note is to describe numerical methods for the analysis of American options in the jump-diffusion model.

2 The Framework

2.1 The jump-diffusion model

We consider a financial market with two assets: one is a non-risky asset, with price S_t^0 at time t governed by the equation $dS_t^0 = rS_t^0 dt$, where r is a positive constant representing the interest rate, the other is a risky asset, with price S_t at time t. Here, $(S_t)_{t \geq 0}$ is assumed to be a random process and is governed by the following stochastic differential equation:

$$\begin{cases} S_0 & = & y \\ \dfrac{dS_t}{S_{t^-}} & = & \mu dt + \sigma dB_t + d\left(\sum_{j=1}^{N_t} U_j\right) \end{cases} \tag{2.1}$$

where y is the spot price at time 0, $(N_t)_{t \geq 0}$ is a Poisson process with parameter λ, and $(U_j)_{j \geq 1}$ is a sequence of square integrable i.i.d. random variables, with values in $]-1, \infty[$ (since the price of a financial asset should be positive). μ, σ are two constants with $\sigma > 0$. The parameter λ of the Poisson process accounts for the frequency of jumps and the random variable U_j accounts for the relative amplitude of jumps. The financial model described above can be interpreted in a more intuitive way as follows: the underlying stock price in this model allows a discontinuous path, with jump times controlled by a Poisson process, it is modeled by a geometric Brownian motion (the Black–Scholes model) between two jump times, and it can leap a random value at jump times.

Furthermore, we assume that the processes $(B_t)_{t \geq 0}$, $(N_t)_{t \geq 0}$, $(U_j)_{j \geq 1}$ are independent. We denote by $\mathcal{F}_t$ the σ-field generated by random variables B_s, N_s, $U_j \mathbf{1}_{\{j \leq N_s\}}$ for $s \leq t$ and $j \geq 1$.

2.2 Pricing formulae

Unlike the Black–Scholes case, the market modeled by Merton's (1976) jump-diffusion model is not complete. Therefore the classical methodology for pricing American options (cf. Bensoussan 1984, Karatzas 1988) cannot be applied. In fact, even for European options, the derivation of pricing formulae requires additional assumptions. If we consider a European option with maturity T, and pay-off h, where h is a positive square integrable $\mathcal{F}_T$-measurable random variable, a 'fair price' cannot be derived by constructing a replicating portfolio in an incomplete market. Various approaches can be used to derive pricing formulae:

(a) Merton (1976) assumes that the risk associated with jumps can be diversified away.

(b) In some cases, the market may be embedded in a complete market. In particular when the U_j's are deterministic, the market can be completed by adding another asset (see Jeanblanc-Picque & Pontier 1990, Mulinacci 1994, Shirakawa 1990).

(c) The risk minimizing approach of Föllmer, Sondermann and Schweizer (see Bouleau & Lamberton 1989, Colwell & Elliott 1993, Föllmer &

Schweizer 1990, Föllmer & Sondermann 1986, Pham & Touzi 1994, Schweizer 1991, Schweizer 1994) can be used to select a pricing equivalent martingale measure.

All these approaches lead to pricing formulae involving the computation of the following conditional expectation:

$$V_t = \mathbf{E}^*(e^{-r(T-t)}h|\mathcal{F}_t) \tag{2.2}$$

under some equivalent martingale measure $\mathbf{P}^*$ (see also El Karoui & Quenez (1995) for the definition of a price range in an incomplete market).

We focus our discussions on the last approach (c) and try to apply results obtained by Föllmer, Sondermann and Schweizer to the jump-diffusion model. Generally, the theory consists in defining a concept of risk, choosing a price and a hedge such that this risk is minimized.

- *Martingale measure case*: $\mathbf{P}^* = \mathbf{P}$
  Föllmer & Sondermann (1986) have shown that when the price of a risky asset is a square integrable martingale, a trading strategy minimizing the quadratic risk can be constructed, and its value can be expressed as a conditional expectation of the discounted pay-off value. Normally, this value will be taken as the corresponding European option price. We suppose that μ, λ, r and the expectation of U_1 satisfy the following relation: $\mu = r - \lambda \mathbf{E} U_1$. This equality yields that the discounted price $(e^{-rt}S_t)_{t\geq 0}$ is a square integrable martingale with respect to the filtration $(\mathcal{F}_t)$. Clearly, this assumption allows us to use the notion of risk-minimizing trading strategy in the sense of Föllmer and Sondermann, thus it yields formula (2.2) with $\mathbf{P}^* = \mathbf{P}$.

  Note that Merton (1976) did not suppose $\mu = r - \lambda \mathbf{E} U_1$, but his price formula can be written as the conditional expectation with respect to a probability $\mathbf{P}^*$ equivalent to the initial probability $\mathbf{P}$ of the model, such that, under $\mathbf{P}^*$, S_t is still a solution of the same type of equation as (2.1), but with $\mu = r - \lambda \mathbf{E} U_1$. So, from the computational point of view, we come back to the same case as above.

- *General case*: $\mathbf{P}^* = \hat{\mathbf{P}}$
  If the equation $\mu = r - \lambda \mathbf{E} U_1$ is not satisfied, then the discounted price is only a semi-martingale. Generally, there is no trading strategy minimizing the quadratic risk. But Föllmer & Schweizer (1990), Schweizer (1991) have introduced definitions of local risk-minimizing trading strategy and minimal probability. They have shown that when a minimal probability $\hat{\mathbf{P}}$ exists, the value of such a strategy at time t can be expressed as a conditional expectation under this minimal probability $\hat{\mathbf{P}}$. Note that Colwell & Elliott (1993) have introduced a 'pricing

measure' which coincides also with the minimal measure of Föllmer and Schweizer.

Let us recall briefly some results of Schweizer. Consider a probability space $(\Omega, \mathcal{F}, \mathbf{P})$ with a filtration $\mathcal{F} = (\mathcal{F}_t)_{0 \leq t \leq T}$ satisfying the usual conditions. Let the discounted price process $\tilde{S}_t$ be a semi-martingale, $\tilde{S}_t = S_0 + M_t + A_t$ where M_t is a square integrable martingale, A_t is a process with finite variation. A_t is assumed to be continuous and absolutely continuous with respect to $< M_t >$. Let $\alpha_t = dA_t / d < M_t >$. Schweizer defines a minimal probability $\hat{\mathbf{P}}$ in the following way:

(i) $\hat{\mathbf{P}}$ is a probability equivalent to $\mathbf{P}$ and $\frac{d\hat{\mathbf{P}}}{d\mathbf{P}} \in \mathcal{L}^2(\Omega, \mathcal{F}, \mathbf{P})$.

(ii) $\tilde{S}_t$ is a martingale under $\hat{\mathbf{P}}$.

(iii) $\mathbf{P} = \hat{\mathbf{P}}$ on $\mathcal{F}_0$.

(iv) Any square integrable martingale orthogonal to M_t under $\mathbf{P}$ is still a martingale under $\hat{\mathbf{P}}$.

Furthermore, he has shown that $\hat{\mathbf{P}}$ exists if and only if the solution of the SDE: $G_t = 1 - \int_0^t G_{s-} \alpha_s dM_s$ is a strictly positive square integrable martingale under $\mathbf{P}$. In this case, $\hat{\mathbf{P}}$ is given by:

$$\left. \frac{d\hat{\mathbf{P}}}{d\mathbf{P}} \right|_{\mathcal{F}_t} = G_t. \tag{2.3}$$

Here two problems arise naturally in the jump-diffusion model:

- Is there a minimal probability $\hat{\mathbf{P}}$ in this model with jumps?
- What is the distribution of the random process S_t under this probability $\hat{\mathbf{P}}$, when it exists? In other words, is S_t still the solution of a SDE of the same form: $dS_t/S_{t-} = \hat{\mu}dt + \hat{\sigma}dW_t + d(\sum_{j=1}^{\hat{N}_t} \hat{U}_j)$ where W_t, $\hat{N}_t$ and $\hat{U}_j$ have the same properties under $\hat{\mathbf{P}}$ as B_t, N_t and U_j under $\mathbf{P}$?

The following two propositions allow us to solve the above problems. Their proofs can be found in Zhang (1994, section 1.2).

Proposition 2.1 *Assume that the support of U_j is $]-1, +\infty[$. Then*

(1) the existence of the minimal probability $\hat{\mathbf{P}}$ is equivalent to the condition: $-1 \leq \eta \leq 0$ with $\eta = \mu + \lambda \mathbf{E} U_1 - r/\sigma^2 + \lambda \mathbf{E} U_1^2$.

(2) the minimal probability $\hat{\mathbf{P}}$ is given by (2.3) with

$$G_t = \exp(-\eta\sigma B_t - \frac{1}{2}\eta^2\sigma^2 t) \prod_{j=1}^{N_t} (1 - \eta U_j) e^{\eta\lambda t \mathbf{E} U_1}.$$

In order to solve the second problem, let us denote: $W_t = B_t + \eta\sigma t$. By using Ito's formula, $\tilde{S}_t$ satisfies the following SDE:

$$d\tilde{S}_t = \tilde{S}_{t^-}\lambda(\eta \mathbf{E}U_1^2 - \mathbf{E}U_1)dt + \sigma\tilde{S}_{t^-}dW_t + \tilde{S}_{t^-}d(\sum_{j=1}^{N_t} U_j).$$

Proposition 2.2 *Assume $\tilde{\mathbf{P}}$ is a probability under which*

(1) W_t is a Brownian motion,

(2) N_t is a Poisson process with parameter $\hat{\lambda} = \lambda(1 - \eta\mathbf{E}U_1)$,

(3) U_j are i.i.d, with distributions $\tilde{\mathbf{P}}_{U_1}(dx) = \frac{1-\eta x}{1-\eta \mathbf{E}U_1}\mathbf{P}_{U_1}(dx)$,

(4) $(W_t)_{t\geq 0}$, $(N_t)_{t\geq 0}$, $(U_j)_{j\geq 1}$ are independent.

Then we have: $\hat{\mathbf{P}}|_{\mathcal{F}_T} = \tilde{\mathbf{P}}|_{\mathcal{F}_T}$.

Let us come back to the discussion of formula (2.2). Since we are only interested in the effective computation of pricing formulae, we will assume throughout this note that $\mathbf{P}^* = \mathbf{P}$ and $\mu = r - \lambda\mathbf{E}(U_1)$. In other words, we will be working under a *given* martingale measure $\mathbf{P}$ and take it as a pricing measure.

Now, we consider an American option with maturity T, allowing a profit $h(S_t)$ when it is exercised at time t. By analogy with the European case, we take the quantity $v(t, S_t)$ with

$$v(t,y) = \sup_{\tau\in\mathcal{T}_{t,T}} \mathbf{E}(e^{-r(\tau-t)}h(S_\tau^{t,y})) \tag{2.4}$$

as the American option price at time t. Here, $\mathcal{T}_{t,T}$ is the set of all stopping times with values in $[t,T]$ and $(S_s^{t,y})_{s\geq t}$ is a random process defined by: $S_s^{t,y} = ye^{(\mu-\frac{\sigma^2}{2})(s-t)+\sigma(B_s-B_t)}\prod_{j=N_t+1}^{N_s}(1+U_j)$. This approach is consistent (up to some changes of the parameters) with Merton's formulae as well as with the minimal martingale measure approach of Föllmer and Schweizer. In addition, if the model turns out to be complete (either because there are no jumps or by addition of other assets as in Jeanblanc-Picque & Pontier (1990) and Pham (1995)), the above formula is consistent with that of Bensoussan (1984) and Karatzas (1988).

3 Valuation of American Options

As mentioned in the introduction, numerical methods for computing the price of American options are necessary but sometimes present considerable difficulties due to the possibility of early exercise. Various approximation approaches are possible, such as lattice methods, analytical methods, finite difference methods ... Among the vast literature on numerical methods for American option pricing, we mention Bensoussan & Lions (1978), Broadie & Detemple (1996), Carr & Faguet (1994), Jaillet *et al.* (1990), MacMillan (1986), Myneni (1992) in the Black–Scholes setting and Amin (1993), Mulinacci (1994), Pham (1995), Zhang (1993, 1995) in the jump-diffusion model. Amin (1993) developed a discrete time model to value options when the underlying process follows a jump-diffusion process. Mulinacci (1994) proposed a probabilistic approximation method when jump sizes are deterministic. Pham (1995) obtained a decomposition of the American put option value as the sum of its corresponding European put price and the early exercise premium. The latter can be written in the form of complicated integral terms. Zhang (1995) derived some quasi-explicit formulae under special assumptions. In this section, we will discuss the valuation of an American option in this jump-diffusion model by finite differences. It is based on a variational inequality.

We focus our discussions on the case of American put options, since the price of an American call on a non-dividend-paying stock coincides with the corresponding European call's price, for the same reason as in the Black–Scholes model (cf. Merton 1973).

3.1 Variational inequality

The relationship between an optimal stopping problem and variational inequalities has been developed by Bensoussan & Lions (1978, 1982) and Friedman. These methods have been applied to American option pricing in diffusion models in Jaillet *et al.* (1990). The main difference between the Jaillet *et al.* (1990) method and ours is that we have to deal with a variational inequality involving a non-local integro-differential operator due to the jump terms. This implies difficulties for the mathematical analysis as well as for the approximation. Numerical methods based on the viscosity solution approach (see Barles *et al.* 1994, 1995, Fitzpatrick & Fleming 1991, Tourin & Zariphopoulou (1994), Pham 1995) could be applied to our setting. The advantage of the classical variational methods is the convergence results for the first order derivatives (see theorem 3.6 below) and the regularity results for the value function (see theorem 3.2).

We now introduce the variational inequality relevant to our problem. We first set: $X_t = \log(S_t)$, and denote: $Z_j = \log(1+U_j)$, $x = \log y$, $\psi(x) = h(e^x)$, $u^*(t,x) = \sup_{\tau \in \mathcal{T}_{t,T}} \mathbf{E}(e^{-r(\tau-t)}\psi(X_\tau^{t,x}))$, with

$$X_\tau^{t,x} = x + \left(\mu - \frac{\sigma^2}{2}\right)(\tau - t) + \sigma(B_\tau - B_t) + \sum_{j=N_t+1}^{N_\tau} Z_j,$$

so that $u^*(t,x) = u^*(t,\log y) = v(t,y)$. Formally, u^* solves the following parabolic inequality:

$$\begin{cases} u(T,\cdot) = \psi(\cdot) \\ \max\left\{\dfrac{\partial u}{\partial t} + Au + Bu, \psi - u\right\} = 0 \quad \text{in } [0,\mathrm{T}] \times \mathbf{R} \end{cases} \tag{3.1}$$

where $Au = \frac{\sigma^2}{2}\frac{\partial^2 u}{\partial x^2} + (\mu - \frac{\sigma^2}{2})\frac{\partial u}{\partial x} - ru$, $Bu = \lambda(\int u(t, x+z)\nu(dz) - u(t,x))$ and ν is the distribution of Z_1.

In order to give the variational form of problem (3.1), we introduce some weighted Sobolev spaces: for $\alpha > 0$,

$$\begin{aligned} H_\alpha &= L^2(\mathbf{R},\, e^{-\alpha|x|}dx) \\ V_\alpha &= \{f \in H_\alpha |\ \ f' \in H_\alpha\} \\ W^{m,p,\alpha}(\mathbf{R}) &= \{f \in L^p(\mathbf{R},\, e^{-\alpha|x|}dx) \mid \text{for } j \leq m,\ \ f^{(j)} \in L^p(\mathbf{R}, e^{-\alpha|x|}dx)\}. \end{aligned}$$

We denote by $(\cdot,\cdot)_\alpha$ the inner product on H_α. We define also:

$$\begin{aligned} a^\alpha(u,v) &= \frac{\sigma^2}{2}\int \frac{\partial u}{\partial x}\frac{\partial v}{\partial x} e^{-\alpha|x|}dx + r\int uve^{-\alpha|x|}dx \\ &\quad - \int\left(\frac{\sigma^2\alpha}{2}\mathrm{sgn}(x) + \left(\mu - \frac{\sigma^2}{2}\right)\right)\frac{\partial u}{\partial x} ve^{-\alpha|x|}dx \\ b^\alpha(u,v) &= -\int (Bu)\cdot ve^{-\alpha|x|}dx. \end{aligned}$$

It is not difficult to show that $a^\alpha(\cdot,\cdot)$ is a bilinear functional on V_α and $b^\alpha(\cdot,\cdot)$ is a bilinear functional on H_α. We introduce another hypothesis:

(H): $\mathbf{E}\, e^{\alpha|Z_1|} < \infty$ for all α.

Theorem 3.1 below states that the price function u^* equals the unique solution of the variational inequality (3.2). This theorem can be proved by using the methods described in Bensoussan & Lions (1978, 1982) and Jaillet *et al.* (1990) (see Zhang 1994, Chapter 3 for details).

Theorem 3.1 *Assume $\psi \in V_\alpha$, and* (**H**) *is satisfied.*

1. *There is one and only one function u satisfying $u \in L^2([0,T]; V_\alpha)$ and $\frac{\partial u}{\partial t} \in L^2([0,T]; H_\alpha)$ such that:*

$$\begin{cases} u(T,x) = \psi(x) \\ u \geq \psi \ \text{ a.e. in } [0,T] \times \mathbf{R}, \ \text{ and} \\ \forall v \in V_\alpha, \ \ v \geq \psi, \ \text{ we have:} \\ -(\frac{\partial u}{\partial t}, v-u)_\alpha + a^\alpha(u, v-u) + b^\alpha(u, v-u) \geq 0. \end{cases} \tag{3.2}$$

2. *The function u^* is equal to the solution of (3.2).*

According to Theorem 3.1, we have:

$$u(t,x) = u^*(t,x)$$
$$= \sup_{\tau \in \mathcal{T}_{t,T}} \mathbf{E}\left(e^{-r(\tau - t)}\psi(x) + \left(\mu - \frac{\sigma^2}{2}\right)(\tau - t) + \sigma(B_\tau - B_t) + \sum_{j=N_t+1}^{N_\tau} Z_j \right).$$

From the above expression, it is easy to see that the function u has the following properties: (i) If $x \to \psi(\log x)$ is convex, then the function $x \to u(t, \log x)$ is convex on $]0,\infty[$, for all $t \in [0,T]$; (ii) If $x \to \psi(\log x)$ is decreasing (resp. increasing), then the function $x \to u(t, \log x)$ is decreasing (resp. increasing) on $]0,\infty[$, for all $t \in [0,T]$. In addition, theorem 3.2 below states some regularity results for the price function u. More precisely, it accounts for the smooth linking of the price function u and the pay-off function ψ on the free boundary (the so-called smooth fit property).

Theorem 3.2 *Suppose that* (**H**) *is satisfied. If ψ is a Lipschitz function and $x \mapsto \psi(\log x)$ is convex, then the solution u of (3.2) admits partial derivatives $\partial u/\partial t$, $\partial u/\partial x$, $\partial^2 u/\partial x^2$ locally bounded on* $[0,T[\times\mathbf{R}$.

Furthermore, the function $\partial u/\partial x$ is continuous on $[0,T[\times\mathbf{R}$.

The proof of the above theorem, which can be found in Zhang (1994), is based on martingale techniques.

It is easy to see that, in the case of an American put, ψ, which equals $(K - e^x)_+$, satisfies the assumptions of theorem 3.2.

3.2 Finite difference method

This section deals with the numerical analysis of variational inequality (3.2). Since this variational inequality is defined on the whole real line $\mathbf{R}$, we need first to localize it in an interval $[-l, l]$ where l is a constant. The localized problem is then solved by the finite difference method.

3.2.1 Localization

To localize the variational inequality (3.2), we introduce:

$$T_l^{t,x} = \inf\{s > t;\ |X_s^{t,x}| > l\}.$$

and we denote:

$$u_l{}^*(t,x) = \sup_{\tau \in \mathcal{T}_{t,T}} \mathbf{E}\left[e^{-r(\tau \wedge T_l^{t,x} - t)}\right.$$

The following proposition (see Zhang 1994 for proof) implies that, to calculate the price u^*, it suffices to compute the value of the function $u_l{}^*$.

Proposition 3.3 *Assume that $\psi(x) \in V_\alpha$, and (**H**) is satisfied. Then u_l^* converges uniformly to u^* on any compact set, as l goes to infinity, i.e:*

$$\forall R > 0, \qquad \lim_{l\to\infty} \sup_{(t,x)\in[0,T]\times[-R,+R]} |u^*(t,x) - u_l^*(t,x)| = 0.$$

Remark 3.4 *More precisely, we have the following estimate:*

$$|u^*(t,x) - u_l{}^*(t,x)| \leq C\left(2e^{-\frac{(l-R-|\mu-\frac{\sigma^2}{2}|T)^2}{4\sigma^2 T}} + e^{-\frac{l-R-|\mu-\frac{\sigma^2}{2}|T}{2}} \cdot e^{\lambda T(\mathbf{E}e^{|Z_1|}-1)}\right)^{1/2}$$

where C is a constant independent of l. In practice, for a given precision ϵ, a suitable value for l can be found by using the above estimate.

Now, suppose that l is chosen so that $u_l{}^*$ and u^* are close enough. For calculating the function $u_l{}^*$, we introduce the following notations:

$$\begin{aligned} \Omega_l &=]-l,\, l\,[, \\ \partial\Omega_l &= \{-l,\, +l\}, \\ H_l &= L^2(\Omega_l), \\ V_l &= \{f \in H_l\,, \frac{\partial f}{\partial x} \in H_l\}, \\ W^{2,2}(\Omega_l) &= \{f \in V_l,\, \frac{\partial^2 f}{\partial x^2} \in H_l\}. \end{aligned}$$

We denote by u_l, v_l functions defined on Ω_l. Let:

$$a(u_l, v_l) = \int_{\Omega_l} \frac{\sigma^2}{2}\frac{\partial u_l}{\partial x}\frac{\partial v_l}{\partial x}dx + \int_{\Omega_l} r u_l v_l dx - \int_{\Omega_l}\left(\mu - \frac{\sigma^2}{2}\right)\frac{\partial u_l}{\partial x} v_l dx.$$

Since the integro-differential operator is not local, we prefer to introduce an operator $\tilde{B}^l$ given by

$$\tilde{B}^l u_l(t,x) = \lambda[\int_{z,z+x\in\Omega_l} u_l(t,x+z)\nu(dz) - u_l(t,x)]$$

and a function f with

$$f(x) = \lambda\int_{z,z+x\notin\Omega_l} \psi(x+z)\nu(dz).$$

We define $b(u_l, v_l) = -(\tilde{B}^l u_l, v_l)_l$ where $(\cdot,\cdot)_l$ denotes the inner product of H_l. By using the same method as in theorem 3.1, we can show that, if $\psi \in V_\alpha$, then the function $u_l{}^*(t,x)$ is the unique solution of the following variational inequality: find $u_l \in L^2([0,T];\, V_l)$ such that $\partial u_l/\partial t \in L^2([0,T];\, H_l)$ and

$$\left\{\begin{array}{lll} u_l(T,x) = \psi(x) & x \in \Omega_l \\ u_l(t,x) \geq \psi(x) & \text{a. e.} \quad \text{in } [0,T]\times\Omega_l \\ u_l(t,x) = \psi(x) & \text{if } x \in \partial\Omega_l, \text{ and} \\ \forall v_l \in V_l, \quad v_l \geq \psi, & \text{we have:} \\ -\left(\frac{\partial u_l}{\partial t}, v_l - u_l\right)_l + a(u_l, v_l - u_l) + b(u_l, v_l - u_l) - (f, v_l - u_l)_l \geq 0 \end{array}\right. \tag{3.3}$$

3.2.2 Discretization

We discretize now the problem (3.3) by using the finite difference method (FDM). We are interested in the case where the random variable Z_1 admits a density function g (i.e. $\nu(dz) = g(z)dz$). Furthermore, we suppose that this function g is C^1. One important example is when $1+U_j$ follows a log-normal distribution, since we then have a simple formula for the European option's price (cf. Merton 1976).

- **Finite Difference Method**

For the numerical solution of problem (3.3) by the FDM, we introduce a grid of mesh points $(t, x) = (ik, jh)$ where h and k are mesh parameters which are small and supposed tending to zero, and where i, j are integers with $i \geq 0$. We then look for an approximate solution of (3.3) at these points, which will be denoted by u_j^i, by solving a problem in which the derivatives in (3.3) are replaced by the finite difference quotients.

More precisely, at each point in $h\mathbf{Z}$ we associate an interval

$$W_h^0(M) =](m - \frac{1}{2})h, (m + \frac{1}{2})h[.$$

We denote $\Omega_h = \{M \in h\mathbf{Z},\ W_h^0(M) \subset \Omega_l\}$, $m_1 = \inf\{m \mid mh \in \Omega_h\}$, $m_2 = \sup\{m \mid mh \in \Omega_h\}$, and V_h the space generated by χ_h^M where χ_h^M is the indicator function of $W_h^0(M)$ and $M \in \Omega_h$. If $u_h \in V_h$, we can write: $u_h(x) = \sum_{j=m_1}^{m_2} u_j \mathbf{1}_{](j-\frac{1}{2})h,\,(j+\frac{1}{2})h]}(x)$. Furthermore, we denote $\delta\varphi(x) = \frac{1}{h}[\varphi(x + \frac{h}{2}) - \varphi(x - \frac{h}{2})]$.

- **Approximations for ψ, $a(\cdot,\cdot)$, $\tilde{B}^l$ and f**

We approximate ψ by $\psi_h = \sum_{M\in\Omega_h} \psi_h(M)\chi_h^M$ where $\psi_h(M) = \frac{1}{h}\int_{W_h^0(M)} \psi(x)dx$. We denote $\psi_s = \psi_h(sh)$ for $s \in \mathbf{Z}$. In addition, we define, for u_h, $v_h \in V_h$,

$$a_h(u_h, v_h) = \int_{\Omega_l} \frac{\sigma^2}{2}\delta u_h\, \delta v_h dx + \int_{\Omega_l} r u_h v_h dx - \int_{\Omega_l} (\mu - \frac{\sigma^2}{2})\delta u_h v_h dx.$$

Note that $a_h(u_h, v_h) = -(A_h u_h, v_h)_l$, where A_h: $V_h \to V_h$ is the operator defined by $A_h u_h = \sum_{j=m_1}^{m_2} (A_h u_h)_j \mathbf{1}_{](j-\frac{1}{2})h,\,(j+\frac{1}{2})h]}$ with:

$$(A_h u_h)_j = \frac{\sigma^2}{2h^2}(u_{j+1} - 2u_j + u_{j-1}) + (\mu - \frac{\sigma^2}{2})\frac{u_{j+1} - u_{j-1}}{2h} - r u_j. \tag{3.4}$$

Furthermore, we introduce another operator B_h: $V_h \to V_h$ given by $B_h u_h = \sum_{j=m_1}^{m_2} (B_h u_h)_j \mathbf{1}_{](j-\frac{1}{2})h,\,(j+\frac{1}{2})h]}$ with:

$$(B_h u_h)_j = \lambda(\sum_{l=m_1-j}^{m_2-j} u_{l+j} g_l h - u_j) \qquad m_1 \leq j \leq m_2 \quad g_l = g(lh). \tag{3.5}$$

It is easy to see that the index l is between $-(m_2 - m_1)$ and $m_2 - m_1$. So $[g_l]$ is a $2(m_2 - m_1) + 1$ vector. In addition, we approximate f by $f_h = \sum_{j=m_1}^{m_2} f_j \mathbf{1}_{](j-\frac{1}{2})h,(j+\frac{1}{2})h]}$ with:

$$f_j = \lambda \sum_{\substack{l+j>m_2 \\ l+j<m_1}} \psi_{l+j} g_l h \tag{3.6}$$

- **Discrete scheme**

Denoting $N = [\frac{T}{k}]$, the approximation scheme of problem (3.3) can be written as follows: find an approximate solution written in the form:

$$u_{h,k}(t,x) = \sum_{i=0}^{N} u_h^i(x) \mathbf{1}_{](i-1)k,ik]}(t)$$

where $u_h^0, \cdots, u_h^N$ are the elements of V_h satisfying the following :

$$\begin{cases} u_h^N = \psi_h, \\ \forall\, i \in \{0, \cdots, N\},\ u_h^i \geq \psi_h \text{ and } u_h^i(m_j h) \ = \ \psi_h(m_j h) \text{ with } j = 1, 2, \text{ and} \\ \forall\, i \in \{0, \cdots, N\},\ \forall\, v_h \in V_h,\ v_h \geq \psi_h, \quad \text{we have:} \\ \left(\dfrac{u_h^i - u_h^{i-1}}{k} + A_h u_h^{i-\theta} + B_h u_h^{i-\bar{\theta}} + f_h, v_h - u_h^{i-1} \right)_l \leq 0 \end{cases} \tag{3.7}$$

with θ, $\bar{\theta}$ fixed in $[0,1]$ and $u_h^{i-\theta} = u_h^i + \theta(u_h^{i-1} - u_h^i)$, $u_h^{i-\bar{\theta}} = u_h^i + \bar{\theta}(u_h^{i-1} - u_h^i)$. The operators A_h, B_h and the function f_h are defined by (3.4), (3.5) and (3.6) respectively. If u_h^i is given, for any θ belonging to $[0,1]$ and $\bar{\theta}$ belonging to $[0,1]$, the inequality (3.7) has a unique solution. The argument for this is that the operator $(I - \theta k A_h - \bar{\theta} k B_h)$ is coercive for a number k sufficiently small. That yields the above result (cf. Glowinski *et al.* (1976) Chapter I, Theorems 2.1 and 2.2).

The system (3.7) provides a recursive relation between u_h^{i-1} and u_h^i, and can be solved step by step from the terminal condition $u_h^N = \psi_h$. The scheme proposed above is of θ-scheme type with different θ for the differential operator and integro-differential operator.

Obviously, two problems arise. The first is whether the scheme (3.7) is convergent. The second is to propose efficient algorithms to solve this approximate scheme.

Before discussing these two problems, we would like to write the discretization scheme in another equivalent matrix form. We denote by (u, v) the inner product of vectors u and v in $\mathbf{R}^n$; and $u \geq v$ if $\forall\, i \in \ \{1, \ldots, n\}$ $u_i \geq v_i$. We say that a matrix M is coercive if there exists a constant $C > 0$ such that, for all vectors $x \in \mathbf{R}^n$, $(Mx, x) \geq C|x|^2$, where $|x|$ is the Euclidean norm of x. The following proposition states that a variational inequality can be expressed as a linear complementarity problem (cf. Jaillet *et al.* 1990, Proposition 5.1).

Proposition 3.5 *Assume that M is a real square matrix; for all vectors u, q, φ of $\mathbf{R}^n$, the following two systems (3.8), (3.9) are equivalent:*

$$u \geq \varphi, \quad \forall\, v \geq \varphi \quad (Mu, v-u) \geq (q, v-u) \tag{3.8}$$

$$u \geq \varphi, \quad Mu \geq q, \quad (Mu-q, \varphi-u) = 0 \tag{3.9}$$

Let us introduce some notation. Suppose that $u(t,x)$ is a solution of problem (3.3). Let $m = m_2 - m_1 - 1$. We denote u^i, ϕ , F, three $\mathbf{R}^m$-vectors whose components are given respectively by $u^i_j = u(ik, (j+m_1)h)$, $\phi_j = \psi_{j+m_1}$, and $F_j = f_{j+m_1}$, for $1 \leq j \leq m$. With this notation, the above system (3.7) is equivalent to the following, due to Proposition 3.5:

$$u_0^{i-1} = \psi(m_1 h), \quad u_{m+1}^{i-1} = \psi(m_2 h), \quad 1 \leq i \leq N$$

$$\left\{ \begin{array}{l} u^N = \phi \\ \left.\begin{array}{l} Mu^{i-1} \geq Q^i, \quad u^{i-1} \geq \phi \\ (Mu^{i-1} - Q^i, \phi - u^{i-1}) = 0 \end{array}\right\} \quad 1 \leq i \leq N \end{array}\right. \tag{3.10}$$

with $M = (I - k\theta A_h - k\bar{\theta} B_h)$ and $Q^i = (I + k(1-\theta)A_h + k(1-\bar{\theta})B_h)u^i + kF$. Since we take Dirichlet boundary conditions, $(I - k\theta A_h - k\bar{\theta} B_h)$ will be represented by the following $m \times m$ matrix:

$$M = \tilde{M} + G \tag{3.11}$$

with $\tilde{M} = (\tilde{m}_{ij})_{m\times m}$ a tridiagonal matrix whose diagonal entries $\tilde{m}_{ii} = 1 + \theta b$, sub-diagonal entries $\tilde{m}_{i+1,i} = \theta a$ with $i = 1, \cdots, m-1$, over-diagonal entries $\tilde{m}_{i-1,i} = \theta c$ with $i = 2, \cdots, m$, otherwise $\tilde{m}_{i,j} = 0$, and $G = (g_{ij})_{m\times m}$ whose diagonal entries $g_{ii} = \lambda\bar{\theta}k(1 - g_0 h)$ and off-diagonal entries $g_{ij} = -\lambda\bar{\theta}khg_{j-i}$, for $i \neq j$. Here,

$$a = -\frac{k\sigma^2}{2h^2} + \frac{\beta k}{2h}, b = \frac{k\sigma^2}{h^2} + rk, c = -\frac{k\sigma^2}{2h^2} - \frac{\beta k}{2h}$$

with $\beta = \mu - (\sigma^2/2)$.

We denote by q^i the $\mathbf{R}^m$-vector given by

$$\begin{aligned} q^i_j &= u^i_j - (1-\theta)(au^i_{j-1} + bu^i_j + cu^i_{j+1}) \\ &\quad + \lambda k(1-\bar{\theta})\left(\sum_{l+j=0}^{m+1} u^i_{l+j} g_l h - u^i_j\right) + \lambda k \sum_{\substack{l+j>m+1 \\ l+j<0}} \phi_{l+j} g_l h. \end{aligned} \tag{3.12}$$

Then, the right-hand side Q^i of (3.10) is defined as follows: for all vectors q^i, we denote by Q^i, a vector with components:

$$\left\{ \begin{array}{ll} Q^i_1 = q^i_1 - \theta a\psi(m_1 h) + y_1 & \\ Q^i_m = q^i_m - \theta c\psi(m_2 h) + y_m & \\ Q^i_j = q^i_j + y_j & 2 \leq j \leq m-1 \end{array}\right.$$

where the vector $[y_j]$ is given by $y_j = \lambda\bar{\theta}khg_{-j}\psi(m_1h) + \lambda\bar{\theta}khg_{m-j+1}\psi(m_2h)$, for $1 \leq j \leq m$.

Clearly $1 + \theta b > 0$. Suppose that $\beta < \sigma^2/h$; we have $a < 0$, $c < 0$. Put $M = (m_{i,j})_{m\times m}$; it is not difficult to see that $m_{ii} > 0$ and $m_{ij} \leq 0$ if $i \neq j$. Furthermore, it can be shown that, for a number h sufficiently small, the matrix M is row and column strictly diagonally dominant. Thus, M is coercive.

From the definition of the matrix M (3.11), it is easy to see that a fully implicit scheme ($\theta = 1$, $\bar{\theta} = 1$) would lead (because of the integral operator) to a full matrix. In our numerical implementation, we chose $\theta = 1$, $\bar{\theta} = 0$. Therefore, we have $M = \tilde{M}$ which is a tridiagonal matrix (see the section 'Linear Complementarity Problem' below for a detailed discussion).

3.2.3 Convergence theorem

In a general setting, R. Glowinsky, J.L. Lions & R. Trémolières (1976) have studied the approximation schemes for variational inequalities and have proved a convergence theorem under a rather strong coercivity hypothesis. This hypothesis is not satisfied in our model with jumps, even in the model of Black–Scholes. So, the results in Glowinski *et al.* (1976) cannot be applied directly. This section presents a convergence theorem for the first order derivatives, which seems new, even in the case of a diffusion model without jumps (the proof of Jaillet *et al.* (1990) simply refers to Glowinski *et al.* (1976)). For the complete proofs of this theorem, refer to Zhang (1994).

Theorem 3.6 *Suppose* $u_{h,k}(t,x) = \sum_{i=1}^{N} u_h^i(x)\mathbf{1}_{](i-1)k,ik]}(t)$ *is the solution of discrete problem (3.7) and* u *is the solution of (3.3). Assume* $k/h^2 \leq \beta$ *where* β *is a sufficiently small constant. Let* $\theta, \bar{\theta} \in [0,1]$, h, $k \to 0$. *Then,*

- *if* $\psi \in W^{2,2,\alpha}(\mathbf{R})$, $u_{h,k}$ *converges strongly to* u *in* $L^2([0,T];H_\iota)$ *and* $\delta u_{h,k}$ *converges strongly to* $\partial u/\partial x$ *in* $L^2([0,T];H_\iota)$;
- *if* $\psi \in W^{1,2,\alpha}(\mathbf{R})$, $u_{h,k}$ *converges strongly to* u *in* $L^2([0,T];H_\iota)$ *and* $\delta u_{h,k}$ *converges weakly to* $\partial u/\partial x$ *in* $L^2([0,T];H_\iota)$.

3.2.4 Linear complementarity problem

As stated above, in order to calculate a value of an American option in the jump-diffusion model, it suffices to resolve problem (3.10). More precisely, for a given time step, it remains to solve a so-called linear complementarity problem (LCP) which can be written as follows: find vectors $W = (W_i)$ and

$Z = (Z_i)$ in $\mathbf{R}^m$ such that

$$\begin{cases} W = V + MZ & (3.13.1) \\ W \geq 0, \quad Z \geq 0 & (3.13.2) \\ W^T Z = 0 & (3.13.3) \end{cases} \tag{3.13}$$

where $M = (m_{ij}) \in \mathbf{R}^m \times \mathbf{R}^m$, $V = (V_i) \in \mathbf{R}^m$.

There are two main families of algorithms for the LCP: pivoting methods (direct methods) such as Cryer's algorithm and iterative methods (indirect methods) such as the projected SOR algorithm. Under suitable conditions, these pivoting methods are finite, and an exact solution will be obtained. Whereas the iterative methods are convergent in the limit, only an approximate value can be found. For problems of small to medium size, pivoting methods are very efficient. As an alternative to the pivoting methods, iterative schemes have their advantages in solving large-scale LCP. For a complete discussion of these problems, we refer readers to Cottle *et al.* (1992), Murty (1988). Our purpose in this section is to give basic ideas of these two types of methods and to suggest some efficient algorithms considering special properties of our problem (3.10).

As already mentioned above, in practice, we take $\theta = 1$, $\bar{\theta} = 0$ in (3.10). The matrix M defined by (3.11) becomes a tridiagonal Minkowski matrix (M has positive principal minors, positive diagonal entries and non-positive off-diagonal entries). Recall that Cryer's (1983) algorithm is efficient for solving LCP with a tridiagonal Minkowski matrix. That is exactly what we need.

We will present briefly the basic idea of this algorithm. For the detailed version, refer to Cryer (1983).

- **The algorithm of Cryer**

This algorithm is based on a direct method and is a modification of Saigal's (1970) algorithm. The main difference is that the data are scanned in alternating forward and backward passes.

The basic idea of this kind of algorithm is: Choose an initial value which satisfies both (3.13.1) and (3.13.3), maintain the two conditions during all steps and satisfy gradually the non-negative condition given in (3.13.2). The solution of the problem (3.13) is then obtained.

Let us introduce the following notation: $N = \{1, 2, \cdots, m\}$, if $J \subset N$, $|J|$ denotes the number of elements in J; $N \backslash J$ denotes the complement of J with respect to N; $(MZ+V)|J \geq 0$ denotes the system of $|J|$ inequalities obtained by deleting column and row indices not belonging to J. We recall that if M is a Minkowski matrix, so is $M|J$, and $M^{-1} \geq 0$. With this notation, the basic steps of this algorithm can be described as follows:

Step 0:

Choose $Z^0 \geq 0$, *such that* $(MZ^0 + V)|J^0 = 0$, *where* $J^0 = \{i \in N : Z_i^0 > 0\}$. *We can assume that*[1] $J^0 \supset Q = \{i \in N : V_i^0 < 0\}$. *Then, set* $p = 0$.

Step 1:

Denote $W^p = MZ^p + V$, *and* $I^p = \{i \in N| \ W_i^p < 0\}$.

- *If* $I^p = \phi$, *stop.* (W^p, Z^p) *is the solution of the problem.*
- *Otherwise, choose* $i^p \in I^p$ *and set* $J^{p+1} = J^p \bigcup \{i^p\}$. *Compute* Z^{p+1} *such that* $(MZ^{p+1} + V)|J^{p+1} = 0$, *and* $Z^{p+1}|N \backslash J^{p+1} = 0$.

Set $p = p + 1$ *and repeat until* $I^p = \phi$ *or* $p = m$.

The above basic algorithm is valid for any Minkowski matrix. In the particular case where M is a tridiagonal Minkowski matrix, an implementation of this basic method which minimizes the amount of computation can be found in Cryer (1983).

We now conclude our discussion about LCP by presenting an iterative method.

- **The algorithm of the projected SOR**

Similar to the SOR (Successive Over Relaxation) scheme for a certain class of matrix equation, the projected SOR scheme (see Cottle *et al.* 1992) is frequently used to solve large-scale LCP. The basic steps can be described as follows:

Step 0:

Choose $Z^0 \geq 0$. *Then, set* $p = 0$.

Step 1:

Denote $Z^p = (Z_i^p)_{1 \leq i \leq m}$; *form an intermediate vector* $Y^{p+1} = (Y_i^{p+1})_{1 \leq i \leq m}$ *as follows:* $Y_i^{p+1} = -\frac{1}{m_{ii}}(V_i + \sum_{j=1}^{i-1} m_{ij} Z_j^{p+1} + \sum_{j=i}^{m} m_{ij} Z_j^p)$, *define the new vector* Z^{p+1} *by:*

$$Z_i^{p+1} = max(0, Z_i^p + \omega(Y_i^{p+1} - Z_i^p)). \tag{3.14}$$

Set $p = p + 1$ *and repeat until* $|Z^{p+1} - Z^p| < \varepsilon$ *where* ε *is the prescribed precision.*

The word '*projected*' refers to the fact that Z_i^{p+1} is the projection of the vector $Z_i^p + \omega(Y_i^{p+1} - Z_i^p)$ onto the non-negative orthant. The constant ω is often called a relaxation parameter. Assuming that the matrix M is invertible and positive definite, and under conditions $0 < \omega < 2$ and $Z^0 \geq 0$, the above algorithm is convergent. In addition, the convergence can be optimized by

[1] At least, for example, we can take $J^0 = Q$, it is easy to see that, $MZ^0|Q = -V|Q > 0$, then $Z^0|Q > 0$ since M is a Minkowski matrix.

choosing a particular value of $\omega \in (1,2)$ which depends on the matrix. The advantage of the projected SOR for LCP is that it has only one more step (3.14) compared to the SOR scheme for the matrix equation. Therefore, the European and American options can be implemented easily in a single program.

Remark 3.7 *To price a European option on several assets by the finite difference method, a so-called alternating direction method is widely used and often proved to be very efficient. By analogy with the European case, valuation of an American option on several assets may be solved iteratively by solving a sequence of 'one-dimensional' LCP involving a tridiagonal Minkowski matrix (see Lin & Cryer (1985) for a detailed discussion about the alternating direction method for LCP).*

Remark 3.8 *Brennan & Schwartz's* (1977) *algorithm* (*see also Jaillet* et al. 1990) *seems still valid numerically in this jump-diffusion model. But, we were not able to give a rigorous justification as in Jaillet* et al. (1990) *because of the jumps term.*

Remark 3.9 *The advantage of the finite difference method with respect to other approximation methods (such as analytical methods ...) is its flexibility. For example, it can be easily adapted to price many non-standard or exotic options and can provide accuracy to any required degree. In addition, it is worth mentioning that all commonly used lattice methods are special cases of explicit finite difference schemes.*

4 Numerical results

We consider two particular cases: the Mini-Krach model (in which the relative amplitude of jumps is constant with a given negative value, denoted by $-a$, $a \geq 0$), and the log-normal model (in which the random variable $1+U_j$ follows a log-normal distribution), since closed-forms for the European option price exist in both cases (Lamberton & Lapeyre 1992, Merton 1976). Clearly, jumps in the log-normal model could be positive or negative. Hereafter, we will show at first some implied volatility curves (IVC), and then we will compare the jump-diffusion model (in the log-normal case) with the Black–Scholes model in the American case.

- **Implied volatility**

It is not difficult to see that the predictable quadratic variation of $\tilde{S}_t (= e^{-rt} S_t)$ in the jump-diffusion model is given by $\bar{\sigma}^2 \int_0^t \tilde{S}_u^2 du$ with $\bar{\sigma}^2 = \sigma^2 + \lambda \mathbf{E} U_1^2$. Recall that, in the Black–Scholes model, the quadratic variation of $\tilde{S}_t$ equals

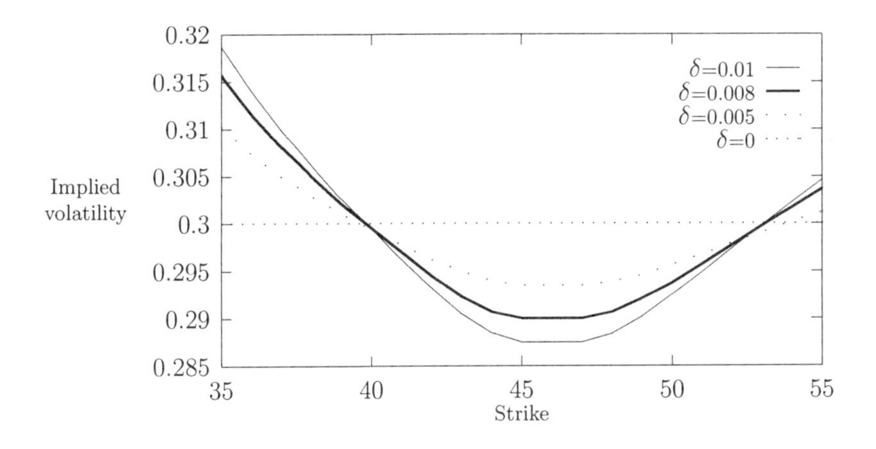

Figure 1: Implied volatility (log-normal model). $\bar{\sigma} = 0.3$, $p = 0.3$, $\mathbf{E}U_1 = 0$

$\sigma^2 \int_0^t \tilde{S}_u^2 du$. Clearly, the presence of jumps modifies the observable volatility of random process $(\tilde{S}_t)_{t\geq 0}$. And the volatility of the Brownian motion (we call the coefficient before the Brownian motion in the SDE the *volatility* of the Brownian motion) is not observable. If $\lambda \mathbf{E}U_1^2 = 0$, the jump-diffusion model reduces to the Black–Scholes model. Let us denote $p = \sigma^2/\bar{\sigma}^2$, $\delta = \mathbf{E}U_1^2$; we get:

$$p + \frac{\lambda}{\bar{\sigma}^2}\delta = 1.$$

Evidently, $\delta = 0$ (i.e. $p = 1$) corresponds to the case without jumps (Black–Scholes model), other cases correspond to the model with jumps.

The implied volatility lets us equalize the price in the Black–Scholes model to the price observed in the market. By changing the value of strike K, we can get a curve of implied volatility as a function of strike. Now, supposing that the 'good' model corresponding to the market is the model with jumps, we calculate the value of the option by using the pricing formula in the jump-diffusion model with a historic volatility $\bar{\sigma} = 0.3$. The remaining parameters are fixed and are chosen in the following way: the interest rate $r = 9\%$, spot $x = 45FF$, and residual maturity $s = T - t = 30/365$. In Figures 1 and 2, we fix values of p and $\bar{\sigma}$, and let the value of δ vary. We find that the implied volatility curve's convexity increases with δ. If we change the value of $\bar{\sigma}$ (or change the value of p), and fix the other parameters, we observe the same phenomenon (see Zhang (1994) Chapter 6). Another interesting thing is that the forms of the implied volatility curves may be changed if the distribution of U_j is different.

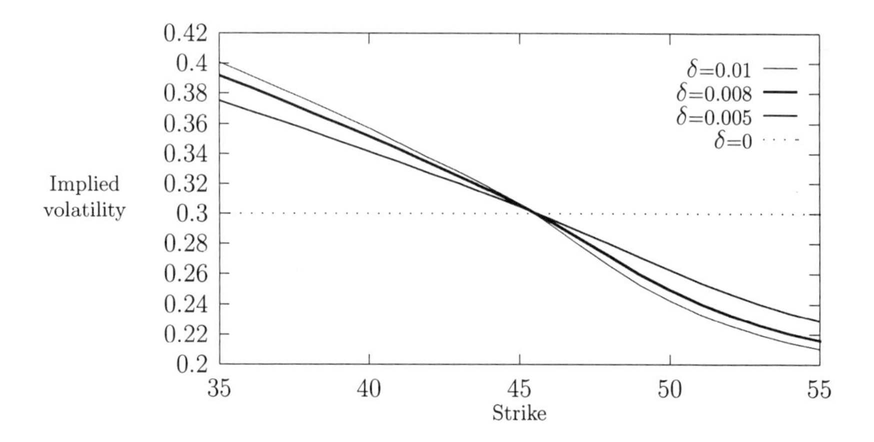

Figure 2: Implied volatility (Mini-Krach model). $\bar{\sigma} = 0.3$, $p = 0.3$

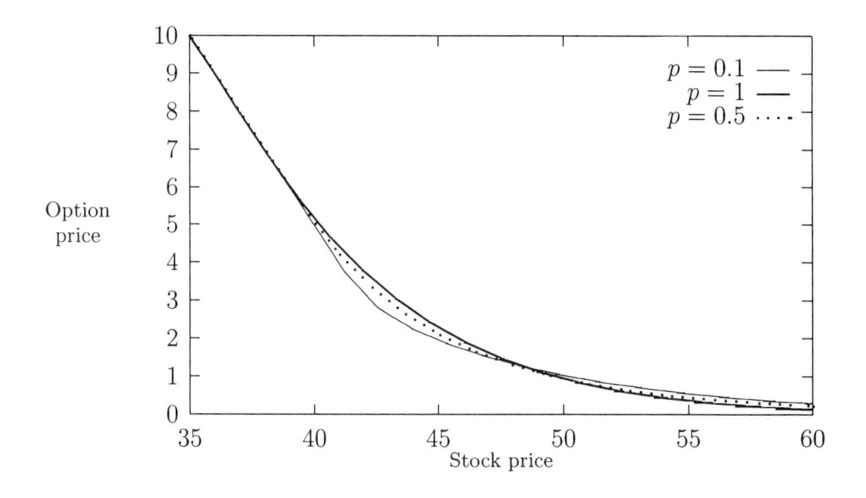

Figure 3: American option. The B-S model and jump-diffusion model (log-normal case). $r = 9\%$, $K = 45$, $T = 1$ year, $\bar{\sigma} = 0.2$, $EU_1 = 0$, $VarU_1 = 0.04$

By using this notion of IVC, we can also provide a complete interpretation for the comparison of the Black–Scholes and jump-diffusion models (see Figure 3). Let us focus our discussion on the log-normal case. Suppose that observed volatility in the market is $\bar{\sigma} = 0.3$. According to Figure 1, when the option is strongly out of the money, the implied volatility σ is greater than the observed volatility $\bar{\sigma}$ in the market. So, if a trader takes $\bar{\sigma}$ as the volatility of Brownian motion, it means that he has taken a weaker volatility. Thus, the price in the Black–Scholes model is less than the price in the model with jumps. On the other hand, when we are at the money, the Brownian volatility σ is weaker than the observed volatility $\bar{\sigma}$. So if a trader takes $\bar{\sigma}$ as the Brownian volatility, he has made an overestimate, which implies the price in the Black–Scholes model is greater than that in the jump-diffusion model.

- **Comparison between Black–Scholes and jump-diffusion models**

Since $\bar{\sigma}$ defined above is the observed volatility in the market, for comparing these two models, we should take $\bar{\sigma}$ as the volatility in both cases. Note that in the jump-diffusion model $\bar{\sigma}^2 = \sigma^2 + \lambda \mathbf{E} U_1^2$ and in the Black–Scholes $\bar{\sigma}^2 = \sigma^2$. The difference of American option prices in the Black–Scholes and jump-diffusion models (log-normal case) are illustrated in Figure 3. When the option is strongly in the money, the existence of jumps is not sufficient for us to 'hope' for the spot to pass over the strike, and we are almost sure that the option will be exercised. We find, both in the Black–Scholes and in the jump-diffusion model, that the value of an American put is near to the pay-off value $K - S$; When the option is strongly out of the money, the option's price in the model with jumps is greater than that in the Black–Scholes model; when the option is at the money, the option's price in the model with jumps is smaller than that in the Black–Scholes model.

Note that we find almost the same phenomena in European case (see Zhang (1994) Chapter 6 for a detailed discussion about numerical results).

Acknowledgments

This work has been done in CERMICS, ENPC. The author is very grateful to her advisor Professor D. LAMBERTON for his instruction and helpful discussions.

References

Amin, K.I. (1993) 'Jump Diffusion Option Valuation in Discrete Time', *Journal of Finance* **48**, 1833–1863.

Ball, A.C. & Torous, W.N. (1985) 'On Jumps in Common Stock Prices and Their Impact on Call Option Pricing', *Journal of Finance* **40**, 155–173.

Barles, G., Daher, C. & Romano, M. (1994) 'Optimal control on the L^∞ norm of a diffusion process', *SIAM Journal of Control and Analysis* **32**, 612–634.

Barles, G., Daher, C. & Romano, M. (1995) 'Convergence of numerical schemes for parabolic equations arising in finance theory', *Math. Models and Methods in Applied Sciences* **5**, 125–143.

Bensoussan, A. (1984) 'On the theory of option', *Acta Applicandae Mathematicae* **2**, 139–158.

Bensoussan, A. & Lions, J.L (1978) *Applications des Inéquations Variationnelles en Contrôle Stochastique*, Dunod.

Bensoussan, A. & Lions, J.L (1982) *Contrôle Impulsionnel et Inéquations Quasi-variationnelles*, Dunod.

Black, F. & Scholes, M. (1973) 'The pricing of options and corporate liabilities', *Journal of Political Economy* **81**, 635–654.

Bouleau, N. & Lamberton, D. (1989) 'Residual risks and hedging strategies in Markovian markets', *Stoch. Proc. and Appl.* **33**, 131–150.

Brennan, M.J. & Schwartz, E.S. (1977) 'The valuation of American put options', *Journal of Finance* **32**, 449–462.

Broadie, M. & Detemple, J. (1996) 'American option valuation: new bounds, approximations, and a comparison of existing methods', *Review of Financial Studies* **9**, No. 4.

Carr, P. & Faguet, D. (1994) 'Fast accurate valuation of American options', *Working paper, Cornell University.*

Colwell, David B. & Elliott, Robert J. (1993) 'Discontinuous asset prices and non-attainable contingent claims', *Mathematical Finance* **3**, 295–308.

Cottle, R.W., Pang, J.S. & Stone, R.E. (1992) 'The Linear Complementarity Problem', Academic Press.

Cox, J.C. & Ross, S.A. (1976) 'The valuation of options for alternative stochastic processes', *Journal of Finance* **3**, 145–166.

Cryer, C.W. (1983) 'The efficient solution of linear complementarity problems for tridiagonal Minkowski matrices', *ACM Trans. Math. Softwave* **9**, 199–214.

El Karoui, N. & Quenez, M.C. (1995) 'Dynamic programming and pricing of contingent claims in an incomplete market', *Siam J. of Control and Opt.* **33**, 29–66.

Fitzpatrick, B.G. & Fleming, W.H. (1991) 'Numerical methods for an optimal investment-consumption model', *Mathematics of Operations Research* **16**, 823–841.

Föllmer, H. & Schweizer, M. (1990) 'Hedging of contingent claims under incomplete information', Davis, M.H.A. & R.J. Elliot (eds.), Gordon and Breach,.

Föllmer, H. & Sondermann, D. (1986) 'Hedging of non-redundant contingent ilaims', In *Contributions to Mathematical Economics in Honor of Gerard Debreu*, Hildebrand, W. & Mas-Colell, A. (eds.), North-Holland.

Glowinski, R., Lions, J.L. & Trémolières, R. (1976) *Analyse Numérique des Inéquations Variationnelles*, Dunod.

Hull, J. & White, A. (1987) 'The pricing of options on assets with stochastic volatilities', *Journal of Finance* **42**, 281–300.

Jaillet, P., Lamberton, D. & Lapeyre, B. (1990) 'Variationnal inequalities and the pricing of American options', *Acta Applicandae Mathematicae* **21**, 263–289.

Jarrow, R. & Rosenfeld, E. (1984) 'Jumps risks and the intertemporal capital asset pricing model', *Journal of Business* **57**, 337–351.

Jeanblanc-Picque, M. & Pontier, M. (1990) 'Optimal portfolio for a small investor in a market model with discontinuous prices', *Applied Mathematics Optimization* **22**, 287–310.

Jorion, P. (1988) 'On jump processes in the Foreign Exchange and Stock Markets', *Review of Financial studies* **4**, 427–445.

Karatzas, I. (1988) 'On the pricing of American options', *Applied Mathematics Optimization* **17**, 37–60.

Lamberton, D. & Lapeyre, B. (1992) *Introduction au Calcul Stochastique Appliqué à la Finance*, Ellipses-Edition Marketing.

Lin, Y. & Cryer, C. W. (1985) 'An alternating direction implicit algorithm for the solution of linear complementarity problems arising from free boundary problems', *Appl. Math. Opt.* **13**, 1–17.

MacMillan, L. (1986) 'Analytic approximation for the American put option', *Advances in Futures and Options Research* **1**, 119–139.

Merton, R.C. (1973) 'Theory of rational option pricing', *Bell J. of Econom. and Management Sci.* **4**, 141–183.

Merton, R.C. (1976) 'Option pricing when underlying stock returns are discontinuous', *J. of Financial Economics* **3**, 125–144.

Mulinacci, S. (1994) 'An approximation of American option prices in a jump-diffusion model', *Preprint.*

Murty, K. (1988) 'Linear Complementarity, Linear and Nonlinear Programming', In *Sigma Series in Applied Mathematics* **3**, Heldermann Verlag.

Myneni, R. (1992) 'The price of the American option', *Annals of Applied Probability* **2**, 1–23.

Pham, H. (1995) *Applications des Méthodes Probabilistes et de Contrôle Stochastique aux Mathématiques Financières*, Thèse de l'Université Paris IX Dauphine.

Pham, H. & Touzi, N. (1994) 'Equilibrium state prices in a stochastic volatility model', *preprint CREST.*

Saigal, R. (1970) 'A note on a special linear complementary problem', *Opsearch* **7**, 175–183.

Schweizer, M. (1991) 'Option hedging for semi-martingales', *Stochastic Processes and their Applications* **37**, 339–363.

Schweizer, M. (1994) 'Approximating random variables by stochastic integrals', *Annals of Probability* **22**, 1536–1575.

Shirakawa, H. (1990) 'Security market Model with Poisson and diffusion type return process', *Institute of Human and Social Sciences, Tokyo Institute of Technology*, preprint.

Tourin, A. & Zariphopoulou, T. (1994) 'Numerical schemes for investment models with singular transactions', in *Computational Economics* **7**, 287–307.

Zhang X.L. (1993) 'Options Américaines et modèles de diffusion avec saut', *Note aux Comptes Rendus de l'Académie des Sciences, Paris, Série I,* **317**, 857–862.

Zhang X.L. (1994) *Méthodes Numériques pour le Calcul des Options Américaine dans des Modèles de Diffusion avec Sauts*, Thèse de doctorat de l'Ecole Nationale des Ponts et Chaussées, Paris.

Zhang X.L. (1995) 'Formules quasi-explicites pour les options Américaines dans un modèle de diffusion avec sauts', *Mathematics and Computers in Simulation* **38**, 151–161.

Some Nonlinear Methods for Studying Far-from-the-money Contingent Claims

E. Fournié, J.M. Lasry and P.L. Lions

1 Introduction

The purpose of most models is to explain the most commonly occurring phenomena. When designing a model and fitting parameter values, large-deviations situations are ignored, at least to begin with, and the success of the model is judged by the way it can explain common situations. Although models designed in this way will have implications for large deviations, those same large-deviations situations may reveal a richer structure not evident in 'typical' behaviour.

We are going to show how, in the framework of diffusion models (the most commonly-used of financial models), one may be led to take as constant quantities whose variable and random characteristics only appear in large motions of the market. We will give a general statement on the information likely to appear in large deviations of the market, and on other latent risks.

This applies naturally to models where volatilities or correlations are supposed constant, even though in reality this is not the case. Our results suggest an explanation of the 'smile' of the implied volatility curve (that is, overvaluation of options far from the money). Furthermore, the results display implied volatility smiles as a quite general phenomenon.

This is the reason why we will start by presenting this particular case, which will give an introduction and an intuition to the general results in Section 3.

Finally, in Section 4, we will show how the asymptotic estimations in Section 3 can be used in order to improve numerical calculation methods for out-of-the-money contingent claims. Our asymptotic results let us considerably accelerate Monte Carlo simulations, and also provide the means to replace ill-conditioned linear PDEs with well conditioned nonlinear PDEs.

2 A First Qualitative Approach Based on an Example

The archetype of financial models is the (celebrated) Black–Scholes model (1973). It contains almost all paradigms providing a framework for the financial models which have been developed since then. The practical use of this model is also typical of adaptations and compromises which have to be made in order to go from a theory to the practical use of models in finance.

Let us briefly recall the main facts concerning the Black–Scholes model (BS).

In the model developed by Black and Scholes to price and hedge options, the volatility of the underlying asset is supposed to be constant. From this hypothesis, Black and Scholes derived an analytical formula giving the value of a European option depending on the volatility parameter σ. But, actually:
(i) the volatility σ is not constant,
(ii) the market does not adopt the value V_{BS} provided by the Black–Scholes model.

However, the practitioners of financial markets widely used the Black – Scholes model up to some modifications not expected in the theory:
(iii) the negotiations preparing the transactions are expressed in terms of implicit volatility rather than in terms of the price P of the option (the implicit volatility σ_{imp} is linked one-to-one to the price by the equation $V_{BS}(\sigma_{\text{imp}}) = P$ where V_{BS} is the value given by the Black–Scholes formula).
(iv) the theoretical delta hedging recommended by the model, i.e. $\Delta(\sigma) = (\partial V_{BS}/\partial S)(\sigma)$ where S is the value of the underlying asset, is used by practitioners, not according to the theory, but by replacing σ by σ_{imp}.
(v) the implicit volatility depends on the option's maturity T and is generally higher far-from-the-money (smile effect).

Thus, the BS model plays a very important role in practice. However, this role is not the one expected by the main hypotheses in the initial BS theory.

One could presume that the compromises and adaptations in the use of BS are anecdotal and specific to this model. But, on the contrary, we wish to show that they provide general indications which can be useful in many other financial models.

The purpose of this paper is to explore [1] what the large deviations of the market reveal about the 'constants' of a model.

When a 'constant' in a diffusion model oscillates between several close values, these oscillations may not be revealed if they are hidden by an averaging or a smoothing. However, large deviations will reveal the existence of these oscillations. This is the meaning of the results presented later on in Section 3.

Let us now give a first intuitive indication of these results from the Black–Scholes model. In order to clarify our ideas, we continue with the equations; we recall the formal framework of the model of Fisher Black and Myron Scholes.

---

[1] With the same purpose – i.e. to study the effect of variations of constants, i.e. phenomena which appear when the constants of a model become random variables – E. Fournié, J. Lebuchoux & N. Touzi (1996) have studied small noise perturbations.

An underlying asset S follows a diffusion law given by the stochastic differential equation below, where W is a standard Brownian motion defined on some probability space $(\Omega, \mathcal{F}, P)$

$$\frac{dS}{S} = b\,dt + \sigma\,dW \tag{2.1}$$

where the volatility σ is supposed to be constant.

The riskless rate is also supposed to be constant. Then the market of the underlying asset S and its derivatives is complete. Under the risk neutral probability Q, we have: $b = r$, the riskless rate, in equation (2.1) and if $C = C(t, S)$ is the value of a contingent claim on S at the date t, we have, according to Itô's lemma

$$dC = \left(\frac{\partial C}{\partial t} + rS\frac{\partial C}{\partial S} + \frac{1}{2}\sigma^2 S^2 \frac{\partial^2 C}{\partial S^2}\right) dt + \sigma S \frac{\partial C}{\partial S}\,d\tilde{W} \tag{2.2}$$

where $\tilde{W}$ is a standard Brownian motion under Q.

Under the risk-neutral probability Q, we have $\mathbf{E}_Q(dC/C) = r\,dt$ and identifying with the previous equation, we get the famous BS equation:

$$-\frac{\partial C}{\partial t} - rS\frac{\partial C}{\partial S} - \frac{1}{2}\sigma^2 S^2 \frac{\partial^2 C}{\partial S^2} + rC = 0.$$

This equation is to be completed by a boundary condition (BC) given by (in the European case) the terminal value of the contingent claim, which specifies completely the derivative product. For instance, for a European call with maturity T and strike K:

$$C(T, S) = \max(S - K, 0).$$

Now, let us suppose that the volatility is not a constant, but on the contrary changes randomly according to a two-state Markov process for simplicity (the general case is studied in Section 3).

Suppose that the transition probability[2] from σ_1 to σ_2 (resp. σ_2 to σ_1) during the infinitesimal time dt is $(c_1/\varepsilon)\,dt$ (resp. $(c_2/\varepsilon)\,dt$) where ε is a 'small' parameter. Taking into account the normal conditions of price observation, what the market will observe if the parameter ε is small, is that, on ordinary days (i.e. 99 days out of 100), the volatility appears constant $\sigma = \bar{\sigma}$. We give in Section 3 a precise statement about this theme. Besides, it can be seen that the value of $\bar{\sigma}$ is linked to the constants $\sigma_1, \sigma_2, c_1, c_2$ by the formula:

$$\bar{\sigma} = \sqrt{\frac{c_2\sigma_1^2 + c_1\sigma_2^2}{c_1 + c_2}}.$$

[2]Here, we speak about real probability. The link with risk-neutral probability will be further discussed in Section 3.

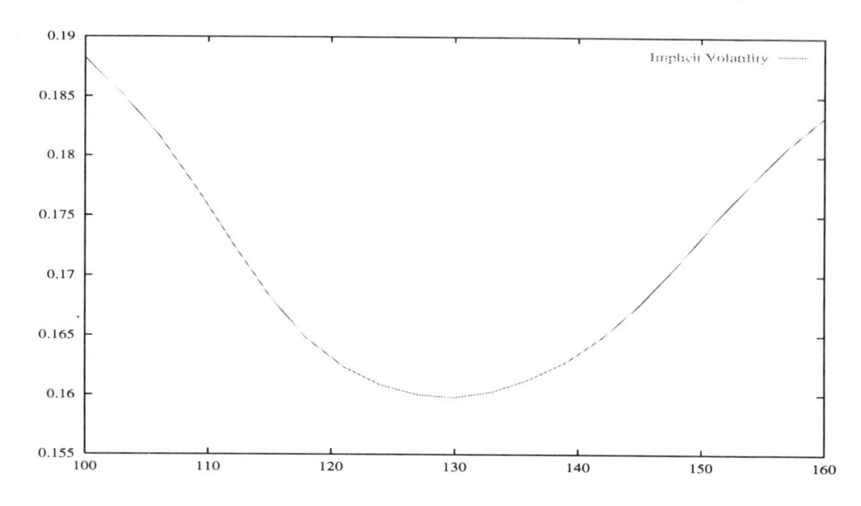

Figure 1: Implicit volatility with $\sigma_1 = 0.15$, $\sigma_2 = 0.2$, $\alpha = 0.75$, $T = 1$ month

The phenomenon is intuitive: if the volatility σ fluctuates quickly between two values σ_1 and σ_2, the observer sees a constant average value $\bar{\sigma}$ (its exact value results from the statistical theorem in Appendix 2).

However, this constant character of σ disappears on days (quite rare) when the price of the underlying asset makes a large deviation. In the next part, we will give an exact and general explanation of the computation of these large deviations. At this point, we would like first to give a intuitive approach.

Let us consider the usual scale of parameters for financial assets, for example $\sigma_1 = 0.12$ and $\sigma_2 = 0.16$. Suppose that we have $c_1 = c_2$, and that ε is 'small' (for instance, $\varepsilon = 0.01$ in this context). Apart from the moments of large deviations, the market 'observes' $\bar{\sigma} = \sqrt{(\sigma_1^2 + \sigma_2^2)/2} = 0.1414$, which implies an option price of $C = C_{\mathrm{BS}}(t, S, \bar{\sigma})$ where C_{BS} is the value resulting from the Black–Scholes model (that is the equation (BS) complete with (BC) boundary conditions with $\sigma = \bar{\sigma}$).

However, if the market perceives the fact that the constant $\bar{\sigma}$ results from a mixture of the two values σ_1 and σ_2, the market can rely on the two values $C_1 = C_{\mathrm{BS}}(t, S, \sigma_1)$ and $C_2 = C_{\mathrm{BS}}(t, S, \sigma_2)$ in the process of formation of equilibrium prices for the options. But, it is easy to notice that C_1/C_2 is close to zero for options far-from-the-money. Actually, far-from-the-money, the value of an option is increasing quickly with the value of volatility.

Suppose that the market prices the option through a combination of C_1 and C_2, $C_{\mathrm{BS}}(S, \sigma_{\mathrm{imp}}) = \alpha\, C_{\mathrm{BS}}(S, \sigma_1) + (1 - \alpha)\, C_{\mathrm{BS}}(S, \sigma_2)$. Then for any value of the parameter α, we find a graph for $\sigma_{\mathrm{imp}}(K)$ (see Figure 1), which shows that the implicit volatility σ_{imp} is close to σ_2 for options far-from-the-money.

In addition, the idea of having an implicit volatility σ_{imp} close to σ_2 far-from-the-money appears as an extra risk at-the-money, leading to an implicit volatility σ_{imp} higher than σ_1 at-the-money.

Thus, even without any specifications of the model, this first analysis results in a graph of the implicit volatility revealing a smile effect.

The model described in the next part is a formalization of this simplified, intuitive approach.

3 Asymptotic Expansion of Value Functions

We present some results and numerical methods which we think are adapted to the study of general phenomena which drive the distortion of financial models, when one retreats from their initial hypotheses, that is the hypotheses adapted to ordinary random events of markets.

Our statements, when applied, show the existence of extra risks which only appear explicitly in the large deviations of the market and are hidden under standard conditions. We will show, by putting together the mathematical results in this part and the qualitative approach in the previous part, that this implies a general characteristic of the smile phenomenon of options.

We will now give in this section mathematical results which provide a precise background to the qualitative suggestions of Section 2. Most financial continuous time models assume that the state dynamics is a diffusion in some subset $\bar{\Omega}$ of $\mathbb{R}^n$. In order to cover very general cases and meaningful examples we will assume that the state is a pair $\left(X^\lambda(t), \lambda(t)\right)$ where $X^\lambda(t)$ follows a diffusion process

$$dX^\lambda \;=\; b^\lambda(X^\lambda)\,dt + \sigma^\lambda(X^\lambda)\,dW$$

where $b^\lambda(.), \sigma^\lambda(.)$ are smooth functions depending on λ and X^λ and where λ is a Markov process with values $\lambda(t)$ in a finite set Λ with transition rate $\{\gamma_{\lambda\mu}; \lambda, \mu \in \Lambda\}$, and W is a standard $\mathbb{R}^m$-valued Brownian motion.

The Markov process will provide a way to investigate risk situations where the correlations between assets or factors can change. This issue is important because such changes in correlations appear on the market and induce a large increase of the probability of the tail events as we will see (and this is intuitive in simple examples – see Section 2).

The above random processes are defined under the risk-neutral probability. But if the risk-neutral probability measure is replaced by the real probability measure, it just changes the drift function b^λ and the Markov transitions $\gamma_{\lambda\mu}$ while σ^λ remains unchanged. As b^λ and $\gamma_{\lambda\mu}$ do not enter the asymptotic expansion results (see the theorem below), the formula will be the same under both probabilities (real and risk-neutral).

The mathematical result reported in this paper is a typical example of the results that can be found in Lasry & Lions (1995).

First we need some technical and natural assumptions and some more notation. We assume for simplicity that Ω is a smooth, bounded open set in $\mathbb{R}^n$ whose boundary is split into two smooth pieces Γ_0, Γ_1. We assume that Γ_0 and Γ_1 are closed in Γ with non-empty interiors in Γ and that $\Gamma_0 \bigcup \Gamma_1 = \Gamma$.

Next we assume that the functions b^λ and σ^λ are Lipschitz on $\bar{\Omega}$ (with values respectively in $\mathbb{R}^n$, and in the space of $n \times m$ matrices, assuming that W is an m-dimensional Brownian motion). We define by $a^\lambda = \frac{1}{2}\sigma^\lambda(\sigma^\lambda)^*$ and we assume that a^λ is positive on $\bar{\Omega}$.

We next introduce τ_i^λ, the first hitting time of Γ_i for $i = 0, 1$, i.e.:

$$\tau_i^\lambda = \inf\{t \geq 0, X^\lambda(t) \in \Gamma_i\}$$

where $X^\lambda(t)$ is the above diffusion process starting at $X^\lambda(0) = x \in \bar{\Omega}$ and $\lambda(0) = \lambda$ (with a small risk of confusing the function $\lambda(.)$ and the constant value λ). We wish to investigate the behaviour for small $t > 0$ of (for instance)[3]

$$P\left(\tau_1^\lambda < \inf(\tau_0^\lambda, t)\right) = u^\lambda(t, x)$$

for $x \in \bar{\Omega}$ and $t \geq 0$ (where $P(F)$ is the probability of event F).

An easy adaptation of classical results yields the following.

Proposition 3.1 *The family (u^λ), $\lambda \in \Lambda$, is the unique smooth $C^{1,2}([0,t] \times \bar{\Omega})$, $t > 0$, solution of*

$$\partial_t u^\lambda - \sum_i b_i^\lambda \partial_i u^\lambda - \sum_{i,j} a_{ij}^\lambda \partial_{ij} u^\lambda + \sum_{\lambda \neq \mu} \gamma_{\lambda\mu}(u^\mu - u^\lambda) \quad = \quad 0$$

with boundary conditions

$$\begin{aligned} u^\lambda &= 0, \text{ on } \Omega, \\ u^\lambda(t,x) &= 0, \text{ on } \Gamma_0, \ t \geq 0 \\ u^\lambda(t,x) &= 1, \text{ on } \Gamma_1, \ t \geq 0. \end{aligned}$$

Before focusing on large deviations, let us investigate briefly what we earlier called fast oscillations of σ leading to a constant value $\bar{\sigma}$. If the transition coefficients $\gamma_{\lambda\mu}$ are large, say $\gamma_{\lambda\mu} = \theta_{\lambda\mu}/\varepsilon$ with small parameter ε, then observers will face a mixture of the $b^\lambda, \sigma^\lambda$ parameters. More precisely, one has the following asymptotic result.

Proposition 3.2 *If the functions (u_ε^λ) are the solutions of the system described in Proposition 3.1 with $\gamma_{\lambda\mu} = \theta_{\lambda\mu}/\varepsilon$ then for all $\lambda \in \Lambda$, $\lim_{\varepsilon \to 0} u_\varepsilon^\lambda = \bar{u}$ where $\bar{u}$ is the solution of*

$$\partial_t \bar{u} - \sum_i \bar{b}_i \partial_i \bar{u} - \sum_{i,j} \bar{a}_{ij} \partial_{ij} \bar{u} \quad = \quad 0$$

[3]The study of $u^\lambda(t,x)$ for small t is, by rescaling, a way to investigate the tail of the probability distribution, the large deviations of the market unexpected in a given amount of time. This rescaling is explained in Appendix 6 – see also Lasry & Lions (1995).

with boundary conditions

$$\begin{aligned}\bar{u}(0,x) &= 0, \text{ on } \Omega,\\ \bar{u}(t,x) &= 0, \text{ on } \Gamma_0,\ t \geq 0\\ \bar{u}(t,x) &= 1, \text{ on } \Gamma_1,\ t \geq 0\end{aligned}$$

and where $\bar{b}_i$ and $\bar{a}_{ij}$ are the expectations of b_i^λ and a_{ij}^λ under the stationary distribution of the Markov process on Λ associated to the transition coefficients $(\theta_{\lambda\mu})_{\lambda,\mu}$.

The proof of this proposition is given in Appendix 2. Introduce τ_i, the first hitting time of Γ_i for $i = 0, 1$, i.e.:

$$\tau_i = \inf\{t \geq 0, X(t) \in \Gamma_i\}$$

where $X(t)$ is the diffusion process defined by $dX = \bar{b}(X)\,dt + \bar{\sigma}(X)\,dW$ with initial state $X(0) = x \in \bar{\Omega}$. Then the above PDE implies that

$$\bar{u}(t,x) = P\left(\tau_1 < \inf(\tau_0, t)\right)$$

for $x \in \bar{\Omega}$ and $t \geq 0$ (where $P(F)$ is as before the probability of event F).

Proposition 3.2 shows that if the Markov process moves fast, an observer will usually only observe the constant $\bar{b}$ and $\bar{\sigma}$, while the richer underlying structure, the Markov process $b^\lambda, \sigma^\lambda, \lambda \in \Lambda$, will be hidden. Our point is that for fixed $t > 0$, if ε is small, the market will only observe $\bar{u}$ and ignore the existence of the Markov process λ , while for fixed ε and $t > 0$ small enough the richer underlying structure will partly appear. This will be expressed by the next theorem and concerns large motions of the market.

So, let us turn now to our main result.

Theorem 3.3 *Let us assume the Markov process (λ_t) satisfies the following condition:*

(T) *for each pair (λ, μ) of elements of Λ $(\lambda \neq \mu)$, there exist an integer k and a sequence $\eta_0, \ldots, \eta_{k+1}$ of elements in Λ such that $\eta_0 = \lambda$, $\eta_{k+1} = \mu$ and $\gamma_{\eta_i\eta_{i+1}} > 0, \forall i = 0, \ldots, k$ (if $\gamma_{\lambda\mu} > 0$, one can just take $k = 0$).*

Then, we have for all $\lambda \in \Lambda$, $-2t \log u^\lambda(t,x) \longrightarrow L^2(x)$ uniformly on compacts subsets of $\Omega \cup \Gamma_0$ when $t \to 0^+$, where L is the unique Lipschitz viscosity solution of

$$\sup_{\lambda\in\Lambda}\left\{\sum_{i,j} a_{ij}^\lambda \partial_i L\, \partial_j L\right\} = 1, \text{ in } \Omega \tag{3.1}$$

$$\geq 1, \text{ on } \Gamma_0 \tag{3.2}$$

$$L = 0, \text{ on } \Gamma_1. \tag{3.3}$$

Moreover, the function L is the solution of the following 'critical path problem'

$$L(x) = \inf\{L(x,y), y \in \Gamma_1\}$$

where

$$L(x,y) = \inf_H \int_0^1 \langle A^{\lambda(t)}(z(t))\, \dot{z}(t), \dot{z}(t)\rangle^{1/2}\, dt$$

where $H \subset C^1$, $H = \{(z(.), \lambda(.)) : z(0) = x, z(1) = y, z(t) \in \Omega \cup \Gamma_0, \forall t \in [0,1], (\lambda_t) \in L^\infty([0,1]; \Lambda)\}$ *and* $A^{\lambda(.)} = (a^{\lambda(.)})^{-1}$.

The proofs of these results, which use nonlinear PDE methods, are given in Lasry & Lions (1995)[4].

Comments on the above results

These results might look quite technical at first sight. Hence we will now give comments and explanations.

(a) The first interesting situation is the case where Λ reduces to one value, i.e. the case where λ is constant. Then the L-equation reduces to:

$$\sum_{i,j} a_{ij}\, \partial_i L\, \partial_j L = 1$$

which, together with the boundary conditions on Γ, means that $L(x)$ is the Riemannian distance from x to Γ_1. We will draw in Appendix 4 some pictures which shows intuitively how and why the Riemannian geodesics appear in this story.

This will also give insight into the role of one special path (the geodesic) and the concentration of probability of the set of paths around this path, i.e. the infinite-dimensional equivalent of the concentration around a point shown in Appendix 1 in an elementary example.

(b) Let us return to the general case where Λ is not reduced to a point. Then the interpretation will be of the same type. Now one has to replace the Riemannian geodesics by paths of minimal length for a metric which will be a minimal mixture of the Riemannian metrics associated with the different values of $\lambda \in \Lambda$. Such a metric is called a Finslerian metric (see Appendix 5 for explanations of how the Finslerian metric is linked to changes of correlations in large motions of the market, and algorithms for fast computation of this metric).

The financial interpretation of this optimization problem is that, far-from-the-money (outside the main range), the order of magnitude of the risk depends on the length of a critical path. This path is the solution of a worst-case

[4]The study of the convergence of the conditional probability on the path set has been intensively studied by several authors by using probabilistic large deviations techniques (see Varadhan (1967), Molchanov (1975), Kiefer (1976), Kanai (1977), Azencott (1984)) in the case where Λ is reduced to one point.

'optimization' problem. This worst-sided optimization will include the choice, all along the path, of volatility matrix σ^λ which at each point makes easiest the local motion towards the risky domain.

(c) Fat tails
We now investigate what the above asymptotic results tell us about the tails of the probability distribution. Remember the above argument (see the comments on Proposition 3.2) that the complex Markov structure might be hidden and that the market may only know the 'average' correlation matrix $\bar{\sigma}$ given in Proposition 3.2. Then the expected distribution tails will be computed using $\bar{\sigma}$. But from Proposition 3.2 we know that these tails are described by the asymptotic (when $t \to 0$) of the function $\bar{u}(t,x) = P(\tau_1 < \inf(\tau_0, t))$.

Then applying the asymptotic result of the theorem to $\bar{u}$ (by reducing, in the above theorem, the set $\{\sigma^\lambda\}$ to a one point set $\{\bar{\sigma}\}$), we have

$$\lim_{t \to 0} -2t \log \bar{u}(t,x) = \bar{L}^2(x)$$

where $\bar{L}(.)$ is the unique Lipschitz viscosity solution of

$$\sum_{i,j} \bar{a}_{ij}\, \partial_i \bar{L}\, \partial_j \bar{L} = 1 \text{ in } \Omega, \quad \geq 1 \text{ on } \Gamma_0, \quad L = 0 \text{ on } \Gamma_1.$$

Hence, if the market believes that a good model for σ is $\sigma = \text{constant} = \bar{\sigma}$, the expected probability tails will be given by the asymptotic of $\bar{u}$, while at the same time the real process for σ is given by the Markov structure, and thus the tails are described by the asymptotic of u^λ given in Theorem 3.3

$$\lim_{t \to 0} -2t \log u^\lambda(t,x) = L^2(x).$$

Now, we have under very general assumptions the strict inequality $\bar{L}(x) < L(x)$ (at least, as soon as σ is not constant, i.e. as soon as Λ is not reduced to one point – see the proof in Appendix 3). Hence it is almost always true that

$$\lim_{t \to 0} \frac{u(t,x)}{\bar{u}(t,x)} = +\infty$$

which means that (asymptotically) the tails are infinitely fatter than expected (i.e. expected if one believes in the constant value σ). Hence the phenomenon qualitatively discussed in Section 2 holds in this quite general setting, with the same type of implication: a constant value of σ might be a good approximation except for the tails, which will be much fatter than computed out by taking constant value $\bar{\sigma}$ of σ. This may come from a more complex structure, namely Markov randomness on σ, which will reveal itself in large market deviations, in the tails of the probability distribution. In this framework large deviations will always have (asymptotically) a larger probability than expected from constant σ, inducing extra premium and smile as in the discussion of Section 2. Note that the asymptotic result in the above theorem does not depend on $\gamma_{\lambda\mu}$, and therefore large deviations will only reveal part of the underlying Markov structure.

4 Numerical issues

Another use of the asymptotic given in the main theorem of Section 3 is to provide a means of speeding up computations of far-from-the-money contingent claims.

This is the point we are going to explain in this section. We will first explain this issue in the case of Monte Carlo simulations, then we will turn briefly to the case of PDEs.

4.1 Asymptotic might help to reduce Monte Carlo simulation computing time by several orders of magnitude

The idea underlying Monte Carlo acceleration is to inject some information into the system in order to perform a well controlled simulation of trajectories. If we have some information about the nature of the unknown we are trying to compute in a Monte Carlo scheme (a particular expectation), the residual variance conditional to this information should be smaller, and so the simulation scheme, according to the central limit theorem, should converge faster to its asymptotic (the value of the expectation).

This point is largely developed in the paper of Fournié, Lasry & Touzi (1997) who propose to inject the information given by different asymptotics (small noise, large deviations) into the Monte Carlo scheme; such information which is often very easy to get.

The classical methods used in the framework of diffusion processes in order to perform a variance reduction are importance sampling and the technique of control variates (see Newton 1994).

The principle of importance sampling is the change of probability measure. One simulates a process different from the initial one, but whose distribution Q on the Wiener space is equivalent to the distribution P of the basic process. Then one corrects the expectation by the Radon–Nikodym density $Z = dP/dQ$ of the two measures given by the Girsanov exponential.

More precisely, let $(\Omega, \mathcal{F}, P^0, (\mathcal{F}_t), (W_t^0))$ be an $\mathbb{R}^m$-valued standard Wiener process and consider the following diffusion in $\mathbb{R}^n$:

$$\begin{aligned} dX_t^0 &= b(t, X_t^0)\, dt + \sigma(t, X_t^0)\, dW_t^0, \ P^0 - a.s. \\ X_0^0 &= x \end{aligned} \tag{4.1}$$

Let $\Phi : \mathbb{R}^n \to \mathbb{R}$ and set $u(t,x) = \mathbf{E}_{P^0}[\Phi(X_T^0) | X_t^0 = x]$. One wants to compute

$$u(0,x) = \mathbf{E}_{P^0}[\Phi(X_T^0)].$$

Now, let $h \in L^2_{\mathcal{F}_t}([0,T]; \mathbb{R}^n)$ and consider

$$Z_t^h = \exp\left\{ \int_0^t \langle h_s, dW_s^0 \rangle + \frac{1}{2} \int_0^t |h_s|^2\, ds \right\}.$$

If $\mathbf{E}_{P^0}\left[1/Z_T^h\right] = 1$, by using Girsanov's theorem, we can define a new probability measure P^h with density

$$\mathbf{E}_{P^0}\left[\frac{dP^h}{dP^0}\bigg|\,\mathcal{F}_t\right] = \frac{1}{Z_t^h}$$

and a P^h-standard Wiener process W^h, $W_t^h = W_t^0 + \int_0^t h_s\,ds, \forall t \in [0,T]$. As an immediate consequence, we have under P^h

$$\begin{aligned} dX_t^h &= \left(b(t,X_t^h) - \sigma(t,X_t^h)h_t\right)\,dt + \sigma(t,X_t^h)\,dW_t^h, \quad X_0^h = x \\ Z_t^h &= \exp\left\{\int_0^t \langle h_s, dW_s^h\rangle - \frac{1}{2}\int_0^t |h_s|^2\,ds\right\}. \end{aligned}$$

Moreover, we have

$$u(0,x) = \mathbf{E}_{P^0}[\Phi(X_T^0)] = \mathbf{E}_{P^h}[\Phi(X_T^h)Z_T^h],$$

hence are led to solve the $(n+1)$-dimensional stochastic system in (X^h, Z^h), instead of the n-dimensional initial SDE in X^0, to get (in a Monte Carlo scheme) the desired expectation $\mathbf{E}_{P^0}[\Phi(X_T^0)]$. This might seem at first a bad methodology.

Now, let us set

$$h(t,x) = -\frac{1}{u(t,x)}\,\sigma^*(t,x)\,\nabla u(t,x);$$

then Itô's formula implies

$$u(0,x) = \Phi(X_T^h)\,Z_T^h,\ P^h - a.s.$$

Thus, $h = (-1/u)\,\sigma^*\,\nabla u$ provides the perfect change of probability. But we do not know the function u (and thus the function h). We don't need to compute complete knowledge of u; we just want its value at point x, at time 0. In actual fact, a good approximation $\hat{u}$ of u may provide an $\hat{h} = (-1/\hat{u})\,\sigma^*\,\nabla\hat{u}$ such that:

$$V_{P^{\hat{h}}}\left[\Phi(X_T^{\hat{h}})\,Z_T^{\hat{h}}\right] \ll V_{P^0}\left[\Phi(X_T^0)\right].$$

Such a reduction of variance may lead to an important acceleration of Monte Carlo simulation schemes.

If the problem is to compute price u and hedge ∇u of an out-of-the-money contingent claim giving $\Phi(X_T)$ at time T, a useful approximation of $\hat{u}$ of u is given by large deviations results. For example, if $\Phi = \mathbb{1}_B$ where B is a compact set far from the point x, according to Theorem 3.3 one can set $\hat{u} = \alpha(t,x)\exp(L^2(x)/(T-t))$, where $L(x)$ is the Riemannian distance from the point x to the set B and $\alpha(t,x)$ is a normalization parameter. But this requires a previous computation of the function L by any nonlinear technique. In the numerical experiments we performed, the reduction of the simulations required to get a given accuracy widely makes up for the loss of time due to the computation of L. Fournié, Lasry & Touzi obtained variance reductions of order 100 to 400 in applying this method to the pricing of a far-from-the-money option in the framework of stochastic volatility models (see Fournié, Lasry & Touzi (1997) for detailed numerical experiments).

We sketch here the basic idea of another method for accelerating Monte Carlo schemes that we mentioned at the begining of this section: the method of control variates. We consider the same problem as before, i.e. compute the expectation $u(0,x) = \mathbf{E}_P[\Phi(X_T)]$, where $u(t,x) = \mathbf{E}_{P^0}[\Phi(X_T^0) \mid X_t^0 = x]$, for the process

$$dX_t = b(t, X_t)\,dt + \sigma(t, X_t)\,dW_t, \quad X_0 = 0, \; P - a.s.$$

From Itô's lemma, we have:

$$\Phi(X_T) = u(t,x) + \int_t^T (\partial_t u + L_t u)(s, X_s)\,ds + \int_t^T (\sigma^* \nabla u)(s, X_s)\,.\,dW_s.$$

But $(u(t, X_t))$ is a martingale and so we get the following representation (which is the simplest case of the Clark–Haussmann–Ocone formula – see Nualart (1995))

$$\Phi(X_T) = \mathbf{E}_P[\Phi(X_T)] + \int_0^T (\sigma^* \nabla u)(s, X_s)\,.\,dW_s.$$

So, the random variable

$$Y^u = \Phi(X_T) - \int_0^T (\sigma^* \nabla u)(s, X_s)\,.\,dW_s$$

is P-$a.s.$ the constant value $\mathbf{E}_P[\Phi(X_T)]$. Thus

$$\int_0^T (\sigma^* \nabla u)(s, X_s)\,.\,dW_s$$

is the perfect control variate.

Let $\bar{u}$ be an approximation of u and consider $Y^{\bar{u}}$; this has mean $\mathbf{E}_P[\Phi(X_T)]$ and variance

$$V_P[Y^{\bar{u}}] = \mathbf{E}_P\left[\int_0^T |\sigma^*(\nabla \bar{u} - \nabla u)|^2\,ds\right].$$

If $\bar{u}$ is a good approximation of u, it may be that $V_P[Y^{\bar{u}}] \ll V_P[\Phi(X_T)]$, thus providing an important acceleration in a Monte Carlo simulation scheme. This is the principle of variance reduction by control variates (see Newton (1994) for more details about this technique).

4.2 Asymptotic suggests nonlinear transformations to replace ill-conditioned linear PDEs by well conditioned nonlinear PDEs

Another way to look at this change of unknowns is to present some general formal computations that will explain both this log transform and the usefulness of large deviations distances. Let us begin by giving a PDE analogue of the classical Girsanov transform. Let us assume that u solves

$$\partial_t u - \sum_i b_i\, \partial_i u - \sum_{i,j} a_{ij}\, \partial_{ij} u \;=\; 0.$$

Then, let w be a 'weight function' and let U denote the following function

$$U \;=\; e^w u.$$

A straightforward computation shows that U should solve

$$\begin{aligned}\partial_t U - \sum_i b_i\, \partial_i U - \sum_{i,j} a_{ij}\, \partial_{ij} U + 2\sum_{i,j} a_{ij}\, \partial_i w\, \partial_j U - \\ - \left[\partial_t w - \sum_i b_i\, \partial_i w - \sum_{i,j} a_{ij}\, \partial_{ij} w + \sum_{i,j} a_{ij}\, \partial_i w\, \partial_j w\right] U \;=\; 0.\end{aligned}$$

At this stage, in order to avoid (for the reasons mentioned above) looking for an unknown with large deviations (from 1 to 10^{-4}), it is natural to use w in order to equalize the value of U. Our extreme case is to make $U = 1$, in which case we look for $w = -\log u$, and we are back to the change of unknowns suggested above. On the other hand, if we replace a_{ij} by εa_{ij} (where ε is a small parameter) and set $b_i = 0$ for the sake of simplicity, a natural choice for w which corresponds to the ideas in the preceding section for Monte Carlo simulation is then $w = L^2/4\varepsilon t$, where L solves

$$\sum_{i,j} a_{ij}\, \partial_i L\, \partial_j L \;=\; 1.$$

Indeed, we then find

$$\partial_t U - \varepsilon \left\{ \sum_{i,j} a_{ij}\, \partial_{ij} U - \sum_{i,j} a_{ij}\, \partial_i U\, \partial_j w - \sum_{i,j} a_{ij}\, \partial_{ij} w\, U \right\} \;=\; 0$$

and so U is better behaved in terms of the range of values.

Appendix 1
Introduction to the ideas and results of large deviations theory through an elementary example

The theory of large deviations has been intensively developed over recent decades. The tools and results we present here are mainly inspired by this theory. Nevertheless it is not necessary to have a good knowledge of large deviations to understand this article but a few general ideas about it are useful. So, let us give a brief but significant taste of large deviations theory through a simple example.

The main general result of large deviations theory could be called the 'conditional law of large numbers'[5].

Recall that the ordinary, classical law of large numbers gives very precise results about frequencies in independently randomly repeated experiments, namely: relative frequencies converge to the *a priori* probabilities.

Large deviations theory says that, in a very general context, conditional on given constraints or information, the relative frequencies converge to some *different* numbers, which we refer to as the *a posteriori* probabilities.

Moreover, the theory shows how these numbers can be computed through a minimization algorithm. More precisely the *a posteriori* probability is the point in the space of all probabilities which minimizes a functional (called the information function) given the constraints.

Let us examine this conditional law of large numbers through a very simple example.

Throw a dice n times. Call f_i for $i = 1, \ldots, 6$, the frequency of number i, and let x be the mean value of the n random values. Hence $x = (f_1 + 2f_2 + 3f_3 + 4f_4 + 5f_5 + 6f_6)$ and by the (ordinary) law of large numbers, x tends to 3.5 while all frequencies tend to 1/6.

Now, conditional on $x \geq 4$, the *conditional law of large numbers*, i.e. large deviations theory, tells us that the frequencies tend to numbers that are no longer equal to 1/6: $f_1 \to 0.103$, $f_2 \to 0.122$, $f_3 \to 0.146$, $f_4 \to 0.174$, $f_5 \to 0.207$, $f_6 \to 0.246$.

This information might be used for hedging. For example asymptotically one can hedge an option on x at strike price 4, by holding (for example) 5.6 options on $f_5 - f_1$ at strike price 0.104 (5.6 by evaluation of the sensitivity)[6].

[5]Actually while this name is not commonly used (surprisingly no name is commonly used for this result), we propose to adopt it, as it fits well the behavior under consideration in large deviations theory.

[6]If one replaces $x \geq 4$ by $x \geq 4.2$ one finds $f_1 \to 0.081$, $f_2 \to 0.104$, $f_3 \to 0.134$, $f_4 \to 0.172$, $f_5 \to 0.221$, $f_6 \to 0.284$.

One could use other combinations of f_i. For example, asymptotically, 5 options on $f_2 - f_1$ at strike price 0.019 also hedges the option on x at strike price 4. But of course one can feel that options on $f_6 - f_1$ provide a more robust hedge than options on $f_2 - f_1$ for less large values of n. The conditional law of large numbers is a kind of 'certainty principle' which tells you that, conditional on information that drives you outside the expected range (such as $x \geq 4$ in the previous example), the conditional probability will concentrate in the close neighborhood of a very specific event (frequencies will be very near some given values, etc.).

Moreover this specific event can be computed through the minimization of a functional – the (Shannon) information function or negentropy (entropy with negative sign). This general feature (concentration of the conditional probability in the neighborhood of a point) of large deviations theory will help us to understand the critical path interpretation of our PDE results: conditional on large motion, the probability on the set of trajectories of the diffusion process will concentrate around the critical path.

Complementary remarks

It is interesting to look briefly at a toy proof of the 'conditional law of large numbers' in the above example. Using the same notation, combinatorics tells us that:

$$p(n_1, \ldots, n_6) = n!(n_1! \ldots n_6!)^{-1} 6^{-n}$$

where $p(n_1, \ldots, n_6)$ is the probability that the numbers $1, \ldots, 6$ appear n_1, $\ldots, n_6$ times (resp.). Using Stirling's formula $k! = k^k e^{-k} \sqrt{2\pi k}(1 + \varepsilon_k)$, one finds:

$$p(n_1, \ldots, n_6) = n^n (n_1^{n_1} \ldots n_6^{n_6})^{-1} 6^{-n} q$$

where $q = \sqrt{2\pi n}(\sqrt{2\pi n_1} \ldots \sqrt{2\pi n_6})^{-1}(1 + \varepsilon)$. As $f_i = n_i n^{-1}$, one has:

$$-\frac{1}{n} \log p(n_1, \ldots, n_6) \simeq H(f_1, \ldots, f_6)$$

where H is the function $H(f_1, \ldots, f_6) = \sum f_i \log(f_i/6)$.

This is a special case of the Shannon information function, or negentropy,

$$H(q, r) = \Sigma q_i \log(q_i/r_i)$$

where r is the *a priori* probability and q the *a posteriori* probability.

Hence $p \simeq ce^{-nH}$ where c is a 'relatively' constant term. This means that when n is large p is concentrated where H is minimal. When there are no constraints the minimizing point is $q = r$ ($f_i = 1/6$ in the above example); this proves the ordinary law of large numbers. But if there is a constraint (like $f_1 + 2f_2 + 3f_3 + 4f_4 + 5f_5 + 6f_6 \geq 4$), the probability will be concentrated around the point which minimizes H under the constraint (or the constraints, if there are many). Hence the numbers given previously are the solutions of the optimization problem

$$\min \sum_i f_i \log(f_i/6)$$
$$\sum_i i f_i \geq 4, \; \sum_i f_i = 1.$$

All these ideas, concepts and computations are similar and still hold in very general random contexts including diffusion processes but need more sophisticated mathematical treatment.

Appendix 2
Averaging and diffusion processes

In this appendix, we prove in particular Proposition 3.2. Let us first recall some notation. Let Λ be a finite set, let f^λ, b^λ and σ^λ be Lipschitz functions on $\mathbb{R}^n$ with values respectively in $\mathbb{R}$, in $\mathbb{R}^n$ or in the space of $n \times m$ symmetric matrices, for all $\lambda \in \Lambda$. We define $a^\lambda = \frac{1}{2}\sigma^\lambda(\sigma^\lambda)^*$ for all $\lambda \in \Lambda$.

Let Ω be a smooth, bounded open set in $\mathbb{R}^n$ – we could as well take $\Omega = \mathbb{R}^n$ in which case no boundary conditions are necessary. We assume for the sake of simplicity that a^λ is positive definite on $\bar{\Omega}$, for each $\lambda \in \Lambda$ – similar results can be obtained without this assumption but dropping it requires some technical considerations from the theory of viscosity solutions.

We consider the solution u_ε^λ of

$$\begin{aligned} \partial_t u_\varepsilon^\lambda - \sum_i b_i^\lambda \, \partial_i u_\varepsilon^\lambda - \sum_{i,j} a_{ij}^\lambda \, \partial_{ij} u_\varepsilon^\lambda + \frac{1}{\varepsilon} \sum_{\lambda \neq \mu} \theta_{\lambda\mu} \, (u_\varepsilon^\mu - u_\varepsilon^\lambda) \; &= \; f^\lambda, \\ u_\varepsilon^\lambda \Big|_{t=0} = u_0 \;\text{ in } \bar{\Omega}, \;\; u_\varepsilon^\lambda = 0 \;\text{ on } \partial\Omega \times (0, \infty) & \end{aligned} \tag{A.1}$$

where u_0 is continuous on $\bar{\Omega}$, vanishes on $\partial\Omega$, $\varepsilon > 0$ and $\theta_{\lambda\mu} > 0$, for all $\lambda \neq \mu \in \Lambda$ (for example). The functions u_ε^λ are of class C^2 in $x \in \Omega$ and C^1 in t for $t \geq 0$.

We could just as well consider more general situations where f^λ, b^λ and σ^λ depend on t but the arguments are identical and we skip this straightforward extension.

As is well known from elementary (Markov) matrix theory, there exists a unique $(\pi_\lambda)_{\lambda \in \Lambda}$ satisfying

$$\pi_\lambda > 0, \; \forall \lambda \in \Lambda; \; \sum_{\lambda \in \Lambda} \pi_\lambda = 1; \; \sum_{\lambda \in \Lambda} \pi_\lambda \, \theta_{\lambda\mu} = 0, \; \forall \mu \in \Lambda \tag{A.2}$$

where we agree that $\theta_{\lambda\lambda} = -\sum_{\lambda \neq \mu} \theta_{\lambda\mu}$. The probability π_λ is obviously nothing but the invariant measure of the Markov 'switching process'.

We may now reformulate Proposition 3.2 as follows.

Proposition A.2 *As ε goes to 0_+, u_ε^λ converges (uniformly on $\bar\Omega \times [0,T]$, for all $T < \infty$) for all $\lambda \in \Lambda$ to the solution $\bar u$ of*

$$\partial_t \bar u - \sum_i \bar b_i\, \partial_i \bar u - \sum_{i,j} \bar a_{ij}\, \partial_{ij} \bar u = \bar f,$$
$$\bar u\big|_{t=0} = u_0 \text{ in } \bar\Omega, \quad \bar u = 0 \text{ on } \partial\Omega \times (0,\infty) \tag{A.3}$$

where $\bar a = \sum_{\lambda\in\Lambda} \pi_\lambda\, a^\lambda$, $\bar b = \sum_{\lambda\in\Lambda} \pi_\lambda\, b^\lambda$ and $\bar f = \sum_{\lambda\in\Lambda} \pi_\lambda\, f^\lambda$.

Remarks As mentioned above, this result admits various extensions to the case when $\Omega = \mathbb{R}^n$, or when the data f^λ, b^λ and σ^λ depend on t, or when u_0 depends on λ. In the last case, the limiting initial condition becomes: $\bar u\big|_{t=0} = \sum_{\lambda\in\Lambda} \pi_\lambda\, u_0^\lambda$ in $\bar\Omega$; and the convergence is uniform on $\bar\Omega \times [\delta, T]$ for all $0 < \delta < T < \infty$. Also, we can deal with this situation when a^λ is no longer assumed to be positive definite (and $\Omega = \mathbb{R}^n$ to avoid technical considerations on the boundary conditions).

Proof We briefly sketch one possible proof which relies upon the theory of viscosity solutions. We introduce for $\lambda \in \Lambda$, $x \in \bar\Omega$, $t \geq 0$

$$\begin{aligned} \bar u^\lambda(t,x) &= \limsup\{u_\varepsilon^\lambda(s,y),\ s \geq 0, y \in \bar\Omega, s \to t, y \to x, \varepsilon \to 0_+\} \\ \underline{u}^\lambda(t,x) &= \liminf\{u_\varepsilon^\lambda(s,y),\ s \geq 0, y \in \bar\Omega, s \to t, y \to x, \varepsilon \to 0_+\}. \end{aligned}$$

These limits are finite since an easy application of the maximum principle shows that $|\bar u^\lambda| \leq \max_{\bar\Omega} |u_0| + t \max_{\bar\Omega} |f^\lambda|$. We only have to show that $\bar u^\lambda \leq \bar u \leq \underline{u}^\lambda$ for all $\lambda \in \Lambda$. To this end, we essentially only need to show that $\bar u^\lambda$ (resp. $\underline{u}^\lambda$) is a viscosity subsolution (resp. supersolution) of the averaged equation (A.3). By symmetry, it suffices to study $\bar u^\lambda$.

In order to do so, we first observe (see also Lasry & Lions (1995) for a similar argument) that (A.1) yields easily the fact that $\bar u^\lambda$ is independent of λ since we have (in the viscosity sense):

$$\left(\sum_{\lambda\neq\mu} \theta_{\lambda\mu}\right) \bar u^\lambda \geq \sum_{\lambda\neq\mu} \theta_{\lambda\mu} \bar u^\mu.$$

Next, if ϕ is an arbitrary smooth test function on $\bar\Omega \times [0,\infty)$, we introduce the ψ^λ solution of

$$\sum_{\lambda\neq\mu} \theta_{\lambda\mu}\, \psi^\mu = \sum_i (b_i^\lambda - \bar b_i)\, \partial_i \phi + \sum_{i,j} (a_{ij}^\lambda - \bar a_{ij})\, \partial_{ij} \phi + f^\lambda - \bar f. \tag{A.4}$$

Such a solution exists since $\sum_\lambda \pi_\lambda\, \chi^\lambda = 0$ where χ^λ denotes the right-hand side of (A.4). It is defined up to a constant (i.e. $\psi^\lambda + C$ for all $\lambda \in \Lambda$) and we can normalize it by requesting that $\sum_\mu \psi^\mu = 0$. Then ψ^μ is Lipschitz in x and smooth in t. In fact, by a further layer of approximation that we do not wish to detail, we can always assume that $\bar{f}, \bar{b}, \bar{\sigma}$ are smooth, in which case ψ^μ is smooth in $(t,x) \in \bar{\Omega} \times [0,\infty)$ for all $\mu \in \Lambda$.

We then define $\phi^\lambda_\varepsilon = \phi + \varepsilon\psi^\lambda$ and we observe that we have for all $\lambda \in \Lambda$

$$\begin{aligned}\partial_t\phi^\lambda_\varepsilon - \sum_i b^\lambda_i\,\partial_i\phi^\lambda_\varepsilon - \sum_{i,j} a^\lambda_{ij}\,\partial_{ij}\phi^\lambda_\varepsilon + \frac{1}{\varepsilon}\sum_{\lambda\neq\mu}\theta_{\lambda\mu}\,\phi^\mu_\varepsilon - f^\lambda \\ = \partial_t\phi - \sum_i \bar{b}_i\,\partial_i\phi - \sum_{i,j}\bar{a}_{ij}\,\partial_{ij}\phi - \bar{f} + r^\lambda_\varepsilon\end{aligned}$$

where $|r^\lambda_\varepsilon| \leq C_0\varepsilon$.

Next, if $(t_0,x_0) \in \Omega\times(0,\infty)$ is a strict local maximum point of $\max\limits_{\lambda\in\Lambda}(u^\lambda - \phi)$, we can find, for ε small enough, a local maximum point of $\max\limits_{\lambda\in\Lambda}[u^\lambda_\varepsilon - \phi^\lambda_\varepsilon - C_0\varepsilon t]$ denoted by $(t_\varepsilon, x_\varepsilon)$ such that $x_\varepsilon \to x_0$, $t_\varepsilon \to t_0$ as ε goes to 0_+.

Let λ_1 be the point in Λ maximizing over Λ the quantity $u^\lambda_\varepsilon(t_\varepsilon,x_\varepsilon) - \phi^\lambda_\varepsilon(t_\varepsilon,x_\varepsilon) - C_0\varepsilon t$. We deduce from (A.1) and the maximum principle that we have

$$\begin{aligned}0 &= \left[\partial_t u^{\lambda_1}_\varepsilon - \sum_i b^{\lambda_1}_i\,\partial_i u^{\lambda_1}_\varepsilon - \sum_{i,j} a^{\lambda_1}_{ij}\,\partial_{ij}u^{\lambda_1}_\varepsilon + \frac{1}{\varepsilon}\sum_{\mu\neq\lambda_1}\theta_{\lambda_1\mu}(u^\mu_\varepsilon - u^\lambda_\varepsilon)\right](t_\varepsilon,x_\varepsilon) \\ &\geq \left[\partial_t\phi^{\lambda_1}_\varepsilon - \sum_i b^{\lambda_1}_i\,\partial_i\phi^{\lambda_1}_\varepsilon - \sum_{i,j} a^{\lambda_1}_{ij}\,\partial_{ij}\phi^{\lambda_1}_\varepsilon + \frac{1}{\varepsilon}\sum_{\mu\neq\lambda_1}\theta_{\lambda_1\mu}\,\phi^\mu_\varepsilon\right](t_\varepsilon,x_\varepsilon) + C_0\varepsilon \\ &\geq \left[\partial_t\phi - \sum_i \bar{b}_i\,\partial_i\phi - \sum_{i,j}\bar{a}_{ij}\,\partial_{ij}\phi\right](t_\varepsilon,x_\varepsilon).\end{aligned}$$

We conclude the proof of the Proposition upon letting ε goes to 0_+. □

We wish to conclude this appendix with an extension to the case when Λ is the closure of a smooth bounded open set in $\mathbb{R}^p$ or a periodic torus $\Lambda = \prod_{i=1}^p(\mathbb{R}/T_i\mathbf{Z})$ (with $T_1,\ldots,T_p > 0$ fixed). We then consider a coupled process (X_t,λ_t) a solution of

$$dX_t = b(X_t,\lambda_t)\,dt + \sigma(X_t,\lambda_t)\,dW, \qquad X_0 = x,$$

while λ_t is either a reflecting diffusion process on Λ or a diffusion process on Λ given by

$$\varepsilon d\lambda_t = \beta(X_t,\lambda_t)\,dt + \gamma(X_t,\lambda_t)\,dB, \qquad \lambda_0 = \lambda \in \Lambda,$$

where W_t, B_t are independent Wiener processes (in dimensions m, r resp.), $\varepsilon > 0$, σ, b, γ, β are Lipschitz on $\bar{\Omega}\times\Lambda$ with values respectively in the space of $n\times m$ matrices, $\mathbb{R}^n$, the space of $p\times r$ matrices, and $\mathbb{R}^p$. We assume furthermore that $\alpha = \frac{1}{2}\gamma\gamma^*$ is positive definite on $\bar{\Omega}\times\Lambda$ and that $a = \frac{1}{2}\sigma\sigma^*$ is also positive definite on $\bar{\Omega}\times\Lambda$ (the latter restriction is not really necessary but the first one is in order to ensure the existence and uniqueness of an invariant measure). The analogue of (A.1) is now

$$\begin{aligned}\frac{\partial u_\varepsilon}{\partial t} - \sum_i b_i \frac{\partial u_\varepsilon}{\partial x_i} - \sum_{i,j} a_{ij} \frac{\partial^2 u_\varepsilon}{\partial x_i \partial x_j} & \\ + \frac{1}{\varepsilon}\left[- \sum_{i=1}^p \beta_i \frac{\partial u_\varepsilon}{\partial \lambda_i} - \sum_{i,j=1}^p \alpha_{ij} \frac{\partial^2 u_\varepsilon}{\partial \lambda_i \partial \lambda_j}\right] &= f, \\ u_\varepsilon|_{t=0} = u_0 \text{ in } \bar{\Omega} \times A, \quad u_\varepsilon = 0 \text{ on } \partial\Omega \times A \times (0,\infty) & \end{aligned} \tag{A.5}$$

with Neumann's boundary conditions or periodic boundary conditions on ∂A. Of course, in the periodic case, we need to assume that all the data are periodic in λ.

As is well known, for each $x \in \bar{\Omega}$, there exists a unique probability density on Λ, $\pi(x,\lambda)$ satisfying

$$-\sum_{i=1}^p \frac{\partial}{\partial \lambda_i}(\beta_i(x,\lambda)\pi(x,\lambda)) + \sum_{i,j=1}^p \frac{\partial^2}{\partial \lambda_i \partial \lambda_j}(\alpha_{ij}(x,\lambda)\pi(x,\lambda)) = 0$$

with the appropriate boundary conditions (periodic in the periodic case, ...). In fact, $\pi \in C^\alpha(\bar{\Omega} \times A)$ for all $\alpha \in (0,1)$, $\pi > 0$ on $\bar{\Omega} \times A$ and, of course, $\int_A \pi(x,\lambda)\, d\lambda = 1$ for all $x \in \bar{\Omega}$.

Then a relatively straightforward extension of Proposition 3.2 is given by the following result – which can be proved along the same lines as in the proof above.

Theorem A.5 *As ε goes to 0_+, u_ε converges (uniformly on $\bar{\Omega} \times A \times [0,T]$ for all $T \in (0,\infty)$) to the solution $\bar{u}(t,x)$ of*

$$\begin{aligned}\partial_t \bar{u} - \sum_i \bar{b}_i\, \partial_i \bar{u} - \sum_{i,j} \bar{a}_{ij}\, \partial_{ij}\bar{u} &= \bar{f}, \text{ in } \Omega \times (0,\infty) \\ \bar{u}|_{t=0} = u_0 \text{ in } \bar{\Omega}, \quad \bar{u} = 0 \text{ on } \partial\Omega \times (0,\infty) & \end{aligned} \tag{A.6}$$

where $\bar{a} = \int_A \bar{a}(x,\lambda)\,\pi(x,\lambda)\,d\lambda$, $\bar{b} = \int_A \bar{b}(x,\lambda)\,\pi(x,\lambda)\,d\lambda$, and $\bar{f} = \int_A \bar{f}(x,\lambda)$ $\pi(x,\lambda)\,d\lambda$.

Appendix 3

In this appendix, we prove that $\bar{L} > L$ on $\Omega \times \Gamma_0$ (with the notation of Section 3) under very natural conditions on $(a^\lambda)_{\lambda \in \Lambda}$. Let us recall that L is the unique solution of

$$\max_{\lambda \in \Lambda}\left(\sum_{i,j=1}^n a_{ij}^\lambda\, \partial_i L\, \partial_j L\right) = 1 \text{ in } \Omega, \geq 1 \text{ on } \Gamma_0, \; L = 0 \text{ on } \Gamma_1$$

while $\bar{L}$ solves

$$\sum_{i,j=1}^{n} \bar{a}_{ij}\, \partial_i \bar{L}\, \partial_j \bar{L} \;=\; 1 \text{ in } \Omega, \geq 1 \text{ on } \Gamma_0, \; \bar{L} \;=\; 0 \text{ on } \Gamma_1$$

where $\bar{a} = \sum_{\lambda \in \Lambda} a^{\lambda}\, \pi_{\lambda}$, $\pi_{\lambda} > 0$ for all $\lambda \in \Lambda$, $\sum_{\lambda \in \Lambda} \pi_{\lambda} = 1$.

Our main condition is that, for all $\xi \in S^{n-1}$ (i.e. $\xi \in \mathbb{R}^n$, $|\xi| = 1$), $\sum_{i,j=1}^{n} a_{ij}^{\lambda}\, \xi_i\, \xi_j$ is not independent of λ. Note that, if $n = 1$, this simply means that a^{λ} is not independent of λ (which amounts in some sense to requesting that Λ does not reduce to a single point!).

This condition immediately implies the existence of some $M > 1$ such that we have for all $\xi \in S^{n-1}$ and for all $x \in \bar{\Omega}$

$$\max_{\lambda \in \Lambda} \left(\sum_{i,j=1}^{n} a_{ij}^{\lambda}(x)\, \xi_i\, \xi_j \right) \;\geq\; M^2 \left(\sum_{i,j=1}^{n} \bar{a}_{ij}(x)\, \xi_i\, \xi_j \right).$$

By homogeneity, this inequality clearly holds for all $\xi \in \mathbb{R}^n$.

Next, we claim that $\bar{L} \geq ML$ on $\bar{\Omega}$. Indeed, it suffices to observe that we have

$$\max_{\lambda \in \Lambda} \left(\sum_{i,j=1}^{n} a_{ij}^{\lambda}(x)\, \frac{\partial}{\partial x_i} \left(\frac{\bar{L}}{M} \right) \frac{\partial}{\partial x_j} \left(\frac{\bar{L}}{M} \right) \right) \;\geq\; 1 \text{ in } \Omega \times \Gamma_0, \; \frac{\bar{L}}{M} \;=\; 0 \text{ on } \Gamma_1.$$

Standard comparison results for (Lipschitz) viscosity solutions of such problems immediately imply our claim.

Since $L > 0$ on $\Omega \times \Gamma_0$, we deduce our initial claim, namely $\bar{L} > L$ on $\Omega \times \Gamma_0$.

Appendix 4
Intuitive approach to Finslerian geodesic paths

Let us try in this appendix to give an intuitive qualitative approach to our Riemannian (and Finslerian) geometric story. We will do that through a simple example, and will treat it in a heuristic style.

We consider the case of a financial asset, or an index, whose volatility is stochastic. The underlying dynamic is supposed to be

$$\begin{aligned} \frac{dS}{S} &= a\, dt + \sigma\, dW_1 \\ \frac{d\sigma}{\sigma} &= b\, dt + \gamma\, dW_2 \end{aligned}$$

but we will mostly discuss the example in a qualitative style. We want to describe the behavior of the asset price S, and its volatility σ, in the case of large deviations in a short amount of time. More precisely, we want to describe a large motion by cutting it into a succession of smaller ones.

So we denote by (S_0, σ_0) the initial values of price and volatility, and we look to the range of 'possible' motions after 1, 2, ... small periods of time. A possible motion is something inside n standard deviations (choose a value $n = 2$ for example). Of course, in a rigorous discussion one should justify this way of computing: justification is possible as a side consequence of our main theorem, but the point in this appendix is not to be tough on technical questions; the point here is to give insights.

During the first small period of time Δt the probability of reaching a point $(S, \sigma) = (S_0 + \Delta S_0, \sigma_0 + \Delta\sigma_0)$ is given by[7]

$$\begin{aligned} &\frac{1}{2\pi} \exp\left[-\{(\Delta W_1)^2 + (\Delta W_2)^2\}/(2\Delta t)\right] \\ = &\frac{1}{2\pi} \exp\left[-\frac{1}{2\Delta t}\left\{\frac{(\Delta S_0 - a\Delta t)^2}{\sigma_0^2} + \frac{(\Delta\sigma_0 - b\Delta t)^2}{\gamma^2}\right\}\right] \end{aligned}$$

which is, for small Δt, equivalent to

$$\frac{1}{2\pi} \exp\left[-\{(\Delta S_0/\sigma_0)^2 + (\Delta\sigma_0/\gamma)^2\}/(2\Delta t)\right].$$

Hence, the motions within n standard deviations are given by

$$\left(\frac{\Delta S_0}{\sigma_0}^2\right) + \left(\frac{\Delta\sigma_0}{\gamma}^2\right) \leq c\Delta_t$$

where c is a constant which depends only on the choosen value for n. In other words, the possible motions are in an ellipsoid around (S_0, σ_0) defined by the above equation (see Figure 2).

Then we will repeat this computation to obtain the range after two periods of time. We compute all the one-period possible motions starting from any point of the range after one period. This gives the possible motions after two periods. One can see in Figure 3 the construction of the second period which is crucial in the understanding. The hull of the different ellipsoids gives a two-periods range which is no longer an ellipsoid (see Figure 4).

Then the construction of the subsequent periods shows the appearance of curves of points at the same distance from (S_0, σ_0). Then, looking at the paths which lead to a given value of $S = S_1$, we get the 'geodesic', the shortest path, i.e. the path which leads to S_1 in the minimum number of periods (see Figure 5).

[7] We suppose that W_1 and W_2 are independent for the sake of simplicity.

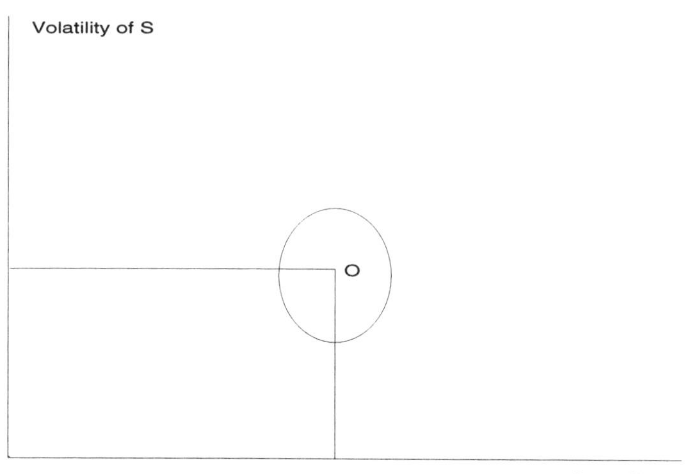

Figure 2: Possible motions within one period of time

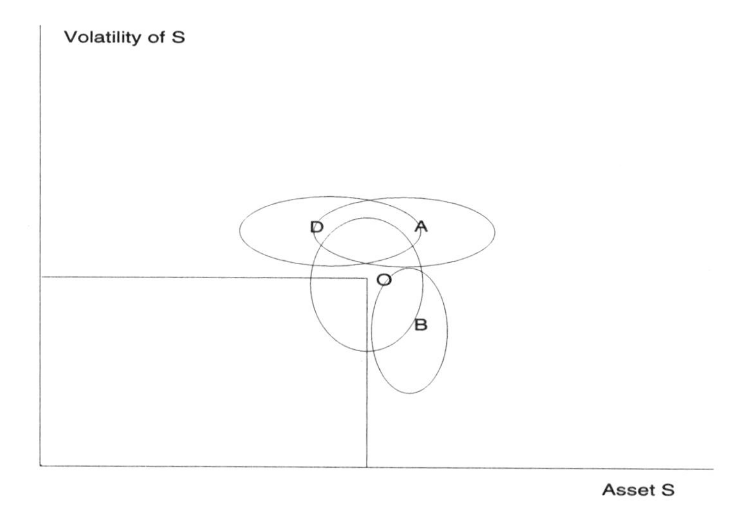

Figure 3: Possible motions during the second period of time. The ellipse centered on B is narrower because the volatility is smaller than in O, while the ellipses centered on A and D are larger because the volatility is bigger

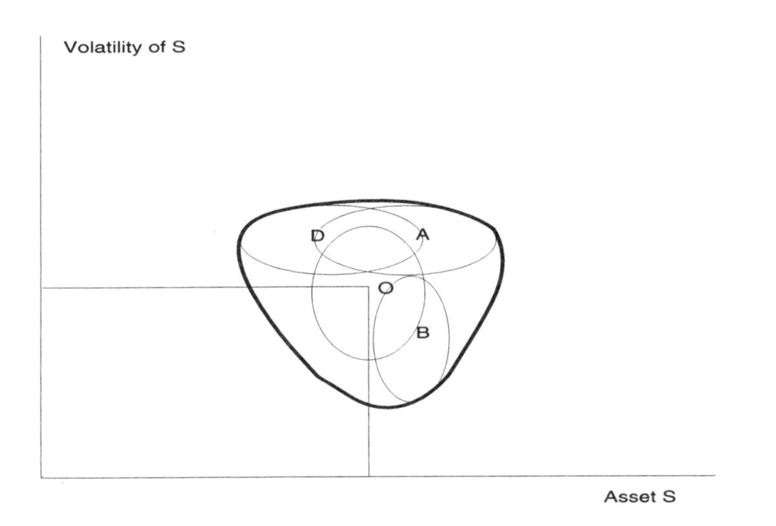

Figure 4: Possible motions within two periods of time, the union of all the ellipses centered on a point of the boundary of the central ellipse

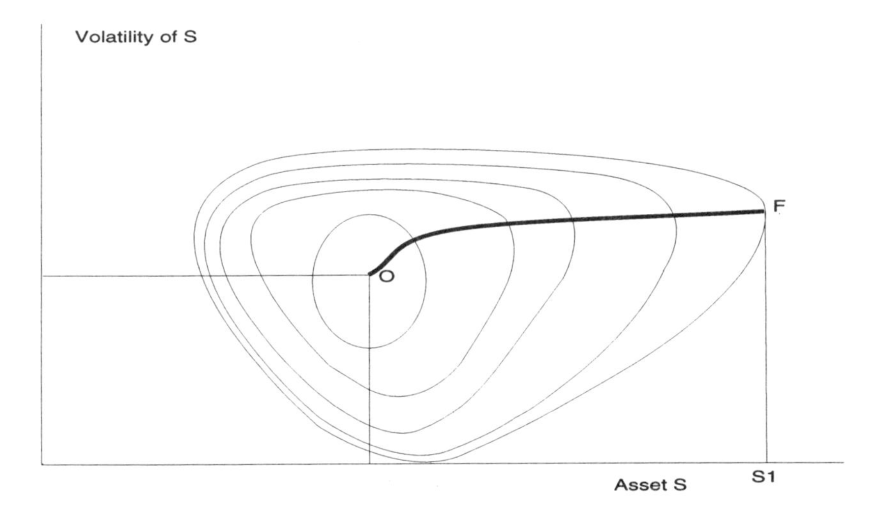

Figure 5: Possible motions within n periods of time. The thick line is the shortest path (geodesic) from O to the half plane defined by inequality $S \geq S_1$. The shape of this geodesic shows that when there is a large rise of S within the model, the move is as follows: first the volatility σ increases quickly while S moves slowly; next S goes up very fast, the volatility staying relatively constant

Appendix 5
Numerical computation of a Finslerian metric

While the Finslerian metric might seem to solve a very technical and nonlinear equation, it is actually easy and fast to compute, as we will explain now on an example. The nonlinear equation (3.1) is only a very simple particular case of the first-order Hamilton–Jacobi equation. This nonlinear equation can be solved by various finite difference methods (see for example Rouy & Tourin (1992)). We suggest a very simple algorithm based on the optimal control interpretation of the function L, i.e. a direct solving of the 'critical path problem' of Theorem 3.3.

We consider, with notation of Theorem 3.3, the Finslerian metric defined by the fields $A^{\lambda(.)} = (a^{\lambda(.)})^{-1}$, and we want to construct an algorithm for solving

$$L(x) = \inf\{L(x,y), y \in \Gamma_1\}$$

where

$$L(x,y) = \inf_H \int_0^1 \langle A^{\lambda(t)}(z(t))\,\dot{z}(t), \dot{z}(t)\rangle^{1/2}\,dt$$

where $H \subset C^1(z)$, $H = \{(z(.),\lambda(.)) : z(0) = x, z(1) = y, z(t) \in \Omega \cup \Gamma_0, \forall t \in [0,1], (\lambda_t) \in L^\infty([0,1];\Lambda)\}$.

We propose an algorithm based on the propagation of the boundary Γ_1 by a direct application of the dynamic programming principle. Such a numerical scheme is not really accurate or robust but it quickly provides sufficiently good approximations of both the required function L and the associated geodesics (the critical paths). We give here its three basic steps.

Step 1 Choose a grid of points in the state space of z (large enough to cover Ω and to avoid unwanted boundary effects)

Step 2 Initialize the area Γ_1 to 0 (boundary condition)

Step 3 Select progressively new sets of points B on the grid and compute, for all $\lambda \in \Lambda$, all the distances $L^\lambda(z,z')$ for the local metric $A^\lambda(z)$ from the point $z \in B$ to the points $z' \in B'$ whose distance $L(z')$ to the set Γ were calculated at the previous step of the algorithm. The distance to the set Γ for any new point $z \in B$ is the minimum over all $z' \in B'$, of $\inf_\lambda L^\lambda(z,z') + L(z')$ and for a given $z_1 \in B$, the point in $z_1' \in B'$ which achieves this minimum is precisely the next point of the geodesic passing through z_1.

We present an application of this algorithm to the following stochastic diffusion model (which is a particularly simple case of the stochastic volatility model – see Fournié, Lasry & Touzi (1997)):

$$\begin{aligned} dS_t &= \sigma_t S_t \left(\sqrt{1-\rho^2}dW_t^1 + \rho\, dW_t^2\right) \\ d\sigma_t &= \gamma\sigma_t\, dW_t^2. \end{aligned} \tag{A.7}$$

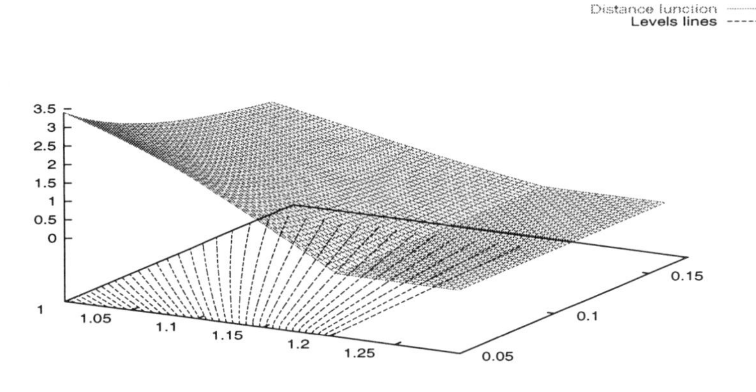

Figure 6: Finslerian metric and level lines

Let us suppose the parameter $(\lambda_t) = (\gamma_t, \rho_t)$ follows an exogeneous pure jump process and takes values in the set of $\Lambda = \{\lambda_i = (\gamma_i, \rho_i), i = 1 \dots n\}$.

The family of metrics $A^\lambda = (a^\lambda)^{-1} = (\sigma^\lambda \sigma^{\lambda *})^{-1}$ associated to this system is

$$a^\lambda = \begin{pmatrix} \sigma^2 S^2 & \rho\sigma^2 S\gamma \\ \rho\sigma^2 S\gamma & \sigma^2\gamma^2 \end{pmatrix},$$

$$A^\lambda = \begin{pmatrix} \dfrac{1}{\sigma^2 S^2(1-\rho^2)} & -\dfrac{\rho}{\sigma^2 S\gamma(1-\rho^2)} \\ -\dfrac{\rho}{\sigma^2 S\gamma(1-\rho^2)} & \dfrac{1}{\sigma^2\gamma^2(1-\rho^2)} \end{pmatrix}.$$

Let $B = \mathbb{R}^2 - \bar{\Omega}$ be an open bounded set. We have to solve the nonlinear equation

$$\sup_\lambda \langle a^\lambda \nabla L, \nabla L\rangle = 1 \text{ on } \Omega, \; L = 0 \text{ on } \bar{B}$$

using the algorithm described above.

In the first following numerical experiment, the parameter $\lambda = (\gamma, \rho)$ is distributed in $\{(\gamma_1, \rho_1) = (0.1, 0.5), (\gamma_2, \rho_2) = (0.6, -0.5)\}$. Figures 6 and 7 present respectively the distance to the set $B = \{S > 1.2\}$ and the associated geodesic paths. Next, we change the distribution of the parameter $\lambda = (\gamma, \rho)$ in $\{(\gamma_1, \rho_1) = (0.1, 0.5), (\gamma_2, \rho_2) = (0.6, 0.5)\}$. Figures 8 and 9 present the corresponding distance and geodesic paths.

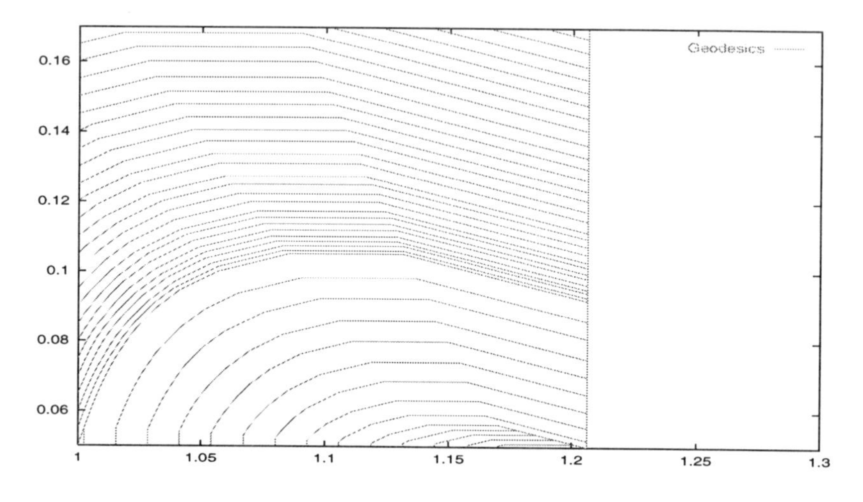

Figure 7: Finslerian geodesic paths

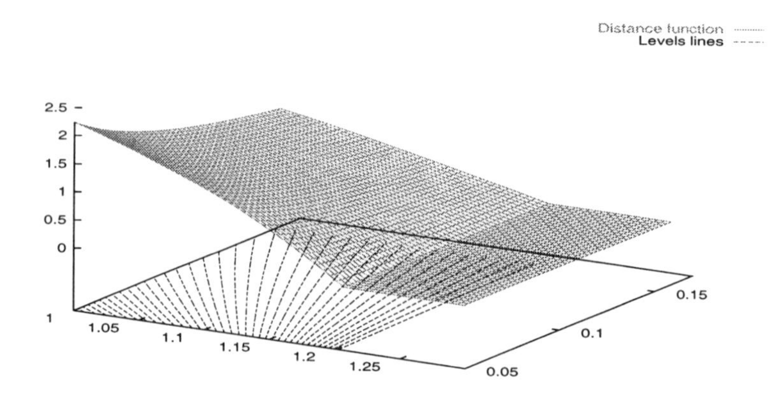

Figure 8: Finslerian metric and level lines

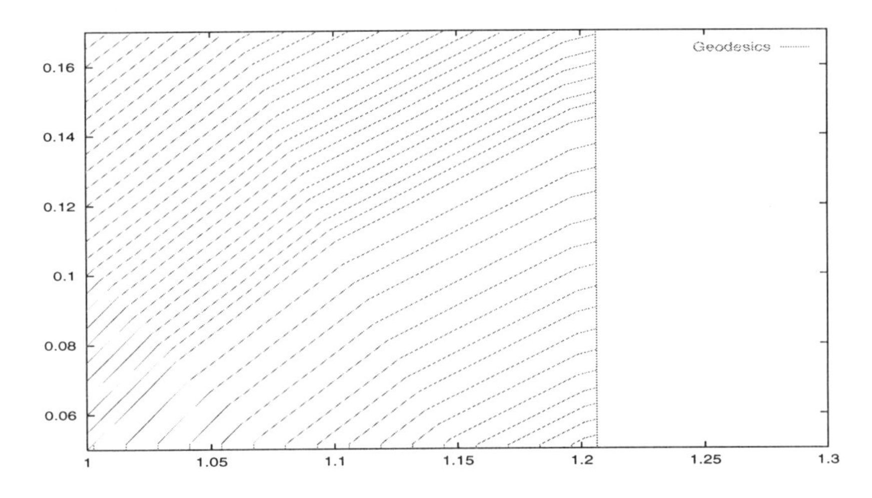

Figure 9: Finslerian geodesic paths

Appendix 6

In this appendix, we explain the relationship between large deviations of the market and small time asymptotics, and we choose to do so with some explicit examples.

We begin with the celebrated Black–Scholes model, namely

$$dS \;=\; bS\,dt + \sigma S\,dW \tag{A.8}$$

where $\sigma > 0$, $b \in \mathbb{R}$. Then, if S_0 is large (a similar argument can be made if S_0 is small), we choose $\varepsilon = 1/\log \bar{S}$ and we set $X = S^\varepsilon$. This change of unknown, which corresponds to the natural scaling associated to log-normal random variables, leads to

$$dX_t \;=\; \varepsilon\left(b - \frac{\varepsilon - 1}{2}\sigma^2\right) X\,dt + \varepsilon\sigma X\,dW, \quad X_0 \;=\; x \tag{A.9}$$

where $x = S_0^\varepsilon = \exp\left(\dfrac{\log S_0}{\log \bar{S}}\right)$, which is of order 1 if $\dfrac{\log S_0}{\log \bar{S}}$ is of order 1. The equation (A.9) corresponds to small time asymptotics and the associated distance function L solves

$$\frac{1}{2}\sigma^2 x^2 \left(\frac{\partial L}{\partial x}\right)^2 \;=\; 1$$

and the associated distance function is $\dfrac{\sqrt{2}}{\sigma}\left|\log \dfrac{x}{y}\right|$.

If we assume now that σ is a Markov process with, say, a finite state space Σ, we are precisely in the structure described in Section 3 (with $\sigma = \lambda!$) and the function L solves

$$\max_{\sigma\in\Sigma}\left[\frac{1}{2}\sigma^2x^2\left(\frac{\partial L}{\partial x}\right)^2\right] = 1$$

i.e.

$$\frac{1}{2}\sigma_{\max}^2x^2\left(\frac{\partial L}{\partial x}\right)^2 = 1$$

where $\sigma_{\max} = \max_{\sigma\in\Sigma}\sigma$. The associated distance is given by $\dfrac{\sqrt{2}}{\sigma_{\max}}\left|\log\dfrac{x}{y}\right|$.

The second example we wish to mention consists of (A.8), (A.9) after scaling, combined with a stochastic differential equation for σ

$$d\sigma = \varepsilon\left\{\beta\,dt + \mu(\sigma)\left[\rho\,dW + \sqrt{1-\rho^2}\,dW^1\right]\right\}$$

where μ, β are given functions of σ (for instance $\mu(\sigma) = \mu\,\sigma$, $\beta(\sigma) = \beta\,\sigma$, $\beta \in \mathbb{R}$, $\mu > 0$), $\rho \in [-1,+1]$ and W^1 is another Wiener process independent of W. The corresponding function L solves

$$\max_{\rho\in[-1,+1]}\left\{\frac{1}{2}\sigma^2x^2\left(\frac{\partial L}{\partial x}\right)^2 + \rho\,\sigma\,\mu\,x\frac{\partial L}{\partial x}\frac{\partial L}{\partial\sigma} + \frac{1}{2}\mu^2x^2\left(\frac{\partial L}{\partial\sigma}\right)^2\right\} = 1.$$

The final remark we wish to make concerns initial or terminal conditions associated with (A.8). A typical example in finance consists of the model (A.8) with the following value

$$V(S_0,t) = \mathbf{E}(S_t - K)_+ \tag{A.10}$$

where $K > 0$ and $t > 0$ - this corresponds to a call. In terms of the process X given by (A.9), (A.10) may be rewritten as

$$V^\varepsilon(t,x) = \mathbf{E}(X_t^{1/\varepsilon} - K)_+. \tag{A.11}$$

From a partial differential equation viewpoint, (A.11) is equivalent to

$$V^\varepsilon\big|_{t=0} = \left(x^{1/\varepsilon} - K\right)_+.$$

Next, we observe that the analysis of Section 3 leads to the change of unknown $v^\varepsilon = -\varepsilon^2\log V^\varepsilon$ and to the convergence of v^ε to the solution v of

$$\frac{\partial v}{\partial t} + \max_{\lambda\in\Lambda}\left[a_\lambda(x)\left(\frac{\partial v}{\partial x}\right)^2\right] = 0$$

with the following initial condition: $v = \infty$ if $x < 1$, $v = 0$ if $x > 1$. The validity of this asymptotic limit and the solution of this initial-value problem are obtained by the same tools and methods as those described in Section 3. The solution is given by

$$v(t,x) = \inf\left\{\frac{L^2(x,y)}{4t},\ y > 1\right\}.$$

In other words, large deviations asymptotics do not see the difference between various terminal functions like $(x-K)_+$ or $a\mathbf{1}_{(x \geq K)}$ $(a > 0)$, ...

Acknowledgements

Parts of this work come within the scope of the GMC research program on Monte Carlo based numerical methods at Caisse Autonome de Refinancement, and benefit from interactions with other work undertaken in this program by Jérôme Lebuchoux, Nizar Touzi and the authors. We are grateful for various numerical experiments from A. Tourin and her helpful comments.

Some Comments about the Bibliography

Varadhan (1967), Molchanov (1975), Kiefer (1976), Kanai (1977), Azencott (1982) and Friedlin & Wentzell (1984) are the classical results on large deviations for stochastic processes. Lasry & Lions (1995) is a recent extension of these results whose proofs are obtained by using the viscosity solution theory due to M.G. Crandall and P.L. Lions (1983).

Fleming & Soner (1992) and Barles (1994) are reference books for the theory of viscosity solutions of the Hamilton–Jacobi equations – see also Lions (1982). Rouy & Tourin (1992) is concerned with the numerical solution of the Hamilton–Jacobi equations, similar to those in this article, arising in image processing.

Milshtein (1988), Wagner (1988), Newton (1994), Fournié, Lasry & Touzi (1997) consider some methods of variance reduction diffusion processes, namely importance sampling and control variates – see also GMC (1996) for other methods of variance reduction for diffusions.

Hull & White (1987), Stein & Stein (1991), Scott (1991), Heston (1993), Avellaneda, Levy & Parès (1995), Renault & Touzi (1996), Fournié, Lasry & Touzi (1997) and Fournié, Lebuchoux & Touzi (1996) give good insights into the theory and the utility of stochastic volatility models for financial assets, and also into the pricing and hedging of options in such models.

References

Avellaneda, M., Levy, A. & Parès (1995), 'Pricing and hedging derivative securities in markets with uncertain volatilities', *preprint.*

Azencott, R. (1982) 'Densités des diffusions en temps petit: développement asympotiques, partie 1', *Lect. Notes in Math.*, Springer Verlag.

Barles G. (1994), *Solutions de viscosité des équations de Hamilton-Jacobi*, Mathématiques et Applications **17**, Springer Verlag.

Black, F. & Scholes, M. (1973), 'The pricing of options and corporate liabilities', *J. of Pol. Econ.* **81**, 637–659.

Crandall, M.G. & Lions, P.L. (1983), 'Viscosity solutions of Hamilton–Jacobi equations', *Trans. Amer. Math. Soc.* **277**, 1–42.

Fleming, W.H. & Soner, M. (1992), *Controlled Markov Processes and Viscosity Solutions*, Springer Verlag.

Fournié, E., Lasry, J.M. & Touzi, N. (1997), 'Monte Carlo methods in stochastic volatility models', this volume.

Fournié, E., Lebuchoux, J. & Touzi, N. (1996), 'Small noise expansion and importance sampling', *preprint.*

Friedlin, M.I. & Wentzell, A.D. (1984), *Random Perturbations of Dynamical Systems*, Springer Verlag.

GMC (1996), *Cahier de la CAR* **10**.

Heston, S.L. (1993), 'A closed form solution for options with stochastic volatility', *Review of Financial Studies* **6**, 293–320.

Hull, J. & White, A. (1987), 'The pricing of options on assets with stochastic volatilities', *Journal of Finance* **3**, 281–300.

Kanai (1977), 'Short time asymptotics for fundamental solutions of partial differential equations', *Communications in PDE* **2**, 781–830.

Kiefer, Y.I. (1976), 'Transition density of diffusion processes with small diffusion', *Theory of Probability and Applications* **21**, 513–522.

Lasry, J.M. & Lions, P.L. (1995), 'Grandes déviations pour des processus de diffusion couplés par un processus de sauts', *C. R. Acad. Sci. Paris* **321**, série 1.

Lions, P.L. (1982), *Generalized Solutions of Hamilton-Jacobi Equations*, Pitman.

Milshtein, G.N. (1988), *The Numerical Integration of Stochastic Differential Equations*, Urals University Press, Sverdlovsk (in Russian).

Molchanov, S. (1975), 'Diffusion processes and Riemannian geometry', *Russ. Math. Surveys* **30**, 1–63.

Newton, N.J. (1994), 'Variance reduction for simulated diffusions', *SIAM Journal on Applied Mathematics* **54**, 1780–1805.

Nualart, D. (1995), *Malliavin Calculus and Related Topics*, Probability and its applications, Springer Verlag.

Renault, E. & Touzi, N. (1996), 'Option hedging and implied volatilities in a stochastic volatility model', *Mathematical Finance* **6**, 279–302.

Rouy, E. & Tourin, A. (1992), 'A viscosity solutions approach to shape-from-shading', *SIAM J. Numer. Anal.* **29**, 867–884.

Scott, L. (1991), 'Random variance option pricing', *Advances in Futures and Options Research* **5**, 113–135.

Stein, E.M. & Stein, J.C. (1991), 'Stock price distribution with stochastic volatility', *Review of Financial Studies* **4**, 727–752.

Varadhan, S. (1967), 'Diffusion processes in small time interval', *Communications in Pure and Applied Mathematics* **20**, 659–685.

Wagner, W. (1988), 'Monte Carlo evaluation of functional of stochastic differential equations – variance reduction and numerical examples', *Stoch. Anal. Appl.* **6**, 447–468.

Monte Carlo Methods for Stochastic Volatility Models

E. Fournié, J.M. Lasry and N. Touzi

1 Introduction

The fundamental theorem of valuation by arbitrage reduces the pricing of any European contingent claim to the computation of the expectation of the discounted final reward under some probability measure, equivalent to the initial probability measure, under which the primitive assets processes are martingales. For some simple models and simple contingent claims one can hope to derive a closed form solution of such an expectation as in the seminal work by Black & Scholes (1973). However empirical work pointed out for a long time that such simple models do not fit financial asset price data. Therefore, it is important to develop some computational methods that can handle more complicated models.

Monte Carlo methods appear as a natural tool for such computations since they can deal with models involving many state variables and are well suited for the computation of path-dependent expectations which is the usual case in finance.

In this article, we present an application of Monte Carlo methods for the valuation of contingent claims in stochastic volatility models. In such models the primitive risky asset price process is driven by a bivariate diffusion. Therefore, even for expectations depending only on the terminal value of the process, deterministic methods based on the discretization of the partial differential equation satisfied by the expectation to be computed are time-consuming. For some path-dependent expectations, deterministic methods can still be used by introducing a new state variable and therefore the dimensionality of the problem is increased (see Barles, Daher & Romano 1990).

Recent developments in Monte Carlo simulations motivated an important interest in these methods. Variance reduction techniques extended to diffusions by Newton (1994) allow Monte Carlo methods to be very competitive with deterministic methods even for low dimensional problems as in the case of our application.

In this article we focus on the importance sampling technique. Intuitively, this technique consists in driving the process (via a change in the drift coefficient) to the region of high contribution to the required expectation; Girsanov's theorem lets us put the right weight in order to compensate for the variation in the drift coefficent. The change in the drift coefficient is obtained through a preliminary approximation of the required expectation. We suggest two preliminary approximations of the expectation. The first is the large deviations approximation which appears to be very well suited to the computation of far from the money options. The second approximation is based on a small noise expansion of the stochastic volatility model around the simple Gaussian Black–Scholes model. In our application, the latter approximation also improves very significantly the precision of the Monte Carlo estimator.

The article is organized as follows. Section 1 is devoted to modeling questions. We introduce stochastic volatility models as a natural extension of the Black–Scholes model, which lets us account for its deficiencies, and discuss the problem raised by market incompleteness. For completeness, Section 2 recalls the importance sampling variance reduction technique in Monte Carlo methods. Section 3 presents an application of the large deviations accelerated Monte Carlo estimator which involves the computation of the Riemannian metric characterized as the solution of a nonlinear differential equation. Finally, Section 4 presents an application of the small noise expansion accelerated Monte Carlo estimator.

2 Stochastic Volatility Models

2.1 The Black–Scholes model

The Black–Scholes (1973) model is a major contribution to modern financial theory based on the arbitrage theory for the valuation of contingent claims. However Black and Scholes illustrated their arbitrage argument in a very simple model.

(i) the primitive asset price process $\{S_t,\ 0 \leq t \leq T\}$ is driven by a geometric Brownian motion

$$\frac{dS_t}{S_t} = \mu dt + \sigma dW_t, \qquad 0 \leq t \leq T \quad a.s.$$

which provides an explicit formula for the European option price,

(ii) the market is complete in the sense that the primitive risky asset together with the non-risky asset, whose price process dynamics is described by

$$\frac{d\beta_t}{\beta_t} = r dt, \qquad 0 \leq t \leq T$$

are sufficient to hedge any risk and therefore the option appears as a redundant asset. In order to simplify the presentation we shall always assume the instantaneous interest rate r to be constant.

These last features of the Black–Scholes model contradict empirical observations as well as economic intuition. First, it is well known by now that financial asset data present a Leptokurtic effect (fat tails in the distribution); such an effect can be produced by ARCH models (Engle 1982) and their various extensions. Second, from an economic point of view, if contingent claims were really redundant assets then there is no reason to introduce them on the market for trading. Therefore in order to justify the large trading volumes in contingent claims observed in practice, it seems necessary to start with an initial economy where only the primitive risky and non-risky assets are traded, which is incomplete so that trading in the additional contingent claim can increase the welfare of the agents (see Hart 1975).

Another important empirical observation which is in contradiction with the Black–Scholes models is the *smile* effect: implied volatilities from observed option prices[1] are U-shaped as a function of the exercise price of the option.

2.2 Stochastic volatility models, the valuation problem

Many models have been suggested in the financial literature in order to account for the defficiencies of the Black–Scholes model. In this article we focus on stochastic volatility models (SVM) first studied by Hull & White (1987) and Scott (1987). The primitive risky asset price process dynamics are described by the following bivariate diffusion process

$$\frac{dS_t}{S_t} = \mu_t dt + \sigma(Y_t) dW_t^1 \tag{2.1}$$

$$dY_t = \tilde{\eta}_t dt + \gamma_t dW_t^2 \tag{2.2}$$

$$d\langle W^1, W^2\rangle_t = \rho_t dt, \tag{2.3}$$

where μ_t, $\tilde{\eta}_t$, γ_t and ρ_t are adapted to the natural filtration $\{\mathcal{F}_t,\ 0 \le t \le T\}$ of the Brownian motions $\{W_t^1,\ 0 \le t \le T\}$ and $\{W_t^2,\ 0 \le t \le T\}$ defined on the probability space $(\Omega, \mathcal{F}, P)$ and satisfy the usual regularity conditions for the existence of a unique strong solution to the last bivariate diffusion.

From recent developments in arbitrage valuation theory (see e.g. Harrison & Kreps (1979) and the more recent paper by Delbaen & Schachermayer (1994)), the absence of arbitrage opportunities is equivalent to the existence of a probability measure Q equivalent to the original probability measure P under which the discounted risky asset price process $\{S_t e^{-rt},\ 0 \le t \le T\}$ is a martingale. Such probability measures are called *equivalent martingale measures* (emm). When the market is complete there is a unique emm for the asset price process $\{S_t,\ 0 \le t \le T\}$. Since the contingent claim price process must also satisfy the martingale property under the emm Q, the time

[1] Implied volatilities are computed as follows. The Black–Scholes model provides a explicit formula of an option price depending on the volatility parameter σ. The implied volatility parameter is the value of σ which equates the Black–Scholes price to the observed price of the option.

t price of the contingent claim is characterized as the expectation under Q of its terminal payoff conditional on the information set $\mathcal{F}_t$ up to time t.

An important problem raised by stochastic volatility models is that the market is incomplete (there are two sources of risk, W^1 and W^2, and only one traded risky asset) and therefore there exists an infinite set of emm characterized as follows. Any probability measure equivalent to P is characterized by a continuous version of its density process with respect to P which can be written, from the integral form of martingale representation (see Karatzas & Shreve 1988), as

$$M_t^{(\lambda,\nu)} = \left.\frac{dQ^{(\lambda,\nu)}}{dP}\right|_{\mathcal{F}_t} = \mathcal{E}\left(-\int_0^t \lambda_u dW_u^1 - \int_0^t \nu_u dW_u^2\right), \quad 0 \le t \le T \;\; a.s.$$

where $\mathcal{E}(.)$ denotes the Doleans-Dade exponential martingale and $\{(\lambda_t, \nu_t), 0 \le t \le T\}$ is adapted to $\{\mathcal{F}_t,\, 0 \le t \le T\}$ and satisfies the condition $\int_0^T (\lambda_u^2 + \nu_u^2) du < \infty$ a.s. Next, requiring the underlying risky asset price process to be a martingale under $Q^{(\lambda,\nu)}$ provides the following restriction

$$\lambda_t \sqrt{1-\rho_t^2} + \nu_t \rho_t = \frac{\mu_t - r}{\sigma_t}, \qquad 0 \le t \le T \quad a.s. \tag{2.4}$$

The last restriction does not suffice to fix the processes λ and ν. Any process $\{(\lambda_t, \nu_t), 0 \le t \le T\}$ satisfying the last restriction (together with the integrability conditions) is called *a risk premia process* and defines an emm $Q^{(\lambda,\nu)}$. Since there is an infinite set of risk premia processes there is an infinite set of emm for the risky asset price process $\{S_t,\, 0 \le t \le T\}$.

Therefore the absence of arbitrage does not provide a unique valuation rule as in a complete market framework and further assumptions should be added in order to select a particular emm from among all admissible ones. Such assumptions usually have an economic interpretation in terms of risk compensation. For example, Föllmer & Schweizer (1991) suggested a pricing rule induced by the minimization of the quadratic error between the hedging portfolio and the contingent claim at maturity. Under suitable conditions this pricing rule is shown to be associated to a particular emm called the minimal martingale measure[2]. The risk premia processes associated to the minimal martingale measure are such that the orthogonal risk to the observed one is non-compensated. Another valuation rule in incomplete markets has been suggested recently by Davis (1994) and is based on utility arguments. The pricing rule turns out to be associated to the minimax martingale measure as defined in Karatzas, Lehoczky & Shreve (1991). More generally Pham & Touzi (1993) provide a characterization of emm which is consistent with an equilibrium model where agents are allowed to trade in the risky asset (in

[2]In a recent paper, Gouriéroux, Laurent & Pham (1995) showed that the emm associated to the minimum quadratic hedge portfolio is the minimax martingale associated to a quadratic utility function, as defined in Karatzas, Lehoczky & Shreve (1991).

one unit supply), the non-risky one (in zero net supply) and some contingent claim (in zero net supply).

2.3 Some stylized facts

Stochastic volatility models can be regarded as the continuous-time limit of discrete-time EGARCH models (see Nelson (1990) and Dassios (1993)). Therefore they inherit the fat tails property of EGARCH models and produce the required Leptokurtic effect noticed in financial data.

The second stylized fact justified in a stochastic volatility model is the well known smile effect: implied volatilities from observed option prices are U-shaped as a function of the exercise price of the option. Renault & Touzi (1996) proved that the smile effect is a natural consequence of stochastic volatility for a large class of valuation rules induced by the absence of arbitrage argument. More precisely, suppose that the pricing rule is induced by any emm under which the volatility process is Markov; then the implied volatility parameter, obtained by equating the Black–Scholes formula to the option price given by such a pricing rule, exhibits a smile effect.

2.4 Market completion

An important problem raised by complete markets is that contingent claims appear as redundant assets and therefore the trading in contingent claims is not justified from an economic viewpoint. Stochastic volatility models induce incomplete markets and therefore they are not subject to such criticisms.

Now suppose that the contingent claim is introduced in the economy in the sense that it is available for trading by agents. If the option completes the market then the equilibrium in this new economy is Pareto optimal and therefore the global welfare is increased. Such an important question has been raised by Ross (1976) in a one-period model with finite state space. Bajeux & Rochet (1996) examined the consistency of Ross' results both in a multiperiod framework and in continuous-time assuming a stochastic volatility model for the primitive risky asset price process. While the one-period model of Ross is solved by examining the space spanned by the traded assets payoffs at the final date, the dynamic model requires the definition of a pricing rule in order to check the market completion at any time before maturity. Bajeux & Rochet (1992) and Romano & Touzi (1993) proved that any contingent claim with convex payoff function (satisfying a logarithmic growth condition) completes the market for the class of valuation rules induced by any emm under which the volatility process is Markov.

In addition to the justification of trading in contingent claims the last result shows that, in a stochastic volatility model, any contingent claim can be perfectly hedged by the primitive risky and non-risky assets and a European option, say. Such a hedging strategy is has been effectively applied for a long time by practitioners and is known as delta-sigma hedging (see Scott 1991). Thus stochastic volatility models also provide a justification of the practitioner's behavior.

2.5 The numerical computation problem

In this article we fix some emm Q and address the problem of numerical computation of the arbitrage price of a European contingent claim induced by such a choice of the emm. More precisely we consider a European contingent claim $(T, \phi(.))$ which pays $\phi(S_T)$ at the terminal date T. An important example of payoff functions ϕ is given by European call options $\phi(s) = (s-K)^+$ and European put options $\phi(s) = (K-s)^+$. Assuming that the interest rate is zero (without loss of generality), the arbitrage price of the contingent claim $(T, \phi(.))$ induced by the equivalent martingale measure Q is

$$u_t = \mathrm{E}_t^Q \left[\phi(S_T) \right], \qquad 0 \leq t \leq T \quad a.s. \tag{2.5}$$

where $\mathrm{E}_t^Q = E^Q[. \mid \mathcal{F}_t]$ is the conditional expectation operator under the emm Q. In general there is no simple closed form solution for the price function u_t. However, for some particular specifications of the dynamics of the bivariate diffusion $\{(S_t, Y_t),\ 0 \leq t \leq T\}$ under the emm Q, a closed form solution of the density Laplace transform of the process $\{(S_t, Y_t),\ 0 \leq t \leq T\}$ is available and the price u_t can be computed through the inverse Laplace transform, see e.g. Stein & Stein (1991) and Heston (1993).

In this article we do not assume any parametric specification of the process $\{(S_t, Y_t),\ 0 \leq t \leq T\}$ and we suggest the use of Monte Carlo simulations together with the importance sampling variance reduction technique recently developed for diffusions by Newton (1994). As we shall see in the following section, the importance sampling variance reduction technique requires a sharp approximation of the price u_t. We present two types of approximations. The first is the large deviations approximation which appears to be very well adapted for the computation of 'far-from-the-money' options for example. The second one is obtained by a small noise expansion of the stochastic volatility model around the Black–Scholes one, as suggested in Fournié, Lebuchoux & Touzi (1995).

3 Importance Sampling for Diffusions

In this section we recall briefly the importance sampling variance reduction technique for diffusions and we refer to Newton (1994) for more details. Let $\{Z_t, 0 \leq t \leq T\}$ be an m-dimensional stochastic process whose dynamics is characterized by

$$dZ_t = b(Z_t)dt + \Sigma(Z_t)dB_t, \quad Z_0 = z, \tag{3.1}$$

where $\{B_t,\ 0 \le t \le T\}$ is an n-dimensional Brownian motion defined on $(\Omega, \mathcal{F}, P)$, and $b(.)$ and $\Sigma(.)$ satisfy the usual conditions for existence of a unique strong solution to the last stochastic differential equation. Given a real-valued function ϕ (with polynomial growth, say), define the function

$$u(t,z) \quad = \quad \mathrm{E}[\phi(Z_T) \mid Z_t = z]. \tag{3.2}$$

The Monte Carlo approximation technique consists of simulating S independent sample paths $\{Z_t(s),\ 0 \le t \le T\}$ in the distribution of the process $\{Z_t,\ 0 \le t \le T\}$ and approximating the required expectation by the sample mean

$$\hat{u}_S^{(0)}(t,z) \quad = \quad \frac{1}{S}\sum_{s=1}^{S} \phi\left(Z_T(s)\right). \tag{3.3}$$

The consistency of such an estimator is trivially ensured by the strong law of large numbers. Also, since the simulations are drawn independently, the variance of $\hat{u}_S^{(0)}(t,z)$ is simply given by

$$\mathrm{V}[\hat{u}_S^{(0)}(t,z)] \quad = \quad \frac{1}{S}\mathrm{V}[\phi(Z_T) \mid Z_t = z]. \tag{3.4}$$

However this is not the only way to define a Monte Carlo approximation of $u(t,x)$. Let $\{h_t,\ 0 \le t \le T\}$ be any $\mathbb{R}^n$-valued square integrable adapted process and define the process $\{M_t,\ 0 \le t \le T\}$ by

$$M_t \quad = \quad \exp\left[\int_0^t \langle h_s, dB_s\rangle + \frac{1}{2}\int_0^t |h_s|^2 ds\right], \qquad 0 \le t \le T, \quad a.s.$$

where $\langle .,. \rangle$ denotes the canonical inner product in $\mathbb{R}^n$ and $|.|$ the corresponding norm. If $E\left[M_t^{-1}\right] = 1$, then $\{M_t^{-1},\ 0 \le t \le T\}$ is a positive martingale and we can define a new probability measure $Q^{(h)}$ by

$$\left.\frac{dQ^{(h)}}{dP}\right|_{\mathcal{F}_t} \quad = \quad \frac{1}{M_t}, \qquad 0 \le t \le T, \quad a.s.$$

and the expectation of interest can be written relative to the new probability measure $Q^{(h)}$ as

$$u(t,z) \quad = \quad \mathrm{E}^{Q^{(h)}}\left[\phi(Z_T)M_T \mid Z_t = z\right].$$

Now by Girsanov's theorem, the process $\{\tilde{B}_t,\ 0 \le t \le T\}$ defined by $\tilde{B}_t = B_t + \int_0^t h_\tau d\tau$ a.s. is a standard Brownian motion under $Q^{(h)}$. Therefore defining the processes $\{Z_t^{(h)},\ 0 \le t \le T\}$ and $\{M_t^{(h)},\ 0 \le t \le T\}$ by

$$dZ_t^{(h)} \quad = \quad \left(b(t, Z_t^{(h)}) - \sigma(t, Z_t^{(h)})h_t\right) dt + \sigma(t, Z_t^{(h)})dB_t \tag{3.5}$$

$$M_t^{(h)} \quad = \quad \exp\left[\int_0^t \langle h_s, dB_s\rangle - \frac{1}{2}\int_0^t |h_s|^2 ds\right], \tag{3.6}$$

the expectation $u(t,z)$ can be written in

$$u(t,z) \;=\; \mathrm{E}\left[\phi(Z_T^{(h)})M_T^{(h)} \mid Z_t^{(h)} = z\right]. \tag{3.7}$$

The last expression suggests an alternative Monte Carlo estimation of the expectation $u(t,z)$ for any choice of the process $\{h_t,\, 0 \le t \le T\}$

$$\hat{u}_S^{(h)}(t,z) \;=\; \frac{1}{S}\sum_{s=1}^{S} \phi(Z_T^{(h)}(s))M_T^{(h)}(s), \tag{3.8}$$

with variance given by

$$\mathrm{V}[\hat{u}_S^{(h)}(t,z)] \;=\; \frac{1}{S}\mathrm{V}[\phi(Z_T^{(h)})M_T^{(h)} \mid Z_t = z]. \tag{3.9}$$

The variance reduction technique consists of determining a process $\{h_t,\, 0 \le t \le T\}$ which induces a smaller variance for the Monte Carlo estimator $\hat{u}_S^{(h)}(t,z)$ than the initial one $\hat{u}_S^{(0)}(t,z)$. Using standard Itô calculus, it is then easily seen that

$$\begin{aligned}\phi\left(Z_T^{(h)}\right) M_T^{(h)} \;&=\; u(t,x)\\ &+\int_t^T M_\tau^{(h)}\langle \sigma^*(t,Z_\tau^{(h)})u_z(\tau,Z_\tau^{(h)}) + h_\tau u(\tau,Z_\tau^{(h)})\,,\, dB_\tau\rangle \end{aligned}\tag{3.10}$$

where * denotes transposition, and therefore the estimator $\hat{u}^{(h)}(t,z)$ is interesting as soon as

$$\begin{aligned}&\mathrm{E}\left[\int_0^T M_t^{(h)2}\left|\sigma^*(t,Z_t^{(h)})u_z(t,Z_t^{(h)}) + h_t u(t,Z_t^{(h)})\right|^2 dt\right]\\ &\qquad\le\; \mathrm{E}\left[\int_0^T \left|\sigma^*(t,Z_t)u_z(t,Z_t)\right|^2 dt\right].\end{aligned}$$

The choice of the process $\{h_t,\, 0 \le t \le T\}$ is basically motivated by the following remark. If the function u were known (the problem is solved!), then the optimal choice turns out to be

$$\overline{h}_t \;=\; -\frac{1}{u(t,Z_t)}\sigma^*(Z_t)u_z(t,Z_t), \qquad 0 \le t \le T \quad a.s. \tag{3.11}$$

since the left-hand side of the last inequality vanishes, i.e. $\mathrm{V}\left[\hat{u}_S^{(\overline{h})}(t,z)\right] = 0$. Since the function u is unknown, the variance reduction technique can be applied using a process $\{h_t,\, 0 \le t \le T\}$ induced by a sharp approximation of u. In the rest of the article, such a process $\{h_t,\, 0 \le t \le T\}$ will be called an accelerator.

4 Large Deviations Accelerated Monte Carlo Estimator

4.1 The large deviations approximation

In this section we use the large deviations approximation of the required expectation

$$u(t,z) \quad = \quad E_t\left[\phi(Z_T)|Z_t = z\right],$$

in order to derive an accelerator h as described in the previous section. The idea behind the use of large deviations is the following. Suppose that the process Z is the bivariate diffusion described in Subsection 2.2 and that the contingent claim of interest is a deep out-of-the-money European call option, i.e. $\phi(s,y) = (s-K)^+$ with a large exercise price K. Then it is clear that a large proportion of simulated paths end up below the level K and have no contribution to the Monte Carlo estimator. However if one considers the large deviations of the process of interest $\{Z_t = (S_t, Y_t),\ 0 \leq t \leq T\}$ around the deterministic system corresponding to $\Sigma(Z_t) = 0$ then the proportion of simulated paths which end up in the region of interest is increased significantly and therefore we can hope that the variance of the Monte Carlo estimator resulting from such an approximation will be reduced significantly.

There is a large literature on the asymptotic expansion in small time of the transition density of diffusion processes. The large deviations approximation of the transition density consists of the first order of such an expansion. The first fundamental result is due to Varadhan (1967) who proved the large deviations result

$$\lim_{t\downarrow 0} -2t\log p(t,z,y) \quad = \quad d^2(z,y),$$

where $p(t,z,y)$ is the transition probability function starting from z at time 0 and $d(z,y)$ is the Riemannian metric defined by the fields $z \mapsto [\Sigma(z)^*\Sigma(z)]^{-1}$. The last result has been improved by Molchanov (1975) who derived an equivalent of $p(t,z,y)$ as $t\downarrow 0$. Asymptotic expansion results of the transition probability up to any order have been obtained by Kiefer (1976), Kanai (1977).

Similarly, large deviations results have been established for the probability of attaining a given region starting from some initial state z at time 0. Consider the function

$$v(t,z) \quad = \quad \mathrm{P}\left(\tau < t \mid Z_0 = z\right) \quad \text{with} \quad \tau \ = \ \inf\{t \geq 0 | Z_t \in \Gamma\}$$

for some smooth open set Γ in $\mathbb{R}^n$. The function v so defined corresponds to the reward function $\phi(z) = 1$ if $z \in \Gamma$ and otherwise 0. Then we have the large deviations result

$$\lim_{t\downarrow 0} -2t\log v(t,z) \quad = \quad L^2(z), \tag{4.1}$$

where L is the unique Lipschitz viscosity solution of the nonlinear differential equation

$$\begin{aligned} \nabla L^*(z)[\Sigma^*(z)\Sigma(z)]\nabla L(z) &= 1 \quad \text{for } z \in \mathbb{R}^n \setminus \Gamma \\ L &= 0 \quad \text{for } z \in \Gamma \end{aligned} \tag{4.2}$$

where ∇ is the gradient operator. We refer to Fleming & Soner (1992) and Lasry & Lions (1995), for more details on these results[3].

Next we apply this large deviations approximation in order to derive an accelerator for the numerical computation of the expectation

$$u(t,s,y) = E\left[1_{\{S_T \geq K\}} \mid S_t = s, Y_t = y\right] \tag{4.3}$$

for large K, where $\{Z_t = (S_t, Y_t),\ 0 \leq t \leq T\}$ is the stochastic volatility process described in Subsection 2.2. We can interpret $u(t,z)$ as the hedging ratio of a deep out of the money call option. Taking $\Gamma = (K, \infty) \times \mathbb{R}$ in the definition of v, the large deviations approximation suggests the following accelerator

$$h(t,z) = \frac{L(z)}{T-t}\Sigma^*(z)\nabla L(z). \tag{4.4}$$

4.2 Numerical computation of the large deviations approximation

The nonlinear equation (4.2) can be solved by finite difference methods. Fournié, Lasry & Lions (this volume) suggest another numerical scheme based on the optimal control interpretation of the function L. Indeed an important feature of the function L defined in (4.2) is that it is the solution to the critical path problem associated with the Riemannian metric defined by the fields $A(z) = [\Sigma(z)^*\Sigma(z)]$. More precisely,

$$L(x) = \inf_{y \in \Gamma} L(x,y) \text{ with } L(x,y) = \inf_{\omega(.) \in H(x,y)} \int_0^1 \langle A(\omega(t))\dot{\omega}(t), \dot{\omega}(t)\rangle^{1/2} dt$$

where $H(x,y) = \{\omega \in C^1([0,1], \mathbb{R}^n) : \omega(0) = x \text{ and } \omega(1) = y\}$ and $\dot{\omega}$ is the derivative of ω with respect to t. Therefore we used their algorithm based on the propagation of the boundary of Γ by a direct application of the dynamic programming principle; such a numerical scheme provides both the required function L and the associated geodesics (the critical path).

1. Choose a grid of points in the state space of z (large enough to avoid boundary effects);

[3]In Lasry & Lions (1995), the coefficients of the diffusion are also driven by an independent Markov chain.

2. Initialize the area Γ to 0;

3. Select progressively new sets of points (z) on the grid and compute the distance $L(z,z')$ for the local metric $A(z)$ to the points (z') whose distances $L(z')$ to the set Γ were calculated at the previous step; the distance to the set Γ for any new point z is the minimum over (z') of $L(z,z') + L(z')$ and the point in (z') which achieves the minimum is precisely the next point of the geodesic.

The last numerical scheme can be easily extended to the case where the coefcients of the diffusion are driven by an exogeneous Markov chain process (see Fournie, Lasry & Lions, this volume).

4.3 Numerical results

We now present an application of the importance sampling variance reduction technique using the large deviations approximation as accelerator. We consider the stochastic volatility model presented in Subsection 2.2 with the following coefficients of the model

$$\begin{aligned} \sigma(y) &= e^{y} 1_{\{y \le M\}}, \quad \text{with a large constant } M \\ \eta(y) &= k(c-y), \quad \gamma(y) = \gamma \quad \text{and} \quad \rho(y) = \rho. \end{aligned}$$

The model specification is motivated by some empirical work in the financial literature which provide evidence of the fact that the log-volatility is mean reverting (see e.g. Scott (1987) and Stein & Stein (1991) amongst others). The expectation of interest is

$$u(s,y) = P\left[S_T \ge K \mid S_0 = s, Y_0 = y\right]. \tag{4.5}$$

The Riemannian metric associated to the system is given by the inverse of the matrix

$$\Sigma^*(s,y)\Sigma(s,y) = \begin{pmatrix} s^2\sigma^2(y) & \rho s \gamma \sigma^2(y) \\ \rho s \gamma \sigma^2(y) & \gamma^2 \sigma^2(y) \end{pmatrix} \tag{4.6}$$

and the minimum distance to the set $\Gamma = (K,\infty) \times \mathbb{R}$ is computed according to the numerical scheme described in the previous section.

Next we consider the large deviations accelerator

$$h(t,s,y) = \frac{L(s,y)}{T-t} \Sigma^* \nabla L(s,y), \tag{4.7}$$

the associated Girsanov weight M_T^h and the process $\{(S_t^h, Y_t^h),\ 0 \le t \le T\}$ deduced from the original one, $\{(S_t, Y_t),,\ 0 \le t \le T\}$ as described in Section 2.

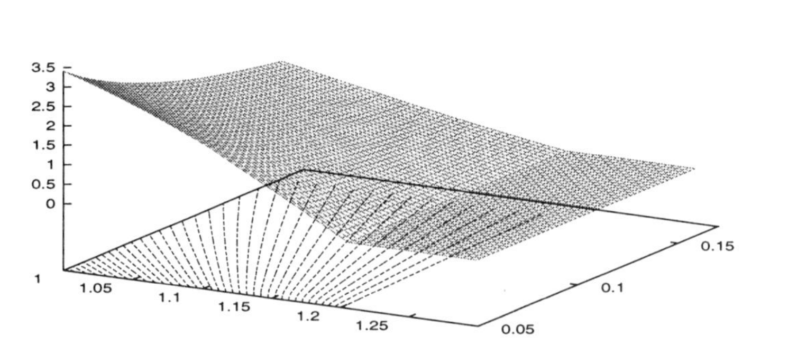

Figure 1: Finslerian metric and level lines

Sample paths of the processes $\{M_t^h,\ 0 \leq t \leq T\}$ and $\{(S_t^h, Y_t^h),\ 0 \leq t \leq T\}$ are simulated according to a classical equally spaced (Δt) Euler approximation (see Pardoux & Talay 1985, Talay 1995) .

Two numerical experiments have been performed. The input parameters of the first one are

$$\begin{aligned} S_0 = 1.1, \quad \sigma_0 = \ln Y_0 = 0.14, \quad \Delta t = 0.03 \\ r = k = c = 0, \quad \gamma = 0.25, \quad \rho = 0.5 \\ T = 30 \text{ days and } K = 1.3. \end{aligned}$$

Figure 1 illustrates the distance to the set $\Gamma = (K, \infty) \times \mathbb{R}$ in the plane $(S, \sigma(y))$. Figure 2 illustrates the sample paths simulated from the original process and those of the controlled one. It shows clearly that controlling the process by its large deviations increases significantly the proportion of sample paths which end up in the region of interest $(K, \infty) \times \mathbb{R}$.

The sample standard deviation of the regular Monte Carlo estimator and the accelerated one are respectively 0.044 and 0.0020. Therefore the importance sampling technique based on the large deviations approximations provides a reduction of the standard deviation by a factor of 22. Figure 3 plots the value of the two estimators – with (solid line) or without variance reduction (dashed line) – as a function of the number of simulated sample paths.

We also performed a second experiment with the following change in the parameters values

$$(r, k, c, \gamma, \rho) = (0.06, 0.5, 0.14, 0.25, 0.5).$$

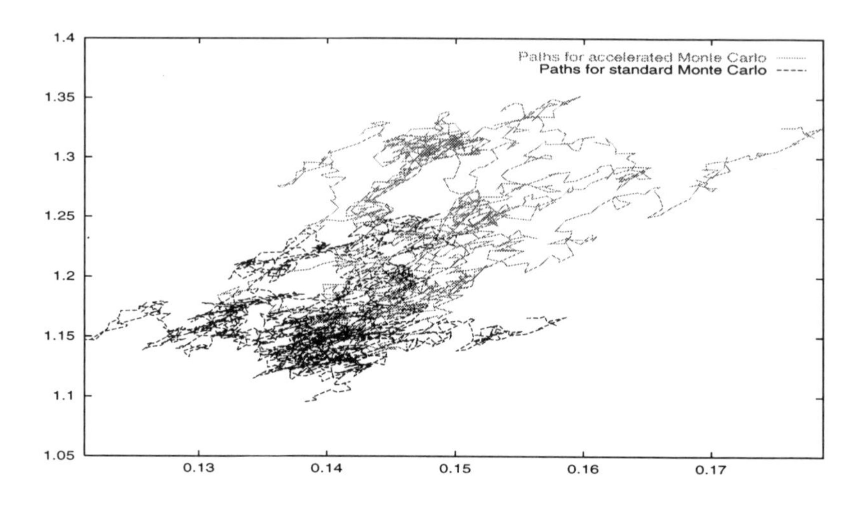

Figure 2: Sample paths of the original and the controlled processes

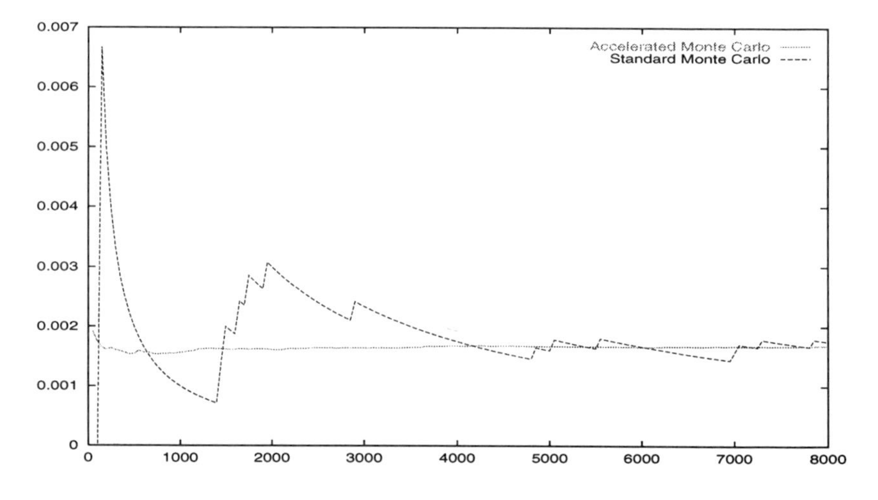

Figure 3: Convergence of Monte Carlo estimators

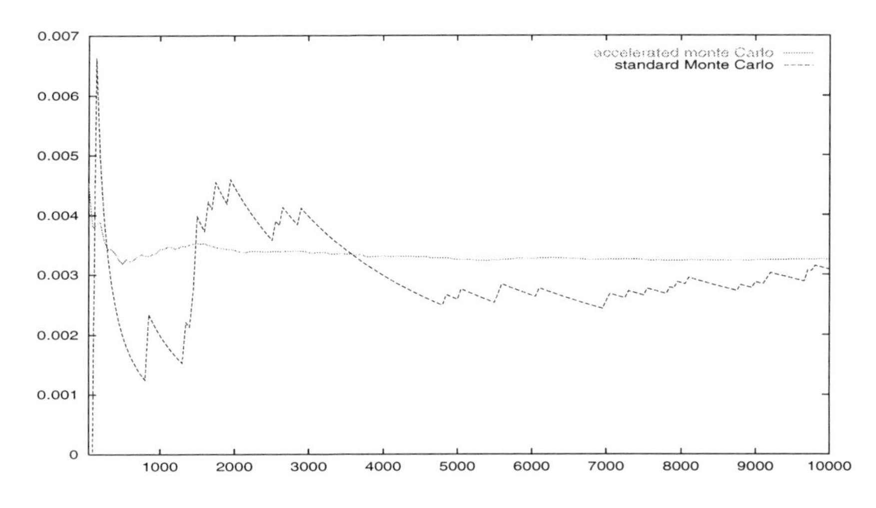

Figure 4: Convergence of Monte Carlo estimators

The sample standard deviations of the two estimators (with or without variance reduction) are respectively 0.0032 and 0.055: the standard deviation is reduced by a factor of 17 in this case. The results of this experiment are illustrated in Figure 4.

5 The Small Noise Expansion Accelerated Monte Carlo Estimator

5.1 The small noise expansion approximation

In this section we suggest an approximation of the process $\{Z_t = (S_t, Y_t), 0 \leq t \leq T\}$ around a Gaussian process since it is easy to derive a closed form solution for the required expectation under the normality assumption. More precisely, we consider again the stochastic volatility model of Subsection 2.2 and we assume that the dynamics of the volatility process depend on a small parameter $\epsilon > 0$ as follows

$$\frac{dS_t^\epsilon}{S_t^\epsilon} = rdt + \sigma(Y_t^\epsilon)d\tilde{W}_t^1 \tag{5.1}$$

$$dY_t^\epsilon = \epsilon\eta(Y_t^\epsilon)dt + \sqrt{\epsilon}\gamma(Y_t^\epsilon)d\tilde{W}_t^2 \tag{5.2}$$

$$\langle dW^1, dW^2\rangle_t = \sqrt{\epsilon}\rho(Y_t^\epsilon)dt, \tag{5.3}$$

where the functions $\sigma(.)$, $\eta(.)$ and $\gamma(.)$ are assumed to be differentiable up to any order with bounded derivatives. As ϵ goes to zero, the process $\{S_t^\epsilon,$

$0 \le t \le T\}$ converges to the process $\{S_t^0,\ 0 \le t \le T\}$ defined by

$$\frac{dS_t^0}{S_t^0} \;=\; rdt + \sigma(Y_0)d\tilde{W}_t^1 \tag{5.4}$$

in the sense that

$$\sup_{0 \le t \le T} E\left|S_t^\epsilon - S_t^0\right| \;\longrightarrow\; 0 \quad \text{as} \quad \epsilon \downarrow 0. \tag{5.5}$$

Therefore a natural approximation of the expectation

$$u_\epsilon(s,y) \;=\; E\left[\phi(S_T^\epsilon) \mid S_0^\epsilon = s\ ,\ Y_0^\epsilon = y\right] \tag{5.6}$$

is obtained by an asymptotic expansion of $u_\epsilon(s,y)$ around the value $\epsilon = 0$ for fixed $(s,y) \in \mathbb{R}_+^* \times \mathbb{R}$. Such asymptotic expansion results have been derived by Azencott (1981) by using the WKB expansion of the perturbed process $\{(S_t^\epsilon, Y_t^\epsilon),\ 0 \le t \le T\}$ in powers of $\sqrt{\epsilon}$. Fournié, Lebuchoux & Touzi (1995) used a partial differential equations approach in order to derive a Taylor expansion of $u_\epsilon(s,y)$ in powers of ϵ. They first proved that $u_\epsilon(s,y)$ is differentiable in $\epsilon = 0$ up to any order and therefore the coefficients of the asymptotic expansion around $\epsilon = 0$ are simply given by the derivatives of $u_\epsilon(s,y)$ with respect to ϵ evaluated at $\epsilon = 0$. Then closed form solutions of the required derivatives are given in terms of the derivatives of $u_0(s,y)$ with respect to s. We refer to Fournié, Lebuchoux & Touzi (1995) for the exact asymptotic expansion result up to any order and we only report the first order which will be used in the application.

$$\begin{aligned}
u_\epsilon(e^x,y) \;=\;& u_0(e^x,y) \\
&+\frac{\epsilon}{2}T^2\Bigg[\eta(y)\left(\mu'(y)\frac{\partial}{\partial x}u_0(e^x,y) + \sigma\sigma'(y)\frac{\partial^2}{\partial x^2}u_0(e^x,y)\right) \\
&\qquad +\sigma\gamma\rho(y)\left(\mu'(y)\frac{\partial^2}{\partial x^2}u_0(e^x,y) + \sigma\sigma'(y)\frac{\partial^3}{\partial x^3}u_0(e^x,y)\right) \\
&\qquad +\frac{1}{2}\gamma^2(y)\left(\mu''(y)\frac{\partial}{\partial x}u_0(e^x,y) + (\sigma\sigma')'(y)\frac{\partial^2}{\partial x^2}u_0(e^x,y)\right)\Bigg] \\
&+\frac{\epsilon}{6}T^3\gamma^2(y)\Bigg[\mu'(y)^2\frac{\partial^2}{\partial x^2}u_0(e^x,y) + 2\mu'\sigma\sigma'(y)\frac{\partial^3}{\partial x^3}u_0(e^x,y) \\
&\qquad\qquad +(\sigma\sigma')^2(y)\frac{\partial^4}{\partial x^4}u_0(e^x,y)\Bigg] \\
&+o(\epsilon)
\end{aligned} \tag{5.7}$$

where $\mu(y) = r - \frac{1}{2}\sigma^2(y)$ is the local mean coefficient of the diffusion $\{X_t = \ln S_t,\ 0 \le t \le T\}$.

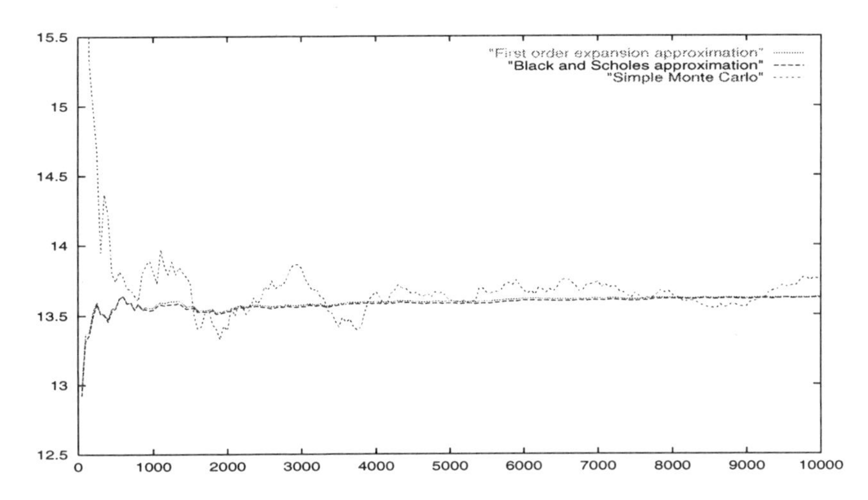

Figure 5: Monte Carlo estimator of at-the-money call option price

5.2 Numerical results

We use the small noise expansion approximation as an accelerator of the Monte Carlo estimator of the European put option price characterized by a terminal payoff function $\phi(s) = (K-s)^+$ where K is a positive constant. The experiments are based on the following coefficients of the model

$$\begin{aligned} \sigma(y) &= e^y 1_{\{y \le M\}}, \quad \text{with a large constant } M \\ \eta(y) &= k(c-y), \quad \gamma(y) = \gamma \quad \text{and} \quad \rho(y) = \rho \end{aligned}$$

with values of the parameters

$$\begin{aligned} r &= 0.1, \quad \sigma_0 = 0.2, \quad S_0 = 100 \\ k &= 0.7, \quad c = -1.6, \quad \gamma = 0.5, \quad \rho = 0.3 \end{aligned}$$

which are very reallistic values observed on the financial markets.

Figure 5 presents the results of our Monte Carlo simulations as a function of the number of simulated sample paths. It is clear that the standard Monte Carlo estimator (without variance reduction techniques) performs very poorly compared to the other two estimators in the sense that its variance is clearly much higher. The two other estimators are obtained by applying the importance sampling procedure with importance sampling function implied by the Black–Scholes approximation (zero-order expansion) and the first-order one. Both estimators converge faster to the true value and exhibit much smaller variances. However, it seems that the important gain in variance is induced by the Black–Scholes approximation. The standard deviations in our experiments are reduced by a factor of 7.3 for an accelerated estimator with zero-order expansion (Black–Scholes) and by a factor of 8.2 with first-order

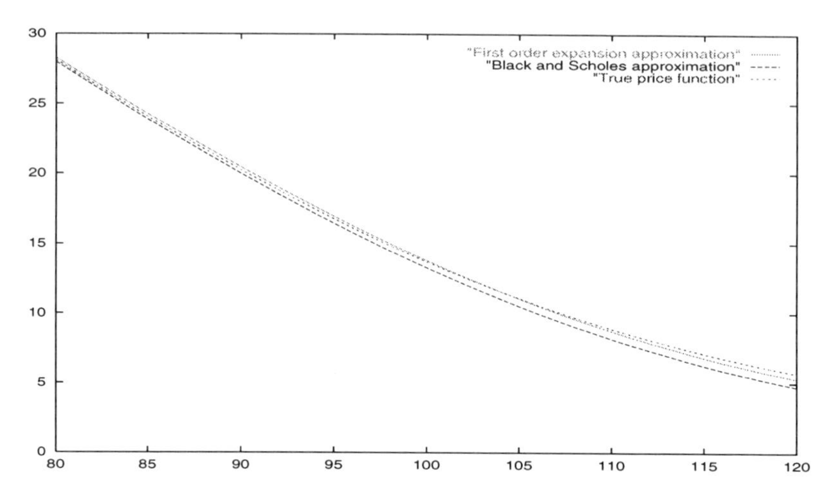

Figure 6: European call option price with varying exercice price

expansion. Notice again that, while the first-order approximation performs strictly better than the Black–Scholes one, the important gain in variance is induced by the simple Black–Scholes approximation.

Figure 6 presents a comparison between the true option price (computed by accelerated Monte Carlo simulations), the Black–Scholes, and the first-order approximations, for various values of the exercise price of the option. Over the range of exercise prices, the first-order approximation of the expectation is closer to the true value of the expectation. The first-order approximation is remarkably close to the true option price for near-from-the money options; as the exercise price gets far-from-the-money the first-order approximation does not perform much better than the Black–Scholes one.

We did not compute higher order approximations but the actual results make us believe that a very sharp approximation of the required expectation can be obtained by a relatively low order approximation.

Acknowledgements

Parts of this work come within the scope of the GMC research program on Monte Carlo based numerical methods at Caisse Autonome de Refinancement, and has benefited from interactions with other work developed in this program by Jérôme Lebuchoux, Pierre-Louis Lions and the authors.

References

Azencott, R. (1981), 'Formules de Taylor stochastiques et développement asymptotique d'intégrales de Feynman', in *Séminaire de Probabilité XVI*, Lecture Notes in Mathematics **921**, 237–285.

Bajeux, I. & Rochet, J.C. (1996), 'Dynamic Spanning: are options an appropriate instrument', *Mathematical Finance* **6**, 1-16.

Barles, G. C., Daher C. & Romano, M. (1990), 'Evaluation of assets with path-dependent cash flows', preprint CEREMADE.

Black, F. & Scholes, M. (1973), 'The pricing of options and corporate liabilities', *Journal of Political Economy* **81**, 637–659.

Dassios, A. (1992), 'Asymptotic expansions to stochastic variance models', Discussion paper, preprint Department of Statistical and Mathematical Sciences, London School of Economics.

Davis, M.H.A. (1994), 'A general option pricing formula', preprint Imperial College London.

Delbaen, F. & Schachermayer, W. (1994), 'A general version of the fundamental theorem of asset pricing ', *Mathematische Annalen* **300**, 463-520.

Engle, R.F. (1982), 'Autoregressive conditional heteroscedasticity with estimates of the variance of UK inflation', *Econometrica* **50**, 987–1008.

Fleming W.H. & Soner H.M. (1992) *Controlled Markov Processes and Viscosity Solutions*, Springer Verlag.

Föllmer, H. & Schweizer, M. (1991), 'Hedging of contingent claims under incomplete information', in *Applied Stochastic Analysis*, M.H.A. Davis and R.J. Elliott (eds.), Gordon and Breach, 389–414.

Fournié, E., Lebuchoux J. & Touzi, N. (1995), 'Small noise expansion and importance sampling', preprint CEREMADE.

Freidlin, M.I. & Wentzell, A.D. (1984), *Random Perturbation of Dynamical Systems*, Springer Verlag.

Friedman, A. (1975), *Stochastic Differential Equations and Applications*, Vol. 1, Academic Press.

Gouriéroux, C., Laurent J.P. & Pham, H. (1995), 'Hedging and numeraire', preprint CREST 9550.

Harrison, J.M. & Kreps, D.M. (1979), 'Martingale and arbitrage in multiperiods securities markets', *Journal of Economic Theory* **20**, 381–408.

Hart, O.D. (1975), 'On the optimality of equilibrium when market structure is incomplete', *Jounal of Economic Theory* **11**, 418–443.

Heston, S.L. (1993), 'A closed-form solution for options with stochastic volatility', *Review of Financial Studies* **6**, 327–343.

Hull, J. & White, A. (1987), 'The pricing of options on assets with stochastic volatilities', *Journal of Finance* **3**, 281–300.

Kanai (1977), 'Short time asymptotics for fundamental solutions of partial differential equations', *Communications in PDE* **2**, 781–830.

Karatzas, I. & Shreve, S.E. (1988) *Brownian Motion and Stochastic Calculus*, Springer Verlag.

Kiefer, Y.I. (1976), 'Transition density of diffusion processes with small diffusion', *Theory of Probability and Applications* **21**, 513–522.

Karatzas, I., Lehoczky, J.P, Shreve, S.E & Xu G.L. (1991), 'Martingale and duality methods for utility maximization in incomplete markets', *SIAM Journal on Control and Optimization* **25**, 1157–86.

Lasry, J.M. & Lions, P.L. (1995), 'Grandes déviations pour des processus de diffusion couplés par un processus de sauts', *CRAS*.

Molchanov, S. (1975), 'Diffusion processes and Riemannian geometry', *Russ. Math. Surveys* **30**, 1–63.

Nelson, D.B. (1990), 'ARCH models as diffusion approximations', *Journal of Econometrics* **45**, 7–38.

Newton, N.J. (1994), 'Variance reduction for simulated diffusions', *SIAM Journal on Applied Mathematics* **54**, 1780–1805.

Pham, H. & Touzi, N. (1993), 'Equilibrium state price processes in a stochastic volatility model', *Mathematical Finance* **6**, 215–236.

Pardoux, E. & Talay, D. (1985), 'Discretisation and simulation of stochastic differential equations', *Acta Appl. Math.* **3**, 23-47.

Renault, E. & Touzi, N. (1996), 'Option hedging and implied volatilities in a stochastic volatility model', *Mathematical Finance* **6**, 279–302.

Romano, M. & Touzi, N. (1993), 'Contingent claims and market completeness in a stochastic volatility model', *Mathematical Finance*, to appear.

Ross, S. (1976), 'Options and efficiency', *Quarterly Journal of Economics* **90**, 75–89.

Scott, L. (1987), 'Option pricing when the variance changes randomly: theory, estimation and application', *Journal of Financial and Quantitative Analysis* **22**, 419–438.

Scott, L. (1991), 'Random variance option pricing', *Advances in Futures and Options Research* **5**, 113–135.

Stein, E.M. & Stein, J.C. (1991), 'Stock price distribution with stochastic volatility: an analytic approach', *Review of Financial Studies* **4**, 727–752.

Talay, D. (1995), 'Simulation of stochastic differential systems', to appear in *Probabilistic Methods in Applied Physics*, P. Kree & W. Wedig (eds.), Lectures Notes in Physics **451**, 54–96, Springer Verlag.

Varadhan, S. (1967), 'Diffusion processes in small time interval', *Communications in Pure and Applied Mathematics* **20**, 659–685.

Dynamic Optimization for a Mixed Portfolio with Transaction Costs

Agnès Sulem

1 Introduction

We consider a model of n risky and one risk-free asset in which the dynamics of the prices of the risky assets are governed by logarithmic Brownian motions (LBM). Risky assets are conventionally called *Stocks* while the risk-free asset is called a *Bond* or *Bank*. Modelling the stock prices by LBM goes back to Black & Scholes (1973). Since the paper by Merton (1971), in which a dynamic optimization consumption/investment model was introduced, there has been a lot of effort expanded to develop and perfect the LBM models.

One of the avenues of research deals with the *transaction cost problem*: Consider an investor who has an initial wealth invested in Stocks and Bonds and who has the ability to transfer funds between the assets. This transfer involves brokerage fees (transaction costs).

In one type of model the proceeds from these operations are used to finance the consumption and the objective is to maximize the cumulative expected utility of consumption over a finite or infinite time horizon (see Davis & Norman 1990, Akian, Menaldi & Sulem 1996, Fleming & Soner 1992, Shreve & Soner 1994, Zariphopoulou 1989). In Akian, Sulem & Séquier (1995), a similar problem is considered but on a finite horizon and without consumption. In Sethi & Taksar (1988) and Akian, Sulem & Taksar (1996), the objective is to maximize the long run average growth of wealth.

Mathematicaly, these problems can be formulated as singular stochastic control problems. Dynamic programming methodology leads to a variational inequality (or dynamic programming equation) for the value function of elliptic, parabolic or ergodic type, that is a nonlinear partial differential equation of second order.

Solution of these equations provides the optimal investment strategy in feedback form which means that the optimal control is a function of the current state of the system.

In general, the value function is not smooth enough to satisfy the dynamic programming equation in the strong sense. Therefore a weak formulation of solutions is necessary: the concept of viscosity (see Lions 1983, and Crandall, Ishii & Lions 1992) is a powerful tool which is well adapted to these equations.

Moreover, except for special cases such as the Merton problem (no transaction costs), dynamic programming equations cannot be solved explicitly and we thus have to turn to numerical solutions.

The aim of this article is to present some of the tools used for dynamic optimisation of portfolios. The article is organized as follows: the mathematical formulation of the problem in terms of the stochastic control problem is given in section 2. Two cases are considered: maximisation of a function of the net wealth on a finite horizon and maximisation of the long run average growth of wealth. We state in Section 3 the variational inequalities satisfied by the value function in both parabolic and ergodic cases. In Section 4, numerical methods are presented to solve the dynamic programming equations: that is, (i) restriction to a bounded domain; (ii) discretisation; (iii) solution of the discretized problem. Finally in the last section numerical results are given in the case of one bank account and two risky assets: optimal transaction policies are obtained and risk-return characteristics are analysed.

2 Formulation of the Problem

Let $(\Omega, \mathcal{F}, P)$ be a fixed complete probability space and $(\mathcal{F}_t)_{t\geq 0}$ a given filtration. We denote by $s_0(t)$ (resp. $s_i(t)$ for $i = 1, \dots, n$) the amount of money in the bank account (resp. in the ith risky asset) at time t and refer to $s(t) = (s_i(t))_{i=0,\dots,n}$ as the investor position at time t. In the absence of control the process $s_0(t)$ grows deterministically at exponential rate r, while $(s_1(t), \dots, s_n(t))$ is governed by a logarithmic Brownian motion. The investment policy is described in terms of two sets of increasing functionals $\mathcal{L}(t) = (\mathcal{L}_1(t), \dots, \mathcal{L}_n(t))$ and $\mathcal{M}(t) = (\mathcal{M}_1(t), \dots, \mathcal{M}_n(t)), i = 1, \dots, n$, representing cumulative purchase and sale of the ith stock at time t. We require that $\mathcal{L}(t)$ and $\mathcal{M}(t)$ are right-continuous, non-decreasing, adapted to $\mathcal{F}_t$ and such that $\mathcal{L}_i(0^-) = \mathcal{M}_i(0^-) = 0$. The transaction processes are singular controls. This formulation allows for instantaneous transactions of finite amounts of stocks: indeed, in contrast to classical control problems in which the displacement of the state due to the control effort is differentiable in time, the singular control models we consider allow this displacement to be discontinuous. We refer to Fleming & Soner (1993).

The dynamics of the system under control is then given by

$$\begin{cases} ds_0(t) &= rs_0(t)dt + \sum_{i=1}^n((1-\mu_i)d\mathcal{M}_i(t) - (1+\lambda_i)d\mathcal{L}_i(t)), \\ ds_i(t) &= \alpha_i s_i(t)dt + \sigma_i s_i(t)dW_i(t) + d\mathcal{L}_i(t) - d\mathcal{M}_i(t), \quad i = 1\dots n \end{cases} \tag{2.1}$$

with initial values $s_i(0^-) = x_i, \quad i = 0, \dots, n$, where $W_i(t)$, $i = 1, \dots, n$, are correlated Wiener processes such that

$$E(W_i(t)W_j(t)) = \rho_{ij}\, t \text{ for } i \neq j.$$

Here λ_i represents the commission for purchasing \$1 worth of the ith stock while μ_i is the commission from the sale of \$1 worth of the ith stock. All the proceeds from sales of stock go into the Bank while all the purchases are financed by withdrawing cash from the Bank.

Define the net wealth of the investor as the amount of money in the bank account resulting from the sale of the risky assets at time t:

$$\rho(t) = s_0(t) + \sum_{i=1}^{n}(1 - \mu_i)s_i(t).$$

Define the admissible region $\mathcal{S} = \mathbf{R}_+^{n+1}$ which corresponds to the absence of borrowing and shortselling. Given an initial endowment x in $\mathcal{S}$, a policy is admissible if $s(t)$ remains in $\mathcal{S}$ during the considered period of time. We denote by $\mathcal{P}$ the set of admissible policies.

We consider the following cases:

- *the finite horizon problem* where the objective is to maximize over all admissible policies the quantity

$$Eu(\rho(T))$$

  where T is some finite horizon, and E denotes expectation, and

- the problem of maximizing *the long run average growth of wealth*

$$\mathcal{J}_x(\mathcal{L}, \mathcal{M}) = \liminf_{T\to\infty} T^{-1}Eu(\rho(T)).$$

  Note that when $\rho(t^-) = 0$, the only admissible policy is to remain at zero. In this case, $\rho(T) = 0$ for all $T \geq t$ and for such a policy $\mathcal{J}_x(\mathcal{L}, \mathcal{M}) = -\infty$.

The function $u(c)$ is a Hyperbolic Absolute Risk Aversion (HARA) utility function defined by

$$u(c) = \frac{c^{1-\gamma}}{1-\gamma}, \quad \gamma \geq 0, \gamma \neq 1 \quad \text{or} \quad u(c) = \log c. \tag{2.2}$$

The coefficient γ is the relative risk aversion. The risk is maximal for $\gamma = 0$ (all the money is put in the asset with the largest rates of return α_i and return variation σ_i) and decreases as γ goes to $+\infty$. The logarithm function corresponds to the limit case $\gamma = 1$.

3 Variational Inequalities

The investor knows the state of the system at each instant of time t. We call this the case of complete information. The best system performance depends on the information available at each time t. Using the dynamic programming method, we can state the equations satisfied by the value function. This approach has the advantage of leading to controls in feedback forms.

- **Parabolic variational inequality**

Define the value function

$$V(t,x) = \sup_{(\mathcal{L},\mathcal{M})\in\mathcal{P}} E[\frac{\rho(T)^{1-\gamma}}{1-\gamma}|s(t) = x]. \tag{3.1}$$

It is shown in Akian, Sulem & Séquier (1995) that $V(t,x)$ satisfies the following parabolic variational inequality:

$$\begin{cases} \max\left\{\dfrac{\partial V}{\partial t} + AV, \max\limits_{1\le i\le n} L_i V, \max\limits_{\substack{1\le i\le n,\\ x_i>0}} M_i V\right\} = 0 \text{ in } \mathcal{S}\setminus\{x_0 = 0\}\times[0,T[, \\ \max\left\{\dfrac{\partial V}{\partial t} + AV, \max\limits_{\substack{1\le i\le n,\\ x_i>0}} M_i V\right\} = 0 \text{ in } \{x_0 = 0\}\times[0,T[, \\ V(T,x) = \dfrac{1}{1-\gamma}\left(x_0 + \displaystyle\sum_{i=1}^{n}(1-\mu_i)x_i\right)^{1-\gamma}, \end{cases} \tag{3.2}$$

where

$$\begin{cases} AV &= \frac{1}{2}\sum_{i=1}^{n} a_{ij}x_i x_j \dfrac{\partial^2 V}{\partial x_i \partial x_j} + \displaystyle\sum_{i=1}^{n} \alpha_i x_i \dfrac{\partial V}{\partial x_i} + r x_0 \dfrac{\partial V}{\partial x_0}, \\ L_i V &= -(1+\lambda_i)\dfrac{\partial V}{\partial x_0} + \dfrac{\partial V}{\partial x_i}, \\ M_i V &= (1-\mu_i)\dfrac{\partial V}{\partial x_0} - \dfrac{\partial V}{\partial x_i}, \end{cases} \tag{3.3}$$

$a_{ii} = {\sigma_i}^2$, and $a_{ij} = \rho_{ij}\sigma_i\sigma_j$ for $i \neq j$.

- **Ergodic variational inequality**

The optimal long run average growth of wealth

$$d = \sup_{(\mathcal{L},\mathcal{M})\in\mathcal{P}} \liminf_{T\to\infty} T^{-1} E[\log(\rho(T)|s(0) = x] \tag{3.4}$$

satisfies

$$\max\{-d + A\tilde{V}, \max_{1\le i\le n} L_i\tilde{V}, \max_{1\le i\le n} M_i\tilde{V}\} = 0 \quad \text{in } \overset{\circ}{\mathcal{S}}$$

(see Akian, Sulem & Taksar 1996).

The quantity $\tilde{V}$ can be interpreted as $\lim_{T\to\infty}\{\mathcal{V}(T,x) - dT\}$ where $\mathcal{V}(T,x) = \sup_{\mathcal{P}} E[\log\rho(T)|s(0) = x]$.

Note that the ergodic problem can be obtained as the limit when the discount factor goes to zero of a discounted problem over a infinite horizon (see Akian, Sulem & Taksar 1996).

Variational inequalities are free boundary problems. The discrete control which selects the equation which satisfies the maximum in (3.2) indicates the optimal policy.

Optimal investment policy At time t, the admissible region $\mathcal{S}$ is divided as follows:

$$\begin{aligned} B_i^{\,t} &= \{x \in \mathcal{S},\ \ L_i V(t,x) = 0\}, \\ S_i^{\,t} &= \{x \in \mathcal{S},\ \ M_i V(t,x) = 0\}, \\ NT_i^{\,t} &= \mathcal{S} \setminus (B_i^{\,t} \cup S_i^{\,t}), \\ NT^t &= \bigcap_{i=1}^{n} NT_i^{\,t}. \end{aligned}$$

NT^t is the no-transaction region at time t. In NT^t the system evolves as a pure diffusion process. Outside NT^t, an instantaneous transaction brings the position to the boundary of NT^t: buy stock i in $B_i^{\,t}$, sell stock i in $S_i^{\,t}$. After the initial transaction, the agent position remains in

$$\overline{NT}^t = \left\{ x \in \mathcal{S}, (\frac{\partial V}{\partial t} + AV)(t,x) = 0 \right\},$$

and further transactions occur only at the boundary (see Davis & Norman 1990).

Existence and uniqueness results for the solution of the variational inequality are obtained in the papers mentioned above. Existence is obtained by stating a weak dynamic programming principle. This principle, due to Bellman in the deterministic case and then extended to stochastic control (see Fleming & Soner 1992) states that 'an optimal policy has the property that, whatever the initial state and control are, the remaining decisions must constitute an optimal policy with regard to the state resulting from the first decision'. Uniqueness of the solution of the parabolic equation is obtained in the class of continuous functions in $\mathcal{S} \times [0,T]$ which satisfy

$$|V(x,t)| \leq C(1 + \|x\|^{1-\gamma}),\ \ \forall x \in \mathcal{S}, t \in [0,T]$$

for some constant $C > 0$.

4 Numerical Methods

4.1 Reduction of the state dimension

The value function V defined by (3.1) has the homothetic property

$$\forall \rho > 0,\ V(t, \rho x) = \rho^{1-\gamma} V(t,x). \tag{4.1}$$

Consider then the new state variables:

$$\begin{cases} \rho &= x_0 + \sum_{i=1}^n (1-\mu_i)x_i \\ y_i &= (1-\mu_i)x_i/\rho, i = 1\dots n \end{cases} \tag{4.2}$$

where ρ represents the net wealth and y_i the fraction invested in stock i. We have

$$V(t,x) = \rho^{1-\gamma}W(t,y) \text{ where}$$

$$W(t,y) = V(t, 1 - \sum_{i=1}^n y_i, \frac{y_1}{1-\mu_1}, \dots, \frac{y_n}{1-\mu_n}). \tag{4.3}$$

The function W satisfies an n-dimensional variational inequality in the simplex $\Delta_n = \{y \in \mathbf{R}^n, y_i \geq 0, \sum_{i=1}^n y_i \leq 1\}$.

The same change of variables for the ergodic problem leads to a variational inequality of the form:

$$\max\left\{\tilde{A}W - d + H(y), \max_{1\leq i\leq n} \tilde{L}_i W, \max_{1\leq i\leq n,\, y_i>0} \tilde{M}_i W\right\} = 0 \text{ in } \Delta_n \tag{4.4}$$

where

$$H(y) = r + \sum_{i=1}^n (\alpha_i - r)y_i - \frac{1}{2}\sum_{i,j=1}^n a_{ij}y_iy_j;$$

here, $\tilde{A}$ is a second order differential operator, and $\tilde{L}$ and $\tilde{M}$ are first order operators.

Remark: We can check that the function W only depends on $\nu = (\nu_i)_{i=1\dots n}$ where $\nu_i = (\lambda_i + \mu_i)/(1-\mu_i)$. Let us denote by $W_{\lambda,\mu}$ the function W in order to express explicitly the dependence of W on the transaction costs; we have:

$$W_{\lambda,\mu}(t,y) = W_{\nu,0}(t,y).$$

The problem can thus be reformulated as if the transaction costs (equal to ν_i) would appear only on purchase and not on sale.

For numerical purposes, we proceed with a technical change of variables which transforms the simplex Δ_n into the cube $[0,1]^n$:

$$\begin{cases} z_1 = y_1 + \dots + y_n, \\ z_i = \dfrac{y_i + \dots + y_n}{y_{i-1} + \dots + y_n} \quad i = 2, \dots, n. \end{cases}$$

4.2 Numerical study of parabolic variational inequality

We have to solve an equation of the form:

$$\begin{cases} \max\left(\dfrac{\partial\phi}{dt} + B^0\phi, \max_{P\in\mathcal{P}(z)} B^P\phi\right) = 0 & \text{in } [0,1]^n \times [0,T[\\ \phi(T,z) = \dfrac{1}{1-\gamma} & \text{in } [0,1]^n \end{cases} \tag{4.5}$$

where B^0 is a second order operator, B^P are first order operators and $\mathcal{P}(z) \subset \mathcal{P} = \{1, 2, \ldots, 2n\}$.

The solution of equation (4.5) leads to the optimal feedback $P(t, z) \in \mathcal{P} \cup \{0\}$. When $P(t, z) = 0$, then $z \in NT^t$, otherwise $z \in B_i^t$ or S_i^t according to the value of $P(t, z)$.

For the numerical study, equation (4.5) is discretized and then solved by means of an iterative method. We use a Crank–Nicholson scheme for the time discretization, with time discretization step $\Delta t = T/M$ ($M \in \mathbf{N}^*$), that is we consider the following approximation:

$$\left(\frac{\partial \phi}{\partial t} + B^0\phi\right)(t, x) \sim \frac{\phi(t + \Delta t, x) - \phi(t, x)}{\Delta t} + \frac{1}{2}(B^0\phi(t + \Delta t, x) + B^0\phi(t, x)).$$

This discretization leads to N elliptic equations with unknowns $\phi(t, .)$ in terms of $\phi(t + \Delta t, .)$, $t = k\Delta t, k = 0, \ldots, M$, which are solved backwards starting from the known function $\phi(T, .)$. Each elliptic equation is solved by using the method described below.

4.3 Numerical study of elliptic variational inequality

We consider equations of the form:

$$\begin{cases} \max\limits_{P \in \mathcal{P}_{ad}} (A^P W + u(P)) = 0 & \text{in } \Omega = [0, 1]^n \setminus \Gamma \\ W = 0 & \text{on } \Gamma \end{cases} \tag{4.6}$$

where A^P is a second order degenerate elliptic operator

$$A^P W(x) = \sum_{i,j=1}^{n} a_{ij}(x, P) \frac{\partial^2 W}{\partial x_i \partial x_j}(x) + \sum_{i=1}^{n} b_i(x, P) \frac{\partial W}{\partial x_i}(x) - \beta(x, P) W(x)$$

with

$$\sum_{i,j=1}^{n} a_{ij}(x, P)\eta_i\eta_j \geq 0, \qquad \beta(x, P) \geq 0 \qquad \forall x \in \Omega,\ \eta \in \mathbf{R}^n,\ P \in \mathcal{P}_{ad}.$$

$\mathcal{P}_{ad}$ is a closed subset of $\mathbf{R}^k$ (which may depend on x) and Γ is a part of the boundary $\partial\Omega$, which consists of faces of the n-cube $[0, 1]^n$. On $\partial\Omega \setminus \Gamma$, the operator A^P is degenerate and no boundary condition is needed.

Let $h = 1/N$ ($N \in \mathbf{N}^*$) denote the finite difference step in each coordinate direction, e_i the unit vector in the ith coordinate direction, and $x = (x_1, \ldots, x_n)$ a point of the uniform grid $\Omega_h = \Omega \cap (h\mathbf{Z})^n$. Equation (4.6) is discretized by a finite difference method. We try to find a discretization such that the operator A_h^P satisfies the discrete maximum principle, i.e.

$$(A_h^P W_h(x) \leq 0 \ \forall x \in \Omega_h) \Rightarrow (W_h(x) \geq 0 \ \forall x \in \Omega_h).$$

To that purpose, since A^P is degenerate, $\partial W/\partial x_i$ is approximated by a one-sided difference approximation according to the sign of the drift term $b_i(x,P)$ (see Kushner 1977). We obtain a system of N_h nonlinear equations in N_h unknowns $\{W_h(x),\ x \in \Omega_h\}$:

$$\max_{P \in \mathcal{P}_{ad}} (A_h^P W_h + u(P))(x) = 0 \quad \forall x \in \Omega_h \tag{4.7}$$

where $N_h = \sharp\Omega_h \sim 1/h^n$. When A_h^P satisfies the discrete maximum principle, then the method is stable (the matrix $(A_h^P)^{-1}$ is not singular) and consistent (i.e. the discretized equation is indeed an approximation of the initial equation). In this case the discretised problem corresponds to the dynamic programming equation associated to a control problem of a Markov chain and we can use classical algorithms: the value iteration method (see Bellman 1971) which consists in a contraction iteration, and the Howard algorithm (see Howard 1960, Bellman 1957, 1971) also named policy iteration.

Another possible approximation procedure is to approximate the original controlled process by an appropriate controlled Markov chain on a finite state space, and also to approximate the original objective function (see Kushner & Dupuis 1992).

In our specific case, because of the degeneracy of the operator A^P at some points of the closed n-cube $\overline{\Omega}$ and the presence of mixed derivatives, A_h^P does not satisfy the discrete maximum principle, and (4.7) may not be stable, even for a small step h. However A_h^P can be written as the sum of a symmetric negative definite operator and an operator which satisfies the discrete maximum principle; we thus infer the stability of A_h^P which is confirmed by numerical experiments.

We have developed an algorithm called a multigrid-Howard algorithm (see Akian, Menaldi & Sulem 1996). It is based on the Howard algorithm which consists of an iteration on the control and value functions (starting from P^0 or W^0):

$$\text{for } k \geq 1, \qquad P^k \in \operatorname*{Argmax}_{P \in \mathcal{P}_h}(A_h^P W^{k-1} + u(P)) \tag{4.8}$$

$$\text{for } k \geq 0, \qquad W^k \text{ is solution of } A_h^{P^k} W + u(P^k) = 0. \tag{4.9}$$

When A_h^P satisfies the discrete maximum principle, the sequence W^k decreases and converges to the solution of (4.7) and the convergence is in general super-linear. The exact computation of step (4.9) is expensive in dimension $n \geq 2$ (the complexity of a direct method is $\mathcal{O}(N_h^{3-2/n})$). We thus use a multigrid method with initial value W^{k-1} to compute W^k . The advantage is that each multigrid iteration takes computing time of $\mathcal{O}(N_h)$ and contracts the error by a factor independent of the discretization step h. For a detailed description of the multigrid algorithm, see for example McCormick (1987), Hackbusch (1985), Hackbusch & Trottenberg (1981).

5 Numerical Results

5.1 Optimal investment policy for the finite horizon problem

As an example for our numerical study, we focus on the domestic asset allocation issue. The riskless rate r is fixed at 6%. The drift α_1 of the first asset (long bond portfolio) is fixed at a level of 9% and its standard deviation σ_1 equal to 7%. The other risky asset simulates the equity market: drift α_2 equal to 11% and volatility σ_2 equal to 20%. The correlation coefficient ρ between the two risky assets is set at 40% and the time horizon T is equal to 1 year.

Consider an investor with a rather low risk aversion: $\gamma = 0.7$. At time $t = 0$, the time remaining before the end of the investment period is long and the investor has a strong incentive to trade on risky assets as long as the transaction costs are not too high (see Table 1). Note that the buy-barrier is very insensitive to the transaction costs. The expenses linked to the trades do not prevent the investor from investing in the risky assets.

transaction costs (equal on both assets)	no transaction region (asset n° 1)	no transaction region (asset n° 2)
0%	$y1 = 17.3\%$	$y2 = 82.7\%$
0.1%	$24.4\% < y1 < 37.5\%$	$96.9\% < y2 < 100\%$
0.5%	$18.8\% < y1 < 43.8\%$	$96.9\% < y2 < 100\%$
0.75%	$18.8\% < y1 < 56.3\%$	$96.9\% < y2 < 100\%$
1%	$18.8\% < y1 < 68.8\%$	$96.9\% < y2 < 100\%$
2%	$18.8\% < y1 < 100\%$	$3.1\% < y2 < 100\%$

Table 1: Sensitivity to transaction costs at the beginning of the investment period

transaction costs (equal on both assets)	no transaction region (asset n° 1)	no transaction region (asset n° 2)
0%	$y1 = 17.3\%$	$y2 = 82.7\%$
0.1%	$24.4\% < y1 < 37.5\%$	$96.9\% < y2 < 100\%$
0.5%	$21.8\% < y1 < 56.3\%$	$15.6\% < y2 < 100\%$
0.75%	$21.8\% < y1 < 78.1\%$	$6.3\% < y2 < 100\%$
1%	$18.8\% < y1 < 93.8\%$	$3.1\% < y2 < 100\%$
2%	$9.4\% < y1 < 100\%$	$0\% < y2 < 100\%$

Table 2: Sensitivity to transaction costs at the end of the investment period

The results are different when the same calculations are performed nearer to the end of the investment period. The time remaining is 20% of the initial period. At this stage of the process, the only case in which it may be sensible to trade on risky assets is when the transaction costs are as low as 0.1% on both assets. As expected, the no-transaction region tends to be larger when the transaction costs increase and when the end of the investment period is closer. Indeed, the investor faces a trade-off between the instantaneous cost of transacting and the expectation of a higher level of final utility if he trades. Given that, the adjustment is all the more efficient when the time remaining is long.

5.2 Risk-return of the portfolio

Given a policy, define the random return of the portfolio at time t as $R(t) = (\rho(t) - \rho)/\rho$ where $\rho = \rho(0)$ is the initial wealth. We are interested in computing the expectation of the return $E(R(T) \mid s(0) = x)$ and the risk measured by the variance of the return $\mathrm{Var}(R(T) \mid s(0) = x)$ for the optimal policy. Given the optimal feedback, define

$$\mathcal{R}(t,x) = E(\rho(T) \mid s(t) = x) \tag{5.1}$$

$$\mathcal{V}(t,x) = E(\rho(T)^2 \mid s(t) = x). \tag{5.2}$$

The functions $\mathcal{R}(t,x)$ and $\mathcal{V}(t,x)$ satisfy the following Kolmogorov equations:

$\forall t \in [0, T[$, $\mathcal{R}(t,x)$ satisfies

$$\begin{cases} \left(\dfrac{\partial \mathcal{R}}{\partial t} + A\mathcal{R}\right)(t,x) = 0 & \text{in } NT^t, \\ L_i \mathcal{R}(t,x) = 0 & \text{in } {B_i}^t, \\ M_i \mathcal{R}(t,x) = 0 & \text{in } {S_i}^t, \end{cases} \tag{5.3}$$

with $\mathcal{R}(T,x) = \rho$ and $\mathcal{V}(t,x)$ satisfies (5.3) with $\mathcal{V}(T,x) = \rho^2$.

We can write:

$$\begin{aligned} E(R(T) \mid s(0) = x) &= \frac{\mathcal{R}(0,x)}{\rho} - 1, \\ \mathrm{Var}(R(T) \mid s(0) = x) &= \frac{\mathcal{V}(0,x)}{\rho^2} - \frac{\mathcal{R}(0,x)^2}{\rho^2}. \end{aligned}$$

The impact of the transaction costs on the financial characteristics of the investment is analysed through graphs in the mean-variance plan. Figure 1 represents the efficient frontiers when the transaction costs are equal respectively to 0 (Merton case) (upper curve) and 1% on both assets. The risk parameter γ varies from 0.7 to 50. When the risk aversion is low enough, the only impact of the transaction costs is a decrease in returns of about 1%.

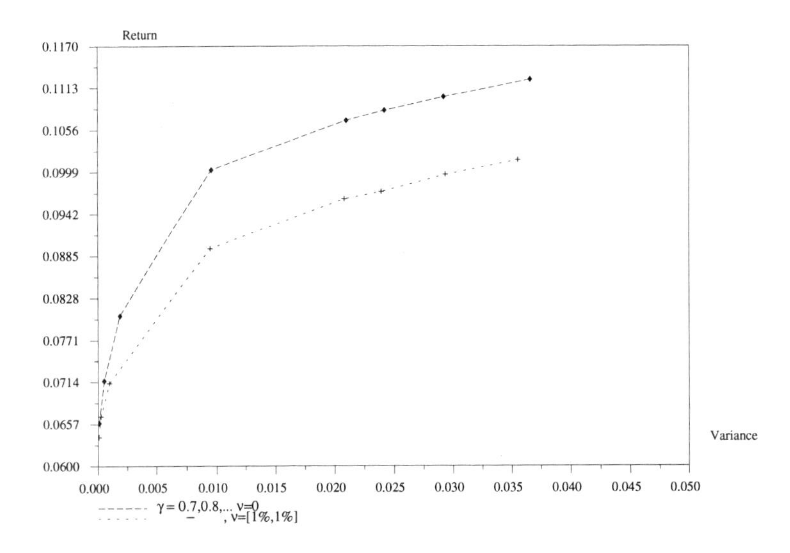

Figure 1: Impact of uniformed transaction costs on the entire efficient frontier

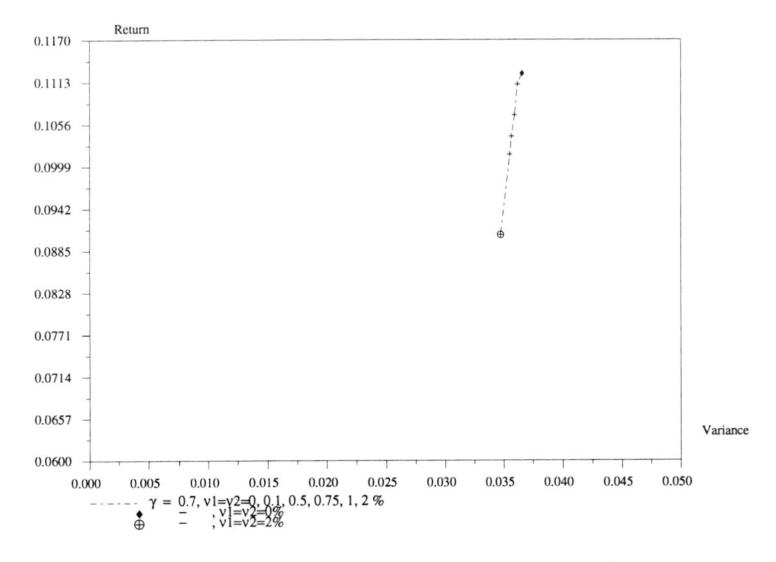

Figure 2: Impact of increasing transaction costs on the strategy

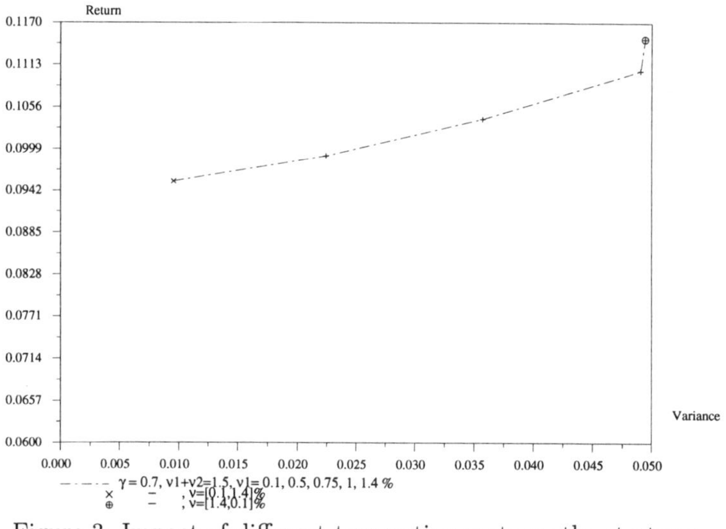

Figure 3: Impact of different transaction costs on the strategy

However, for an investor with such behavior towards risk, the expected returns of the risky assets remain interesting so that he does not modify the risk level of his portfolio. At the left end of the graph, the conclusion is slightly different since risk adverse investors tend to be even less adventurous when they face transaction costs. As expected, the risk adverse investor demands high rewards to take risk. Figure 2 shows how the risk return of the optimal policy is modified when the transaction costs increase from 0 to 2%. The impact of having two different transaction costs for the risky assets is displayed in Figure 3. This test has been performed with a rather low risk aversion $\gamma = 0.7$ and with $\nu_1 + \nu_2 = 1.5$. The transaction cost on bond increases from 0.1 to 1.4 as cost on stock decreases from 1.4 to 0.1 from left to right on the graph. When bonds have much lower transaction costs compared to the stocks, the investor prefers a portfolio with lower risk than the ones he would have implemented with comparable transaction costs on the two assets. On the other hand, if the stocks are much cheaper to buy and sell the investor increases the risk of his portfolio. When the bonds incur high costs there is no point buying them since they no longer provide a satisfactory risk premium over the riskless asset. Then, since the investor considered is not too risk adverse, he prefers to invest on the stock market.

5.3 Optimal investment policy for the long run growth rate

Numerical results are presented in the case of one bank account and two risky assets. They provide the optimal investment strategy and indicate the shape

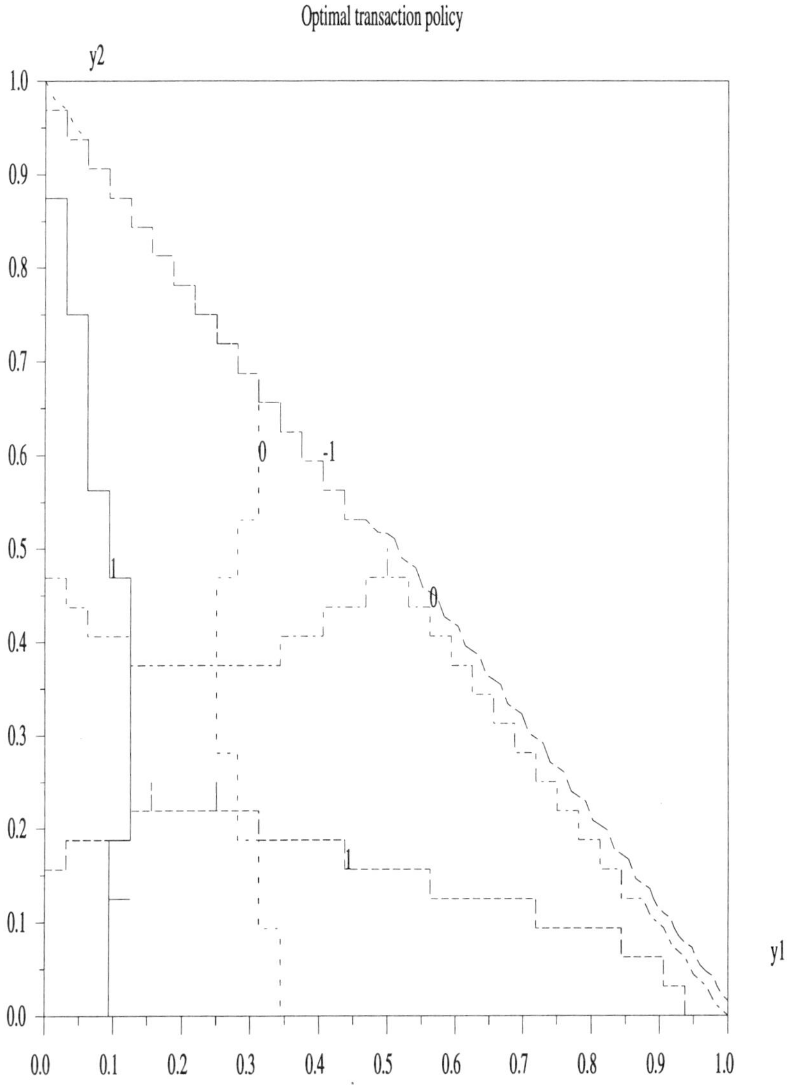

Optimal transaction policy

y2

1.0
0.9
0.8
0.7
0.6
0.5
0.4
0.3
0.2
0.1
0.0

0 -1
1 0
1

0.0 0.1 0.2 0.3 0.4 0.5 0.6 0.7 0.8 0.9 1.0

y1

Figure 4: Optimal investment strategy for the long run average case

of the transaction and no-transaction regions.

The boundaries of the transaction regions are displayed in Figure 4 for $r = 7\%, \alpha = (9\%, 11\%), a_{11} = \sqrt{0.22}, a_{12} = a_{21} = 0.028, a_{22} = \sqrt{0.32}$. The transaction costs are equal to 1% for both assets.

As expected, 9 regions appear: buy (resp. sell) asset i when y_i is below (resp. above) a critical level π_i^- (resp. π_i^+) depending on y_j ($j \neq i$), and no transaction between π_i^- and π_i^+.

After the first transaction, the position of the investor evolves as a diffusion process with reflection on the boundary of NT.

The optimal long run average growth of wealth defined in (3.4) is $d = 0.0785$. It is interesting to compare this with the Merton optimal performance d_m. When the transaction costs are equal to zero, the optimal investment strategy is to keep a constant fraction of total wealth in each risky asset (see Merton 1971, Davis & Norman 1990, Akian, Menaldi & Sulem 1996). Indeed, set $\lambda = \mu = 0$ in equation (4.4). We obtain:

$$\begin{cases} W(y) = \text{ constant} \\ H(y) \geq d_m, \quad \forall y, \end{cases} \tag{5.4}$$

which leads to

$$d_m = -r - \frac{1}{2}(\alpha - r, a^{-1}(\alpha - r))$$

where a is the matrix (a_{ij}). We obtain the optimal proportion

$$\pi^* = \operatorname*{Argmin}_{y \in \mathbb{R}_+^n} H(y) = a^{-1}(\alpha - r)$$

if $\pi^* \in \Delta_n$.

In our numerical example, the Merton optimal performance $d_m = 0.0788$ is of course larger than d. The optimal Merton proportion is $(0.22, 0.33)$ and is located in the no-transaction region.

Concluding remarks

1. The limit of application of the dynamic programming method is the dimension of the system for numerical reasons. We can use fast algorithms like multigrids but we cannot handle high dimension systems: for example a portfolio with hundreds of assets. In this case, we must turn to direct optimisation methods leading to controls in open loop.

2. We have considered here constant volatilities. We can also handle diffusion terms which depend on the state as long as we keep Markovian processes.

References

Akian, M., Menaldi, J.L. & Sulem, A. (1996) 'On an investment-consumption model with transaction costs', *SIAM J. of Control and Optim.* **34** 329–364.

Akian, M., Séquier, P., Sulem, A. & Aboulalaa, A. (1995) 'A finite horizon portfolio selection problem with multi risky assets and transaction costs: the domestic asset allocation example', *Proceedings Association Française de Finance, Bordeaux, June 1995.*

Akian, M., Sulem A. & Séquier P. (1996) 'A finite horizon multidimensional portfolio selection problem with singular transactions', *Proceedings CDC New Orleans* **3** 2193–2198.

Akian, M., Sulem A. & Taksar, M. (1996) 'Dynamic optimisation of a long term growth rate for a mixed portfolio with transaction costs', *working paper.*

Barles G. & Souganidis, P.E. (1991) 'Convergence of approximation schemes for fully nonlinear second-order equations', *Asymptotic Analysis* **4** 271–283.

Bellman, R. (1957) *Dynamic Programming*, Princeton University Press.

Bellman, R. (1971) *Introduction to the Mathematical Theory of Control Processes*, Academic Press.

Black, F. & Scholes, M., (1973) 'The pricing of options and corporate liabilities', *Journal of Political Economy* **81** 635–654.

Chancelier, J.Ph., Gomez, C., Quadrat, J.P. & Sulem, A. (1987) 'Automatic study in stochastic control', in *Stochastic Differential Systems, Stochastic Control Theory and Applications*, W. Fleming & P.L. Lions (eds.), IMA Volumes in Mathematics and its Applications **10**, Springer Verlag, 79–86.

Constantinides, G.M. (1986) 'Capital market equilibrium with transaction costs', *J. of Political Economy* **94** 842–862.

Crandall, M.G., Ishii, H. & Lions, P.L. (1992) 'User's guide to viscosity solutions of second order partial differential equations', *Bull. Amer. Math. Soc.* **27** 1–67.

Davis, M. & Norman, A. (1990) 'Portfolio selection with transaction costs', *Math. Oper. Res.,* **15** 676–713.

Fitzpatrick, B. & Fleming, W.H. (1991) 'Numerical methods for an optimal investment-consumption model', *Math. Oper. Res.* **16** 823–841.

Fleming, W.H. & Soner, H.M. (1993) *Controlled Markov Processes and Viscosity Solutions*, Springer Verlag.

Hackbusch, W. (1985) *Multigrid Methods and Applications*, Springer Verlag.

Hackbusch, W. & Trottenberg U. (eds.) (1981) *Multigrid Methods*, Lecture Notes in Mathematics, **960**, Springer Verlag.

Howard, R.A. (1960) *Dynamic Programming and Markov processes*, MIT Press.

Karatzas, I., Lehoczky, J., Sethi, S. & Shreve, S. (1986) 'Explicit solution of a general consumption/investment problem', *Math. Oper. Res.* **11** 261–294.

Kushner, H.J. (1977) *Probability Methods in Stochastic Control and for Elliptic Equations*, Academic Press.

Kushner, H.J. & Dupuis, P.G. (1993) *Numerical Methods for Stochastic Control Problems in Continuous time*, Applications of Mathematics **24**, Springer Verlag.

Lions, P.L. (1983) 'Optimal control of diffusion processes and Hamilton-Jacobi-Bellman equations', Part 1: 'The dynamic programming principle and applications, *Comm. in Partial Diff. Eq.,* **8** 1101–1174. and Part 2: 'Viscosity solutions and uniqueness', *Comm. in Partial Diff. Eq.* **8** 1229–1276.

Magill, M.J.P. & Constantinides, G.M. (1976) 'Portfolio selection with transaction costs', *J. Econ. Theory* **13** 245–263.

McCormick, S.F. (ed.) (1987) *Multigrid Methods*, Frontiers in Applied Mathematics **5**, SIAM.

Merton, R.C. (1971) 'Optimum consumption and portfolio rules in a continuous time model', *J. Economic Theory* **3** 373–413.

Sethi, S. & Taksar, M. (1988) 'A note on Merton's "Optimum consumption and portfolio rules in continuous-time model" ', *J. Econ. Theory* **46** *2*.

Shreve, S.E. & Soner, H.M. (1994) 'Optimal investment and consumption with transaction costs', *Annals of Applied Probability* **4** 909–692.

Shreve, S.E., Soner, H.M. & Xu, V. (1991) 'Optimal investment and consumption with two bonds and transaction costs', *Mathematical Finance* **1** 53–84.

Taksar, M., Klass, M.J. & Assaf, D. (1988) 'A diffusion model for optimal portfolio selection in the presence of brokerage fees', *Math. Oper. Res.* **13** 277–294.

Zariphopoulou, T. (1992) 'Investment-consumption model with transaction fees and Markov chain parameters', *SIAM J. of Control and Optim.* **30** 613–636.

Imperfect Markets and Backward Stochastic Differential Equations

N. El Karoui and M.C. Quenez

1 Introduction

In this article we are concerned with Backward Stochastic Differential Equations (BSDEs) and with their applications to finance. These equations were first introduced by Bismut (1973) for the linear case and in the general case by Peng & Pardoux (1990). The solution of a BSDE consists of a pair of adapted processes (Y, Z) satisfying:

$$- \, dY_t = f(t, Y_t, Z_t)\, dt \, - \, Z_t^* dW_t \; ; \quad Y_T = \xi, \tag{1.1}$$

where f is called the generator and ξ the terminal condition.

Actually, this kind of equation appears in numerous problems in finance. First, the theory of contingent claim valuation in a complete market studied by Harrison & Pliska (1983), Duffie (1988) among others, can be expressed in terms of BSDEs. A contingent claim is a contract which pays the amount ξ at time T. The basic idea of valuing a contingent claim $\xi \geq 0$ with maturity T is to construct a hedging portfolio, i.e., a portfolio holding some number of shares of traded stocks and some number of shares of a money market, so that the hedging portfolio replicates the claim. At any time, the price of the contingent claim should have the same value as the hedging portfolio if the market is arbitrage free. However, the problem is more complicated since in general an attainable contingent claim can be replicated by an infinite number of hedging portfolios. The arbitrage-free pricing theory imposes some restrictions on the integrability of the hedging portfolios. In general, these assumptions are related to a risk-ajusted probability measure. Using the BSDE theory, we will show that the problem is well posed, that is, there exist a unique price and a unique hedging portfolio, by restricting admissible strategies to square integrable ones under the primitive probability.

In an incomplete market, where only some securities can be traded, it is not always possible to construct an admissible portfolio which replicates exactly ξ, and the price cannot be determined by no-arbitrage arguments. The replication error is called the tracking error. Föllmer & Schweitzer (1990) introduced the notion of local risk-minimizing strategies, for wich the tracking error is a square integrable martingale orthogonal to the basic securities. We show that this pricing rule is in fact a standard valuation in a market where only the traded securities have a return different from the spot rate. The price of a contingent claim is still a solution of a linear BSDE.

In an incomplete market, El Karoui & Quenez (1991) only considered superstrategies, which are strategies with positive tracking error and defined the upper price for each contingent claim ξ as the smallest investment allowing superhedging of ξ. They showed that the upper price is a superstrategy that is not a solution of a BSDE, but an increasing limit of nonlinear BSDEs. Cvitanic & Karatzas (1992) generalized this result to general constraints on the portfolios.

Other nonlinear backward equations were introduced in imperfect markets, for example, for the hedging problem with a higher interest rate for borrowing, or different risk-premiums for the seller and the buyer. In this case, the dynamics of the wealth process are given by a nonlinear convex BSDE. More generally, using the BSDE theory, we develop a nonlinear arbitrage pricing theory in an imperfect market.

In 1989, Duffie and Epstein presented a stochastic differential formulation of a recursive utility for the consumption, as the objective function of an problem of investment-consumption. Recursive utility is an extension of the standard additive utility with the instantaneous utility depending not only on the instantaneous consumption rate c_t but also on the future utility. In their paper, Duffie & Epstein (1992) showed that, under Lipshitz conditions, the recursive utility exists and satisfies the usual properties of standard utilities (concavity with respect to consumption if the BSDE is concave). The BSDE point of view gives a simple formulation of recursive utilities and their properties.

In this article, we first exhibit typical situations in finance where the BSDE point of view is efficient. Then we summarize the results on existence and uniqueness by Peng & Pardoux (1990) and give new (shorter) proofs. We state some a priori estimates of the spread between the solutions of two different BSDEs; existence and uniqueness of the solution of a BSDE follow from a fixed point theorem. One of the most important properties of BSDEs is the comparison theorem which can be obtained under quite general conditions. For example, this theorem gives a sufficient condition for the wealth process to be non-negative and yields the classical properties of utilities. When the BSDE is associated with a classical forward solution of a stochastic differential equation, we show that the solution of the BSDE is related to that of a nonlinear PDE. In the last part, we develop the links between convex BSDEs and optimal control theory. Moreover, this result can be applied to European option pricing. In some constrained cases, the price of a contingent claim is given by the solution of a nonlinear convex BSDE. As the utility function, the price can be written as the maximum of 'ex-post' prices, taken over all changes of 'numeraire' feasible for the wealth; also, the set of controls is bounded and the maximum is attained for an optimal change of numeraire.

2 Backward Differential Equations, a Useful Tool in Finance

2.1 The model

We begin with the typical set-up for continuous-time asset pricing: the basic securities consist of $n+1$ assets. One of them is a non-risky asset (the money market instrument or bond), with price per unit S^0 governed by the equation,

$$dS_t^0 = S_t^0 r_t \, dt, \tag{2.1}$$

where r_t is the short rate. In addition to the bond, n risky securities (the stocks) are continuously traded. The price process S^i for one share of ith stock is modeled by the linear stochastic differential equation,

$$dS_t^i = S_t^i [b_t^i \, dt + \sum_{j=1}^{n} \sigma_t^{i,j} \, dW_t^j], \tag{2.2}$$

driven by a standard n-dimensional Wiener process $W = (W^1, \dots, W^n)^*$, defined on a filtered probability space $(\Omega, (\mathcal{F}_t)_{t\in[0,T]}, \mathbb{P})$. We assume that the filtration $(\mathcal{F}_t)$ is that generated by the Wiener process W. We say that $\mathbb{P}$ is the 'objective' probability measure.

Generally speaking, the coefficients $\sigma^i = (\sigma^{i,j})_{j=1}^n$, b^i and r are assumed to be bounded and predictable processes, with values in $\mathbb{R}^n$, $\mathbb{R}$ and $\mathbb{R}$, respectively. For notational convenience, we will sometimes write $\sigma = [\sigma^{i,j}]$ to denote the random $n \times n$ volatility matrix, and $b = (b^i)$ to denote the stock appreciation rates vector. The assumption that the number of risky securities is equal to the dimension of the underlying noise is introduced for the sake of simplicity.

To ensure the absence of arbitrage opportunities in the market, we assume that there exists an n-dimensional bounded predictable vector process θ, such that

$$b_t - r_t \mathbf{1} = \sigma_t \theta_t, \qquad dt \otimes \mathbb{P} \ a.s,$$

where $\mathbf{1}$ is the vector whose every component is 1. θ is said to be a risk premium vector. We will assume more generally that the square matrix σ_t has a full rank for any fixed $t \in [0,T]$.

Let us consider a small investor, whose actions cannot affect market prices, and who can decide at time $t \in [0,T]$ what amount π_t^i of the wealth V_t to invest in the ith stock, $i = 1, \dots, n$. Of course, his decisions can only be based on the current information $(\mathcal{F}_t)$, i.e., $\pi = (\pi^1, \pi^2, \dots, \pi^n)^*$[1] and $\pi^0 = V - \sum_{i=0}^n \pi^i$ are predictable processes. Following Harrison & Pliska (1981), a strategy is

[1] We denote by σ^* the transpose of the matrix σ

self-financing if the wealth process $V = \sum_{i=0}^{n} \pi^i$ satisfies the equality

$$V_t = V_0 + \int_0^t \sum_{i=1}^{n} \pi_s^i \frac{dS_s^i}{S_s^i},$$

or equivalently, if the wealth process is governed by the linear stochastic differential equation (LSDE)

$$dV_t = r_t V_t dt + \pi_t^*(b_t - r_t \mathbf{1})dt + \pi_t^* \sigma_t dW_t = r_t V_t dt + \pi_t^* \sigma_t [dW_t + \theta_t dt]. \quad (2.3)$$

Such a pair (V, π) is said to be a self-financing trading strategy. Generally, the initial wealth $x = V_0$ is taken as a primitive, and for an initial endowment and portfolio process pair (x, π), there exists a unique wealth process V that is solution of the linear equation (2.3) with initial condition $V_0 = x$. Therefore, there exists a one-to-one correspondence between pairs (x, π) and trading strategies (V, π) which it will be sometimes convenient to use.

2.2 Pricing and hedging contingent claims

A European contingent claim ξ settled at time T is a $\mathcal{F}_T$-measurable random variable. It can be thought of as a contract which pays ξ at maturity T. The pricing of a contingent claim is based on the following principle: if we start with the price of the claim as initial endowment and invest it in the $n+1$ assets, the value of the portfolio at time T must be enough to finance ξ. A contingent claim ξ is said to be attainable if it can be replicated by means of a self-financing trading strategy. In general, an attainable contingent claim may be replicated by an infinite number of self-financing strategies[2]. Using the theory of backward stochastic differential equations, we will show that the pricing problem is well posed, that is the replicating strategy is unique, if we consider only square-integrable replicating strategies. In this framework, the notion of admissible strategy is defined in reference to the objective probability measure $\mathbb{P}$, which is different from most classical presentations of arbitrage pricing theory, where the integrability constraint is related to a risk-adjusted probability measure.

Definition & Theorem 2.1 *A hedging strategy against a contingent claim ξ is a self-financing strategy (V,π) such that $V_T = \xi$ with $\mathbb{E} \int_0^T |\sigma_t^* \pi_t|^2 dt < +\infty$. An attainable square integrable contingent claim ξ is replicated by a unique hedging strategy (V, π).*

Proof Let us give a short proof of the uniqueness property. By linearity, it is clearly equivalent to show that the portfolio equal to 0 is the only hedging

[2]They can be constructed from the suicide strategy (such that $V_0 = 1$, $V_T = 0$) given by Harrison & Pliska (1981)

portfolio with terminal value 0. Let (X, π) be such a hedging square integrable portfolio. Ito's formula applied to $e^{\beta s}|X_s|^2$ between t and T leads to

$$e^{\beta t}|X_t|^2 = 0 - \int_t^T e^{\beta s}|X_s|^2(\beta + 2r_s)ds - \int_t^T e^{\beta s} 2X_s\pi_s^*\sigma_s[dW_s + \theta_s ds] - \int_t^T e^{\beta s}|\sigma_s^*\pi_s|^2 ds.$$

For large β, the quadratic form

$$x^2(\beta + 2r_s) + 2xz^*\sigma_s\theta_s + |\sigma_s^* z|^2 = |\sigma_s^* z + x\theta|^2 + x^2(\beta + 2r_s - |\theta_s|^2)$$

is positive and $e^{\beta t}|X_t|^2$ is a positive local submartingale with terminal value 0. By semimartingale inequalities, the square integrable semimartingale X_t is uniformly bounded in $\mathbb{L}^2$ (i.e., $\sup_{t\in[0,T]}|X_t| \in \mathbb{L}^2$), so by Fatou's lemma $\mathbb{E}[e^{\beta t}|X_t|^2] \leq 0$, which completes the proof. □

According to the result of Pardoux & Peng (1990) (Section 2) on the existence and uniqueness of the solutions of backward stochastic differential equations, we prove that all the square integrable claims are attainable by square integrable replicating strategies.

Theorem 2.2 *Any square integrable contingent claim is attainable; the market is said to be complete. In other words, for any square integrable ξ, there exists a unique pair (X, π) such that $\mathbb{E}\int_0^T |\sigma_t^*\pi_t|^2 dt < +\infty$ and*

$$dX_t = r_t X_t dt + \pi_t^*\sigma_t\theta_t dt + \pi_t^*\sigma_t dW_t; \qquad X_T = \xi. \tag{2.4}$$

X_t is the price of the claim at time t, given by the closed formula,

$$X_t = \mathbb{E}[H_T^t \xi | \mathcal{F}_t], \tag{2.5}$$

where H_s^t is the deflator process, started at time t, such that

$$dH_s^t = -H_s^t[r_s ds + \theta_s^* dW_s]; \qquad H_t^t = 1. \tag{2.6}$$

By Ito's calculus, the process [3] $(H_t X_t, t \in [0, T])$ is a stochastic integral such that $dH_t X_t = H_t[\pi_t^*\sigma_t - X_t\theta_s^*]dW_t$. Classical results about the solutions of the linear stochastic differential equation (2.6) with bounded coefficients give that the martingale H is uniformly bounded in $\mathbb{L}^2$, and that the process $(H_t X_t, t \in [0, T])$ is a uniformly integrable martingale. So,

$$H_t X_t = \mathbb{E}[H_T\xi|\mathcal{F}_t], \text{ or equivalently } X_t = \mathbb{E}[H_T^t\xi|\mathcal{F}_t].$$

The closed form of the deflator process $H_s^t = \exp -[\int_t^s r_u du + \int_t^s \theta_s^* dW_s + 1/2\int_t^s |\theta_u|^2 du]$ leads to the more classical formulation of the price of contingent claim as

$$X_t = \mathbb{E}[e^{-[\int_t^T \theta_u^* dW_u + 1/2\int_t^T |\theta_u|^2 du]} e^{-\int_t^T r_u du}\xi|\mathcal{F}_t] = \mathbb{E}_{\mathbb{Q}}[e^{-\int_t^T r_u du}\xi|\mathcal{F}_t]$$

[3]For notational simplicity we denote the process H_t^0 by H_t

where $e^{-\int_t^T r_u du}$ is the discounted factor over $[t,T]$ and $\mathbb{Q}$ the risk-adjusted probability measure defined by its Radon-Nikodym derivative with respect to $\mathbb{P}$,

$$\frac{d\mathbb{Q}}{d\mathbb{P}} = \exp -[\int_0^T \theta_s^* \, dW_s \; + \frac{1}{2}\int_0^T |\theta_s|^2 \, ds].$$

Notice that $\mathbb{Q}$ is a martingale-measure, that is the discounted wealth processes are $\mathbb{Q}$-local martingales.

Classically, the price is defined by arbitrage, by considering only hedging strategies for which the discounted wealth process is a $\mathbb{Q}$-martingale (Harrison & Pliska 1981) or by considering only hedging strategies such that $\mathbb{E}_{\mathbb{Q}} \int_0^T |\sigma_t^* \pi_t|^2 dt < +\infty$. The major drawback of this point of view is its dependence on the martingale measure $\mathbb{Q}$.

Another approach is proposed by Karatzas & Shreve (1987) who consider trading strategies for which the wealth is bounded from below (in fact positive), instead of satisfying some integrability conditions. These authors prove that the fair price X_t^f at time t of a square integrable positive contingent claim ξ, defined as the smallest endowment which lets us hedge ξ, is still given by $X_t^f = \mathbb{E}[H_T^t \xi | \mathcal{F}_t]$. With a positive hedging strategy (V,π) is associated a positive local martingale $(H_t V_t, t \in [0,T]))$. So, $(H_t V_t, t \in [0,T]))$ is a supermartingale with terminal value $H_T \xi$, which dominates the martingale $(H_t X_t^f, \in [0,T])$. A nice feature of this point of view is that it does not depend on the risk-neutral probability measure. The drawback is that the class of trading strategies is not a linear space.

Notice that it is not easy to deduce directly from the closed form of the price $X_t = \mathbb{E}[H_T^t \xi | \mathcal{F}_t]$, that X is the wealth of a square integrable strategy under the 'objective' probability. The BSDE point of view gives another way to treat this problem by using approximated solutions which have no financial meaning (Section 3).

2.3 Constrained portfolios

Recently, in studying the pricing of contingent claims with constraints on the wealth or portfolio processes, many authors have introduced some nonlinear backward equations for the fair price of claims. We present two different situations, the first relative to a complete market where only some particular securities can be held in hedging strategies, the second where the market is imperfect with different lending and borrowing rates, or different risk premiums for seller or buyer.

2.3.1 Föllmer-Schweizer Hedging Strategy

We consider a complete market modeled as before, where only somes securities, called primary securities, must be held in a hedging portfolio. The seller of a contingent claim cannot have a perfect hedge because of the constraints on the portfolio. Föllmer & Schweizer (1990) introduce the notion of self-financing in mean strategy, which minimize the variance of the tracking error. El Karoui & Quenez (1991) propose to introduce a seller's price, as the smallest investment which is enough to finance the claim with constrained portfolios.

For sake of simplicity, we introduce the following conventional notation: the primary securities are the first j ones, the $j \times n$ volatility matrix of the primary securities is denoted by $\sigma_t^1 = (\sigma_t^{i,k})_{i=1..j,k=1..n}$, and the volatility matrix of the others is denoted by σ_t^2. Let us remark that the matrix σ_t^1 has full rank, since the global matrix σ_t has full rank, and so the matrix $(\sigma_t^1)(\sigma_t^1)^*$ is invertible. The amount of a general portfolio π_t invested in the primary securities is denoted by ${}^1\pi_t$ and the amount in the others is denoted by ${}^2\pi_t$, such that

$$\pi_t^* \sigma_t = ({}^1\pi_t)^* \sigma_t^1 + ({}^2\pi_t)^* \sigma_t^2.$$

So, an admissible hedging portfolio (${}^2\pi_t = 0$) is to be constrained to $\pi_t^* \sigma_t = ({}^1\pi_t)^* \sigma_t^1$, and the admissible wealth is modeled by

$$dV_t = r_t V_t \, dt + ({}^1\pi_t)^* \sigma_t^1 (dW_t + \theta_t dt).$$

Notice that this equation is unchanged if θ_t is replaced by the 'minimal risk premium' θ_t^1 defined as the orthogonal projection of θ_t onto the range of $(\sigma_t^1)^*$ since $\theta_t - \theta_t^1$ belongs to the kernel of σ_t^1. Notice that classical results from linear algebra allow us to give a closed formula for θ_t^1,

$$\theta_t^1 = (\sigma_t^1)^* [(\sigma_t^1)(\sigma_t^1)^*]^{-1} \sigma_t^1 \theta_t. \tag{2.7}$$

In what follows, we suppose the process θ^1 to be bounded.

Given a square integrable contingent claim ξ, there no longer exists an admissible hedging portfolio to finance ξ; in order words, the BSDE

$$dX_t = r_t X_t + ({}^1\pi_t)^* \sigma_t^1 (dW_t + \theta_t dt) \qquad X_T = \xi,$$

can have no solution. So Föllmer & Schweitzer (1990) introduced the notion of self-financing in mean strategies (in short FS-strategies), $(V, {}^1\pi, \phi)$

$$dV_t = r_t V_t + ({}^1\pi_t)^* \sigma_t^1 (dW_t + \theta_t^1 dt) - d\phi_t$$

where, by definition, the tracking error ϕ is a square integrable martingale, orthogonal to the martingale $(\int_0^t \sigma_t^1 dW_t, t \in [0,T])$, and prove the existence of a unique FS-strategy which finances a square integrable contingent claim ξ. Actually, the FS-strategy is simply given by the solution of a linear BSDE.

Theorem 2.3 *The FS-strategy is the hedging strategy (X, ψ) in a market associated with the the 'minimal risk premium' θ^1, that is*

$$dX_t = r_t X_t dt + \psi_t^* \sigma_t \theta_t^1 dt + \psi_t^* \sigma_t dW_t; \quad X_T = \xi. \tag{2.8}$$

More precisely let us denote by (q_t^1, q_t^2) the orthogonal decomposition of $(\sigma_t)^*\psi_t$ onto the range of $(\sigma_t^1)^*$ and by ${}^1\pi_t$ the vector process such that $(\sigma_t^1)^*({}^1\pi_t) = q_t^1$. Then $(X_t, {}^1\pi_t, \phi_t = \int_0^t (q_s^2)^* dW_s, t \in [0,T])$ is the unique FS-strategy associated with ξ.

Proof of Theorem 2.3 Let (X, ψ) be the no-constrained hedging strategy solution of (2.8). Project the vector $q_t = \sigma_t^* \psi_t$ orthogonally onto the range of $(\sigma_t^1)^*$, so

$$q_t = q_t^1 + q_t^2 \qquad \text{where } q_t^1 \in \text{Range}[(\sigma_t^1)^*] \qquad \text{and } q_t^2 \in \text{Ker}(\sigma_t^1).$$

We have an explicit formula for q^1,

$$\begin{aligned} q_t^1 &= (\sigma_t^1)^*[(\sigma_t^1)(\sigma_t^1)^*]^{-1}\sigma_t^1[(\sigma_t^1)^*(\psi_t^1) + (\sigma_t^2)^*\psi_t^2] \\ &= (\sigma_t^1)^*(\psi_t^1 + [(\sigma_t^1)(\sigma_t^1)^*]^{-1}(\sigma_t^2)^*\psi_t^2) = (\sigma_t^1)^*({}^1\pi_t). \end{aligned}$$

Since θ_t^1 is the orthogonal projection of θ_t onto the range of $(\sigma_t^1)^*$,

$$\psi_t^* \sigma_t \theta_t^1 = (q_t^1)^* \theta_t^1 = (q_t^1)^* \theta_t.$$

The martingale $(\phi_t = \int_0^t (q_t^2)^* dW_s, t \in [0,T])$ is orthogonal to $\int_0^t \sigma_s^1 dW_s$ and $(X, {}^1\pi, \phi)$ is a FS-strategy.

Conversely, let $(X, {}^1\pi, \phi)$ be a FS-strategy. Let q^2 be such that $\phi_t = \int_0^t (q_t^2)^* dW_s$ and put $(\sigma_t)^*\psi_t = (\sigma_t^1)^*({}^1\pi_t) + q_t^2$. Then (X, ψ) is a solution of the BSDE (2.8) and it follows from the uniqueness that the FS-strategy is unique. □

Remark 1 The admissible portfolio is given by ${}^1\pi_t = {}^1\psi_t + [(\sigma_t^1)(\sigma_t^1)^*]^{-1} \times \sigma_t^1(\sigma_t^2)^*\psi_t$. In particular, if the matrix $\sigma_t^1(\sigma_t^2)^*$ is the null matrix, or in financial terminology, if the no-primary securities do not introduce supplementary risk on the admissible portfolios, then the FS-strategy consists in holding the amount ${}^1\pi_t = {}^1\psi_t$ in the primary securities. In general the FS-strategy does not depend on the matrix σ^2. In particular the simplest way to calculate the FS-strategy is to complete the 'primary market' by introducing other securities whose volatility matrix satisfies $\sigma_t^1(\sigma_t^2)^* = 0$.

Remark 2 Notice that θ^1 is the risk premium associated with the minimal martingale measure introduced by Föllmer and Schweitzer as a martingale measure such that any $\mathbb{P}$-local martingale, orthogonal to the martingale $\int_0^t \sigma_s^1 dW_s$, is a $\mathbb{Q}^1$-local martingale, where $\mathbb{Q}^1$ is the risk-neutral probability measure associated to the risk premium θ^1, also called the minimal martingale measure. So X_t is the conditional expectation of the discounted contingent claim computed under the minimal martingale measure.

2.3.2 Superhedging in Incomplete Markets

Another point of view is to consider superstrategies instead of the FS-strategies, that is strategies allowed to have a positive tracking error at time T, and more precisely whose tracking error is an adapted (but not necessarily predictable) increasing process. The notation is the same as in the previous example, where we supposed $\sigma_t^1(\sigma_t^2)^*$ is the null matrix.

EL Karoui & Quenez (1991) showed that the upper price process X for a contingent claim ξ, defined as the minimal wealth to invest in superstrategies which replicate ξ, is a superstrategy. This process does not correspond exactly to the solution of a BSDE, but it can be obtained as the increasing limit of a sequence of processes X^k that are solutions of nonlinear BSDEs.

Let ξ be a contingent claim. The upper price for ξ is given by

$$X_0 = \inf\{V_0; \exists(V,\pi,C)\,\text{s.t.}\, V_T = \xi;\ dV_t = r_t V_t dt + ({}^1\pi_t)^*\sigma_t^1[dW_t + \theta_t^1\, dt] - dC_t\}.$$

The problem is to show that the upper price is achieved by a superhedging strategy. This property can be proved by a dual method: EL Karoui & Quenez (1991) show that the upper price is the value function of a control problem, which is necessarily a superstrategy. This point of view will be presented in detail in Section 3.

Another way is to penalize the portfolios for which the constraint $({}^2\pi_t = 0)$ is not satisfied. Recall that the assumption $(\sigma_t^1)^*\sigma_t^2 = 0$ implies that the projection (Proj_t) of the vector $\sigma_t^*\pi_t$ on the kernel of σ_t^1 is $(\sigma_t^2)^*({}^2\pi_t)$. Let us introduce the processes (X^k, π^k) solutions of the following BDSE

$$\begin{aligned}
dX_t^k &= r_t X_t^k\, dt - k||\text{Proj}_t(\sigma_t^*\pi_t^k)||dt + (\pi_t^k)^*\sigma_t[dW_t + \theta_t^1 dt] \\
&= r_t X_t^k\, dt + ({}^1\pi_t^k)^*\sigma_t^1[dW_t + \theta_t^1 dt] - k||(\sigma_t^2)^*({}^2\pi_t)||dt + ({}^2\pi_t^k)^*\sigma_t^2 dW_t \\
X_T^k &= \xi.
\end{aligned}$$

The processes can be viewed as the wealth process of strategies whose tracking error ($\phi_t = \int_0^t k||(\sigma_t s^2)^*({}^2\pi_s)||ds - ({}^2\pi_s^k)^*\sigma_s^2 dW_s$) is a submartingale. The penalizing process given by $k\int_0^t ||\text{Proj}_t(\sigma_t^*\pi_t)||dt$ has an intensity proportional to the length of the non-admissible part of $(\sigma_t^*\pi_t)$. The greater this length, the more the local 'variance' of the non-admissible martingale part ${}^2(\pi_t^k)^*\sigma_t^2 dW_t$ is large and the more expensive the penalty.

By Peng and Pardoux's results on BSDEs, which we recall in Section 3, these equations admit a unique square integrable solution. By the comparison theorem for the BSDE, we see that the sequence of processes X^k is increasing, and by using their interpretation as value functions of a control problem that their limit is a superstrategy. Then it is easy to verify that the limit is the smallest superstrategy against ξ. We shall see in Section 3 that this interpretation as value function of a control problem only results from the convexity of the function Proj_t.

Cvitanic & Karatzas (1992) applied this methodology for very general constraints on the portfolio and gave similar approximations for the upper price of a contingent claim ξ. Bardhan (1993) also studied some problems in this area. More recently, Kramkov (1994) extended this methodology to a general arbitrage-free asset pricing setting in incomplete markets.

2.3.3 Hedging in Imperfect Markets

We present an example of an imperfect market, which takes into account a higher interest rate for borrowing than for lending. As do Bergman (1991), Korn (1992) and Cvitanic & Karatzas (1992) we consider the following problem: the investor is allowed to borrow money at an interest rate $R_t > r_t$, the bond rate. Both processes R and r are supposed adapted and bounded. It is not reasonable to borrow money and to invest money in the bond at the same time. Therefore, we restrict to policies for which the amount borrowed at time t is equal to $(V_t - \sum_{i=1}^{n} \pi_t^i)^-$. Then the square integrable strategy (wealth, portfolio) (V, π) satisfies

$$dV_t = r_t V_t \, dt + \pi_t^* \sigma_t \theta_t \, dt + \pi_t^* \sigma_t dW_t + (R_t - r_t)(V_t - \sum_{i=1}^{n} \pi_t^i)^- \, dt. \tag{2.9}$$

Given an initial investment $V_0 = x$ and a risky portfolio π, there exists a unique solution to this forward stochastic differential equation with Lipschitz coefficients. The upper price of a claim, still defined as the minimal endowment to finance an admissible strategy which guarantees ξ at time T, appears as the solution of the nonlinear backward stochastic differential equation, where the nonlinear term depends on both processes, wealth and portfolio.

Similar equations appear in continuous trading with short sales constraints or transaction costs (Jouini & Kallal 1992, and He & Pearson 1991). Let $\theta^l - \theta^s$ be the difference in excess return between long and short positions in the stock. Then the present value corresponding to the strategy π satisfies

$$dV_t = r_t V_t \, dt + [\pi_t^*]^- \sigma_t [\theta_t^l - \theta_t^s] dt + \pi_t^* \sigma_t (dW_t + \theta_t^l dt). \tag{2.10}$$

In Section 3.3 below, we develop a general pricing theory for contingent claim with respect to convex constrained portfolios.

2.4 Recursive utility

The problem of optimal selection of a portfolio has been extensively studied since the pioneering paper of Merton (1971) (see for example the books of Duffie (1992) or of Karatzas & Shreve (1996)). In the standard portfolio selection problem, a small investor is faced with the problem of maximization of his expected utility from trading and consumption strategies with a given initial wealth. The utility may be concerned with the rate of consumption and (or) the terminal wealth.

Let us consider a small investor who can consume while also investing in a standard complete market. Let us denote by c_t the (positive) consumption rate per time. With obvious notation, the dynamics of the wealth are given by

$$dV_t = r_t V_t \, dt - c_t dt + \pi_t^* \sigma_t [dW_t + \theta_t dt], \qquad V_0 = x. \tag{2.11}$$

The objective of the agent is to maximize the expected utility of consumption over the planning horizon and (or) to maximize the expected utility of the wealth at the end of the period. The agent's preference is measured by a utility function u for consumption and a utility g for terminal wealth, so that the expected utility Y_0 is given by

$$Y_0 = E\left[g(V_T)\, e^{-\beta T} + \int_0^T u(c_s)\, e^{-\beta s} ds\right].$$

To take into account a more realistic situation, Epstein & Zin (1989) in the deterministic case, and Duffie & Epstein (1992) under uncertainty, introduced the notion of recursive utility where the expected utility at time t depends on the consumption path and on the future utility, so that the utility process is defined in a backward recursive way by

$$Y_t = E\left[g(V_T)\, e^{-\beta(T-t)} + \int_t^T f(s, c_s, Y_s)\, e^{-\beta(s-t)} ds \,|\mathcal{F}_t\right],$$

where f is said to be **the generator**.

In the case of information generated by Brownian motion (W_t), the recursive utility is the solution of the BSDE:

$$-\,dY_t = f(t, c_t, Y_t)dt \;-\; Z_t^* dW_t\;; \quad Y_T = g(V_T). \tag{2.12}$$

Examples

Because of their economic motivation, we list the following examples of recursive utilities which are given by Duffie & Epstein (1992).

Standard additive utility The generator of the standard utility is given by:

$$f(c, y) = u(t, c) - \beta\, y.$$

The recursive utility is associated with the standard criterion,

$$Y_t = E\left[g(V_T)\, e^{-\beta(T-t)} + \int_t^T u(c_s)\, e^{-\beta(s-t)} ds \,|\mathcal{F}_t\right].$$

Uzawa utility The generator has the same form as the additive utility, but the discounting rate β depends on the consumption rate c_t: $f(c, y) = u(c) - \beta(c)\, y$. The recursive utility has the same form as in the additive case.

Kreps-Porteus utility Let $0 \neq \rho \leq 1$, $0 \leq \beta$, $0 \neq \alpha \leq 1$. The generator is defined by

$$f(c, y) = \frac{\beta}{\alpha} \frac{c^\rho - y^\rho}{y^{\alpha - 1}}.$$

Properties of Recursive Utilities

Recall that, in general, utilities must satisfy the following classical properties:

1. Monotonicity with respect to the terminal value and to the consumption.

2. Concavity with respect to the consumption.

3. Time consistency: this means that, for any two consumption processes c^1 and c^2 and any time t, if c^1 and c^2 are identical up to time t, and if the continuation of c^1 is preferred to the continuation of c^2 at time t, then c^1 is preferred to c^2 at time 0.

Duffie & Epstein (1989) show that if f is Lipschitz with respect to y, the following equation

$$dY_t = f(t, c_t, Y_t)\, dt \;-\; Z_t^* dW_t\,; \qquad Y_T = Y,$$

has a unique solution. Also, they state that if f is concave with respect to (c, y) and increasing with respect to c, the above properties (1–3) are satisfied[4]

In this article, we will consider a more general class of recursive utilities. In the case of Brownian information, the utility at time t satisfies the BSDE given by:

$$-\; dY_t = f(t, c_t, Y_t, Z_t)\, dt \;-\; Z_t^* dW_t\,; \qquad Y_T = Y. \tag{2.13}$$

Given the path of consumption $(c_t; t \in [0, T])$, and the assumption that the generator is uniformly Lipschitz with respect to (y, z), the existence and uniqueness of recursive utilities (solutions of (2.13)) and the above properties (1–3) are deduced easily from the theory of backward stochastic equations. However, the main contribution of this theory to the concept of recursive utilities is in the interpretation of recursive utilities as value function of a control problem.

3 Backward Stochastic Differential Equations

In this section the main results on BSDE are given and sometimes proved. A more exhaustive exposition can be found in El Karoui, Peng & Quenez (1997).

The first paper concerned with nonlinear BSDE is to our knowledge that of Bismut (1978), where he introduced a nonlinear Ricatti BSDE for which

[4]More recently, Duffie & Singleton (1994) and Duffie & Huang (1994) have also used this type of BSDE to solve some pricing problems.

he showed existence and uniqueness. Pardoux & Peng (1990) were the first to consider general BSDEs. They introduced the notion of an adapted solution of a BSDE on the probability space of Brownian motions as a pair of adapted processes (Y_t, Z_t) satisfying the BDSE and the terminal condition $Y_T = \xi$. Several papers extended their results, in particular Antonelli (1992), Ma, Protter & Yong (1993), Buckdahn (1993), Buckdahn & Pardoux (1993) and of course, Pardoux & Peng (1990, 1992, 1993) and Peng (1990, 1991, 1992a,b, 1993).

In this section, we first state some a priori estimates of the spread between the solutions of two BSDEs, from which we derive existence and uniqueness results. Then we give different properties concerning BSDEs. In particular we study linear BSDEs which are classical in finance, for which we state a comparison theorem.

Firstly fix some notation. On a probability space $(\Omega, \mathcal{F}, \mathbb{P})$ equipped with the filtration $(\mathcal{F}_t)$ of an $\mathbb{R}^n$-valued Brownian motion (W_t) we consider all kinds of spaces of variable or processes:

- $\mathbb{L}_T^{2,d} = \{X \in \mathbb{R}^d;\ \ X \in \mathcal{F}_T\text{-measurable}\ ;\ \ ||X||^2 = \mathbb{E}(|X|^2) < +\infty\}$,
- $\mathbb{H}_T^{2,d} = \{\varphi;\ \varphi_t \in \mathbb{R}^d\ ,\ \text{predictable}\ ,\ ||\varphi||^2 = \mathbb{E}\int_0^T |\varphi_t|^2 dt < +\infty\}$. Such processes are said to be square integrable (for short).
- For $\beta > 0$ and $\varphi \in \mathbb{H}_T^{2,d}$, $||\varphi||_\beta^2$ will denote $\mathbb{E}\int_0^T e^{\beta t}|\varphi_t|^2 dt$. $\mathbb{H}_\beta^{2,d}$ denotes the space $\mathbb{H}_T^2$ endowed with the norm $||.||_\beta$.
- $\mathbb{H}_T^{1,d} = \{\varphi;\ \varphi_t \in \mathbb{R}^d;\ \text{predictable};\quad \mathbb{E}\sqrt{\int_0^T |\varphi_t|^2 dt} < +\infty\}$.

For $x \in \mathbb{R}^d$, $|x|$ will denote its Euclidian norm, and $\langle x, y\rangle$ the inner product. An element $y \in \mathbb{R}^{n\times d}$ will be considered as a $n \times d$ matrix; note that its Euclidean norm is given by : $|y| = \sqrt{\text{trace}(y\,y^*)}$; $\langle y, z\rangle = \text{trace}(yz^*)$.

3.1 Existence and uniqueness of a solution to a BSDE

3.1.1 The Main Result

Consider the BSDE:

$$-\ dY_t = f(t, Y_t, Z_t)\,dt\ -\ Z_t^* dW_t\ ;\quad Y_T = \xi, \tag{3.1}$$

or equivalently

$$Y_t = \xi + \int_t^T f(s, Y_s, Z_s)\,dt\ -\ \int_t^T Z_s^* dW_s, \tag{3.2}$$

where:

- W is an n-dimensional Brownian motion
- $f : \Omega \times \mathbb{R}^+ \times \mathbb{R}^d \times \mathbb{R}^{n\times d} \longrightarrow \mathbb{R}^d$ is $\mathcal{P} \otimes \mathcal{B}^d \otimes \mathcal{B}^{n\times d}$-measurable.
- $\xi \in \mathbb{L}_T^{2,d}$ and $f(.,0,0) \in \mathbb{H}_T^{2,d}$.
- f is uniformly Lipschitz, i.e., there exists $C > 0$ such that

$$|f(t,y_1,z_1) - f(t,y_2,z_2)| \leq C\left(|y_1 - y_2| + |z_1 - z_2|\right) \quad dt \otimes d\mathbb{P} a.s.$$

- Y and Z are $\mathbb{R}^d$ and $\mathbb{R}^{n\times d}$-valued predictable processes and the process Y is continuous.

f is called the *generator* of the BSDE and ξ the *terminal value.* If (f,ξ) satisfies the above assumptions, (f,ξ) are said to be *standard* parameters for the BSDE.

Theorem 3.1 (Pardoux & Peng (1990)) *Given standard parameters (f,ξ), there exists a unique pair $(Y,Z) \in \mathbb{H}_T^{2,d} \times \mathbb{H}_T^{2,n\times d}$ which solves eq. (3.1).*

A proof can be found in Pardoux & Peng (1990). We give here a shorter direct proof, using useful a priori estimates for the solutions.

3.1.2 A Priori Estimates

Proposition 3.2 *Let $((f^i,\xi^i);\ i = 1,2)$ be two standard parameters of the BSDE and $((Y^i,Z^i); i = 1,2)$ be the square integrable associated solutions. Let C be a Lipschitz constant for f^1, and put $\delta Y_t = Y_t^1 - Y_t^2$, $\delta Z_t = Z_t^1 - Z_t^2$ and $\delta_2 f_t = f^1(t,Y_t^2,Z_t^2) - f^2(t,Y_t^2,Z_t^2)$.*
For any $\beta > C(2+C)$, the following a priori estimates hold:

$$\| \delta Y \|_\beta^2 \leq T[e^{\beta T}\mathbb{E}(|\delta Y_T|^2) + \frac{1}{\beta - 2C - C^2} \| \delta_2 f \|_\beta^2] \tag{3.3}$$

$$\| \delta Z \|_\beta^2 \leq (2 + 2C^2 T)e^{\beta T}\mathbb{E}(|\delta Y_T|^2) + \frac{2+2C^2T}{\beta - 2C - C^2} \| \delta_2 f \|_\beta^2 . \tag{3.4}$$

Remark By classical results on the norms of semimartingales, we prove similarly that

$$\mathbb{E}[\sup_{t\leq T}|\delta Y_t|^2] \leq K_T[\mathbb{E}(|\delta Y_T|^2 + \frac{1}{\beta - 2C - C^2} \| \delta_2 f \|_\beta^2],$$

where K_T is a positive constant only depending on T.

Proof of Proposition 3.2 Let $(Y,Z)\in \mathbb{H}_T^{2,d}\times\mathbb{H}_T^{2,n\times d}$ be a solution of (3.1). It follows from well-known inequalities for semimartingales that $\sup_{s\le T}|Y_s|\in \mathbb{L}_T^{2,1}$.

Now consider (Y^1,Z^1) and (Y^2,Z^2), two solutions associated with (f^1,ξ^1) and (f^2,ξ^2), respectively. From Itô's formula applied from $s=t$ to $s=T$ to the semimartingale $e^{\beta t}|\delta Y_t|^2$, it follows that

$$\begin{aligned} e^{\beta t}|\delta Y_t|^2 &= e^{\beta T}|\delta Y_T|^2 + 2\int_t^T e^{\beta s}\langle \delta Y_s,\, f^1(s,Y_s^1,Z_s^1) - f^2(s,Y_s^2,Z_s^2)\rangle ds \\ &- \beta\int_t^T e^{\beta s}|\delta Y_s|^2 ds - \int_t^T e^{\beta s}|\delta Z_s|^2 ds - 2\int_t^T e^{\beta s}\langle \delta Y_s, \delta Z_s^* dW_s\rangle. \end{aligned}$$

Since $\sup_{s\le T}|\delta Y_s|$ belongs to $L_T^{2,1}$, $\delta Z\,\delta Y$ belongs to $\mathbb{H}_T^{1,n}$ and the stochastic integral $\int_t^T e^{\beta s}\langle \delta Y_s, \delta Z_s^* dW_s\rangle$ is $\mathbb{P}$-integrable, with zero expectation. Moreover, from the Lipschitz property of the generator f^1, it follows that

$$\begin{aligned} |f^1(s,Y_s^1,Z_s^1) - f^2(s,Y_s^2,Z_s^2)| &\le |f^1(s,Y_s^1,Z_s^1) - f^1(s,Y_s^2,Z_s^2)| + |\delta_2 f_s| \\ &\le C[|\delta Y_s| + |\delta Z_s|] + |\delta_2 f_s|\,. \end{aligned}$$

The quadratic form $Q(y,z) = -\beta|y|^2 + 2C|y|^2 + 2C|y||z| + 2|\delta_2 f_s||y| - |z|^2$, which appears on the right of the above equality can be reduced to

$$\begin{aligned} Q(y,z) &= -\beta|y|^2 + 2C|y|^2 + C^2|y|^2 + 2|\delta_2 f_s||y| - (|z| - C|y|)^2 \\ &= -\beta_C((|y| - \beta_C^{-1}|\delta_2 f_s|)^2 - (|z| - C|y|)^2 + \beta_C^{-1}|\delta_2 f_s|^2 \end{aligned}$$

where $\beta_C := \beta - 2C - C^2 > 0$ is assumed to be strictly positive.

$$\begin{aligned} \mathbb{E}[e^{\beta t}|\delta Y_t|^2] &+ \beta_C\int_t^T e^{\beta s}(|\delta Y_s| - \beta_C^{-1}|\delta_2 f_s|)^2 ds + \int_t^T e^{\beta s}|\delta Z_s - C\delta Y_s|^2\, ds] \\ &\le \mathbb{E}[e^{\beta T}|\delta Y_T|^2] + \mathbb{E}\int_t^T e^{\beta s}\frac{|\delta_2 f_s|^2}{\beta_C} ds. \end{aligned} \tag{3.5}$$

This inequality leads to estimates of the β-norm of Y in two ways: firstly by integration the right side between 0 and T; secondly by using the inequality $|\delta Y_s|^2 \le 2[(|\delta Y_s| - \beta_C^{-1}|\delta_2 f_s|)^2 + \beta_C^{-2}|\delta_2 f_s|^2]$. The first way gives better estimates for short time T, the second for large T, and large β of course. The control of the norm of the process δZ follows by inequality (3.5). □

Proof of Theorem 3.1 We use a fixed point theorem for the mapping from $\mathbb{H}_T^{2,d}\times\mathbb{H}_T^{2,n\times d}$ into itself, which maps (y,z) onto the solution (Y,Z) of the BSDE with generator $f(t,y_t,z_t)$, i.e.,

$$Y_t = \xi + \int_t^T f(s,y_s,z_s)ds - \int_t^T Z_s^* dW_s\,.$$

Let (y^1,z^1), (y^2,z^2) be two elements of $\mathbb{H}_T^{2,d}\times\mathbb{H}_T^{2,n\times d}$ and let (Y^1,Z^1) and (Y^2,Z^2) be the associated solutions. By Proposition 3.2 applied with $C=0$, we obtain for a 'good' choice of β

$$\|\,\delta Y\,\|_\beta^2 \le \frac{T}{\beta}\mathbb{E}\int_0^T e^{\beta s}|f(s,y_s^1,z_s^1) - f(s,y_s^2,z_s^2)|^2 ds$$

and

$$\| \delta Z \|_\beta^2 \leq \frac{2}{\beta}\mathbb{E}\int_0^T e^{\beta s}|f(s,y_s^1,z_s^1) - f(s,y_s^2,z_s^2)|^2 ds\,.$$

Now since f is Lipschitz with constant C, it is easy to show that

$$\| \delta Y \|_\beta^2 + \| \delta Z \|_\beta^2 \leq \frac{2(2+T)C^2}{\beta}[\| \delta y \|_\beta^2 + \| \delta z \|_\beta^2]. \tag{3.6}$$

Choosing β such that $2(2+T)C^2 < \beta$, we see that this mapping is contracting from $(\mathbb{H}_T^{2,d} \otimes \mathbb{H}_T^{2,n\times d}, \| \cdot \|_\beta + \| \cdot \|_\beta)$ onto itself, and that there exists a fixed point, which is the unique continuous solution of the BSDE. □

Remark: Parametrized BSDE Let $(f(\alpha,.),\xi(\alpha,.)\, \alpha \in \mathbb{R})$ be a family of standard parameters of a BSDE with solutions denoted by (Y^α, Z^α). Using the a priori estimates, we state that the regularity properties such as continuity and differentiability with respect to α of $(Y^\alpha, Z^{\alpha)}, \alpha \in \mathbb{R})$ follow from regularity properties of the generator $f(\alpha,.)$ and terminal condition $\xi(\alpha,.)$. The proof can be found in El Karoui, Peng & Quenez (1997).

3.1.3 Flow of a BSDE

Recall the dependence of the solutions of a BSDE with respect to a terminal condition by the notation $(Y(T,\xi), Z(T,\xi))$. We provide similar regularity results as for the forward SDE.

Proposition 3.3 *Let (Y,Z) be the solution of a BSDE with standard parameters (T,ξ,f). For any stopping time $S \leq T$, and $t \leq S$*

$$Y_t(T,\xi) = Y_t(S, Y_S(T,\xi)), \quad Z_t(T,\xi) = Z_t(S, Y_S(T,\xi)), \ dt \otimes d\mathbb{P}\, a.s.$$

Suppose that the sequence of stopping times S_n (resp. of terminal variables $\xi_n \in \mathcal{F}_{S_n}$) converges a.s. to S (resp. converges in $\mathbb{L}^2$ to $\xi \in \mathcal{F}_S$). Then $(Y(S_n,\xi_n), Z(S_n,\xi_n))$ converges in $\mathbb{H}_\beta^2 \times \mathbb{H}_\beta^2$ to $(Y(S,\xi), Z(S,\xi))$.

Proof By conventional notation, we define the solution of the BSDE with terminal condition (T,ξ) for $t \geq T$ by $(Y_t = \xi, Z_t = 0)$. So if $T' \geq T$, then $(Y_t, Z_t\ ; t \leq T')$ is the unique solution of the BSDE with standard parameters $(T', \xi, f(t,y,z)1_{\{t\leq T\}})$.

Now let $S \leq T$ be a stopping time, and denote by $Y(S,\xi_S)$ the solution of the BSDE with standard parameters $(T,\xi_S, f(t,y,z)1_{\{t\leq S\}})$. Both processes $(Y(S,Y_S), Z(S,Y_S))$ and $(Y_{\wedge S}(T,\xi),\ Z(T,\xi)1_{[0,S]})$ are solutions of the BSDE with parameters $(T, Y_S, f(t,y,z)1_{\{t\leq S\}})$. By uniqueness these processes are equal. The convergence result follows immediately from Proposition 3.2, since the parameters $(T,\xi_n, f(t,y,z)1_{\{t\leq S_n\}})$ satisfy the assumptions of this proposition.

3.2 Comparison theorem

3.2.1 Linear BSDE

In the linear and one dimensional case, Theorem 3.1 specifies the integrability properties of the solution of the standard pricing problem (Theorem 2.1).

Proposition 3.4 *Let (β, γ) be a bounded $(\mathbb{R}, \mathbb{R}^n)$-valued predictable process, φ be an element of $\mathbb{H}_T^{2,1}$, and ξ be an element of $\mathbb{L}_T^{2,1}$. Then the LBSDE*

$$- \; dY_t = [\varphi_t \; + Y_t \beta_t + Z_t^* \gamma_t] \, dt \; - \; Z_t^* dW_t \, ; \qquad Y_T = \xi \tag{3.7}$$

has a unique solution (Y, Z) in $\mathbb{H}_T^{2,1} \times \mathbb{H}_T^{2,n}$ given explicitly by:

$$\Gamma_t Y_t \; = \; \mathbb{E} \left[\xi \Gamma_T + \int_t^T \Gamma_s \varphi_s ds \, | \mathcal{F}_t \right] , \tag{3.8}$$

where Γ_t is the adjoint process defined by the forward LSDE,

$$d\Gamma_t = \Gamma_t [\beta_t \, dt + \gamma_t^* dW_t] \; ; \quad \Gamma_0 = 1. \tag{3.9}$$

In particular if ξ and φ are non-negative, the process (Y_t) is non-negative. If in addition $Y_0 = 0$, then for any t, $Y_t = 0$ a.s., $\xi = 0$ a.s. and $\varphi_t = 0$ $dt \otimes d\mathbb{P}$-a.s.

Proof By Theorem 3.1 there is a unique solution (Y, Z) of the BSDE. By standard calculations (see Section 2.1), we see that $\Gamma_t Y_t + \int_0^t \Gamma_s \varphi_s ds$ is a local martingale. Now, $\sup_{s \leq T} |Y_s|$ and $\sup_{s \leq T} |\Gamma_s|$ belong to $\mathbb{L}_T^{2,1}$ and $\sup_{s \leq T} |Y_s| \times \sup_{s \leq T} |\Gamma_s|$ belongs to $\mathbb{L}^{1,1}$. Therefore the local martingale $(\Gamma_t Y_t + \int_0^t \Gamma_s \varphi_s ds), t \in [0, T]$ is a uniformly integrable martingale, whose t-time value is the $\mathcal{F}_t$-conditional expectation of its terminal value. In particular, if ξ and φ are non-negative, Y_t is also non-negative. If in addition $Y_0 = 0$, the expectation of the non-negative variable $\xi \Gamma_T + \int_0^T \Gamma_s \varphi_s ds$ is equal to 0. So, $\xi = 0$, $\varphi_t = 0$ and $Y_t = 0$ a.s. $dt \otimes d\mathbb{P}$. □

3.2.2 Comparison Theorem

As an immediate consequence of Proposition 3.4, we state in the one dimensional case a comparison theorem (first obtained by Peng (1992a,b).

Theorem 3.5 (Comparison theorem) *Let (f^1, ξ^1), (f^2, ξ^2) be two standard parameters of BSDEs, and (Y^1, Z^1), (Y^2, Z^2) be the associated solutions. We suppose that $\xi^1 \geq \xi^2$ $\mathbb{P}$ a.s., and $\delta_2 f_t = f^1(t, Y_t^2, Z_t^2) - f^2(t, Y_t^2, Z_t^2) \geq 0$ and $dt \times d\mathbb{P}$ a.s. Then we have $Y^1 \geq Y^2$ $\mathbb{P}$ a.s. Moreover the comparison is strict, that is, if in addition $Y_0^1 = Y_0^2$, then $\xi^1 = \xi^2$, $f^1(s, Y_s^1, Z_s^1) = f^2(s, Y_s^2, Z_s^2)$, and $Y_s^1 = Y_s^2$ a.s.*

By applying the comparison theorem to $Y^2 = 0$, we deduce a sufficient condition for non-negativity:

Corollary 3.6 *If $\xi \geq 0$ a.s. and $f(t,0,0) \geq 0$ $dt \times d\mathbb{P}$ a.s., then $Y_t \geq 0$.*

Proof of Theorem 3.5 We use the notation of Proposition 3.2. For the sake of simplicity we suppose that $n = 1$. The pair $(\delta Y, \delta Z)$ is the solution of the following LBSDE:

$$\begin{cases} -d\delta Y_t = [\Delta_y f^1(t)\delta Y_t + \Delta_z f^1(t)\delta Z_t + \delta_2 f_t]dt - \delta Z_t^* dW_t \\ \delta Y_T \;= \xi^1 - \xi^2, \end{cases} \tag{3.10}$$

where

$$\Delta_y f^1(t) = \begin{cases} \dfrac{f^1(t, Y_t^1, Z_t^1) - f^1(t, Y_t^2, Z_t^1)}{Y_t^1 - Y_t^2} & \text{if } Y_t^1 - Y_t^2 \neq 0, \\ 0 & \text{otherwise.} \end{cases}$$

Also,

$$\Delta_z f^1(t) = \begin{cases} \dfrac{f^1(t, Y_t^2, Z_t^1) - f^1(t, Y_t^2, Z_t^2)}{Z_t^1 - Z_t^2} & \text{if } Z_t^1 - Z_t^2 \neq 0, \\ 0 & \text{otherwise.} \end{cases}$$

Now, since by assumption the generator f^1 is uniformly Lipschitz with respect to (y,z), it follows that $\Delta_y f^1$ and $\Delta_z f^1$ are bounded processes. Also by assumption $\delta_2 f_t$ and δY_T are non-negative. Proposition 3.4 gives that the unique solution $(\delta Y, \delta Z)$ of the LBSDE (3.10) is non-negative and satisfies

$$\Gamma_t \delta Y_t = \mathbb{E}\left[(\xi^1 - \xi^2)\Gamma_T + \int_t^T \Gamma_s \delta_2 f_s ds | \mathcal{F}_t\right], \tag{3.11}$$

where Γ is the adjoint process (positive) of the above LBSDE. □

3.3 Markovian case

3.3.1 Forward-Backward Stochastic Differential Equation

In this section we consider the solution of some BSDEs associated with some forward classical SDEs. For example, the forward equation can be the dynamics of some basic securities. For any given $(t,x), \in [0,T] \times \mathbb{R}^p$, consider the following classical Itô stochastic differential equation defined on $[0,T]$,

$$\begin{cases} dS_u = b(u, S_u)du + \sigma(u, S_u)dW_u, & t \leq u \leq T \\ S_u = x, & 0 \leq u \leq t\,. \end{cases} \tag{3.12}$$

We then consider the associated BSDE,

$$-\,dY_u = f(s, S_u^{t,x}, Y_u, Z_u)ds - Z_u^* dW_u\ , \qquad\qquad Y_T = \Psi(S_T^{t,x}). \qquad (3.13)$$

The solutions will be denoted $(S_u^{t,x}, Y_u^{t,x}, Z_u^{t,x},\ 0 \leq u \leq T)$. The system (3.12) and (3.13) is called a forward-backward stochastic differential equation (FBSDE). Here, f (resp. Ψ) is a $\mathbb{R}^d$-valued Borel function defined on $[0,T] \times \mathbb{R}^p \times \mathbb{R}^d \times \mathbb{R}^{n\times d}$ (resp. on $\mathbb{R}^p$), and b (resp. σ) is a $\mathbb{R}^p$-valued (resp. $\mathbb{R}^{p\times n}$-valued) function defined on $[0,T] \times \mathbb{R}^p$. The function f is supposed to be uniformly Lipschitz with respect to (y,z) with Lipschitz constant C, that is,

$$|f(t,x,y_1,z_1) - f(t,x,y_2,z_2)| \leq C[|y_1 - y_2| + |z_1 - z_2|].$$

Moreover, b, σ are supposed to be uniformly Lipschitz with respect to x. Finally, we will suppose that there exists a constant C such that for each (s,x,y,z)

$$|\sigma(x,0)| + |b(x,0)| \leq C(1+|x|);$$
$$|f(s,x,y,z)| + |\Psi(x)| \leq C(1+|x|^p)$$

for real $p \geq 1/2$.

3.3.2 Markov Property of Solutions of Forward-Backward Equations

The main property of the solution of the FBSDE is that Y_u is a deterministic function of $S_u^{t,x}$, which we can see by using the iterative construction of the solution of the BSDE. Notice that if f does not depend on y, z, then the property follows from the Markov property of $S_u^{t,x}$:

$$Y_r = \mathbb{E}[\Psi(S_T^{t,x}) + \int_t^T f(u, S_u^{t,x})du|\mathcal{F}_r] \ = \Phi(r, S_r^{t,x})\,,$$

where

$$\Phi(r,y) = \mathbb{E}[\Psi(S_T^{r,y}) + \int_r^T f(v, S_v^{r,y})dv]\,.$$

The solution of the BSDE is said to be *Markovian.* Furthermore, the process Z associated with Y by the martingale representation theorem is also a deterministic function of $S_u^{t,x}$. This result can be obtained from the paper of Cinlar, Jacod, Protter & Sharpe (1980) on the functional additive martingale of a diffusion process. By using the almost sure convergence of the Picard approximations of the solution of the BSDE, it follows that

Theorem 3.7 *There exist two measurable deterministic functions $u(t,x)$ and $d(t,x)$ such that the solution $(Y_v^{t,x}, Z_v^{t,x})$ of BSDE* (3.13) *is given by*

$$Y_v^{t,x} = u(v, S_v^{t,x})\,;\ \ Z_v^{t,x} = \sigma^*(v, S_v^{t,x})d(v, S_v^{t,x})\,, 0 \leq v \leq T\ dv \otimes d\mathbb{P}\, a.s.$$

3.3.3 BSDE and Partial Differential Equations

Now we study the relation between these forward-backward equations and partial differential equations (PDEs). We first give a generalization of the Feynman-Kac formula stated by Pardoux & Peng (1992). Then we recall that, conversely, under smooth conditions the function $u(t,x) = Y_t^{t,x}$ is a solution in some sense of a PDE.

Proposition 3.8 (Generalization of the Feynman-Kac formula) *Let w be a d-valued function of class $C^{1,2}$ (or smooth enough to be able to apply Itô's formula to $w(v, S_v^{t,x})$) and suppose that there exists a constant C such that, for each (s,x),*

$$|w(s,x)| + |\partial_x w(s,x)\sigma(s,x)| \leq C(1+|x|).$$

Also, w is supposed to be the solution of the following system of quasilinear parabolic partial differential equations:

$$\begin{cases} \partial_t w(t,x) + \mathcal{L}w(t,x) + f(t,x,w(t,x), \partial_x w(t,x)\sigma(t,x)) &= 0 \\ w(T,x) &= \Psi(x)\,. \end{cases} \tag{3.14}$$

where $\partial_x w$ is the gradient of w and $\mathcal{L}_{(t,x)}$ denotes the following second order differential operator:

$$\mathcal{L}_{(t,x)} = \sum_{i,j} a_{ij}(t,x)\partial^2_{x_i x_j} + \sum_i b_i(t,x)\partial_{x_i}; \quad a_{ij} = \frac{1}{2}\left[\sigma\sigma^*\right]_{ij}\,.$$

Then, $w(t,x) = Y_t^{(t,x)}$, where $(Y_s^{t,x}, Z_s^{t,x}, t \leq s \leq T)$ is the unique solution of BSDE (3.13). *Also, $(Y_u^{t,x}, Z_u^{t,x}) = (w(u,S_u), \partial_x v(u,S_u)\sigma(u,S_u)\,, t \leq u \leq T$.*

Proof By applying Itô's formula to $w(u, S_u^{t,x})$ we obtain

$$d\,w(u,S_u) = (\partial_t w(u,S_u) + \mathcal{L}w(u,S_u))\,du \,+\, \partial_x w(u,S_u)\sigma(u,S_u)dW_u.$$

Since w solves (3.14), it follows that

$$\begin{aligned} -d\,w(u,S_u) \;\;&=\;\; f(u,S_u,w(u,S_u), \partial_x w(u,S_u)\sigma(u,S_u))\,du \\ &\qquad - \partial_x w(u,S_u)\sigma(u,S_u)\,dW_u \quad w(T,S_T) = \Psi(P_T). \end{aligned}$$

So $(w(u,S_u), \partial_x w(u,S_u)\sigma(u,S_u); u \in [0,T])$ is the unique solution of BSDE (3.13), and the proof is complete. □

Remark Ma, Protter & Yong (1994) use this point of view to study some more general forward-backward stochastic differential equations of the type, for $t \leq u \leq T$,

$$\begin{aligned} dS_u &= b(u,S_u,Y_u,Z_u)du + \sigma(u,S_u,Y_u)dW_u\,, \\ -dY_u &= f(u,S_u,Y_u,Z_u)du - Z_u^* dW_u, \qquad (3.15) \\ S_t &= x\,, \qquad Y_T = \Psi(S_T^{t,x}). \qquad (3.16) \end{aligned}$$

Now we show that conversely, under some additional assumptions, the solution of the BSDE (3.13) corresponds to the solution of the PDE (3.14). If $d = 1$, we can use the comparison theorem to show that if b, σ, f, Ψ satisfy the assumptions at the beginning of the section and if f, Ψ are supposed to be uniformly continuous with respect to x, then $u(t,x)$ is a viscosity solution of the above PDE (3.14) (see Peng (1992b) and Pardoux & Peng (1992)).

Theorem 3.9 *We suppose that $d = 1$ and that f, Ψ are uniformly continuous with respect to x. Then, the function u defined by $u(t,x) = Y_t^{t,x}$ is a viscosity solution of PDE (3.14).*

Furthermore, if we suppose that, in addition, for each $R > 0$, there exists a continuous function $m_R : \mathbb{R}_+ \to \mathbb{R}_+$ such that $m_R(0) = 0$ and

$$|f(t,x,y,z) - f(t,x',y,z)| \leq m_R(|x - x'|(1 + |z|)), \tag{3.17}$$

for all $t \in [0,T]$, $|x|$, $|x'| \leq R$, $|z| \leq R$, $z \in \mathbb{R}^n$, then u is the unique viscosity solution of PDE (3.14).

4 Concave BSDEs and Control Problems

We are concerned with the solution of a BSDE with respect to a generator that is an infimum of standard generators. In this case, this property can be translated into the solutions, under some mild conditions. In other words,

$$Y_t(\inf f^\alpha) = \operatorname{ess\,inf} Y_t(f^\alpha).$$

This property can be applied to some classical control problems. Under certain conditions, the value function can be characterized as the solution of a (concave) BSDE. From this point of view, the classical properties of the value function can be derived. Then we show that, conversely, the solution (Y, Z) of a BSDE (with $d = 1$) with concave generator may be considered as the value function of a control problem.

4.1 BSDE and optimization

4.1.1 Solution of BSDE as a Minimum or a Minimax

In this section, we are concerned with standard generators f which can be obtained as infima of standard generators f^α. From the comparison theorem, the solution of the BSDE associated with f is less than the infimum of the solutions associated with f^α. The question is to know when equality holds.

Proposition 4.1 *Let (f, ξ) and (f^α, ξ^α) be a family of standard parameters, and let (Y, Z) and (Y^α, Z^α) be the associated solutions. Suppose that there exists a parameter $\overline{\alpha}$ such that*

$$\begin{aligned} f(t, Y_t, Z_t) &= \operatorname{ess\,inf}_\alpha f^\alpha(t, Y_t, Z_t) = f^{\overline{\alpha}}(t, Y_t, Z_t) \qquad dt \otimes d\mathbb{P} a.s., \\ \xi &= \operatorname{ess\,inf}_\alpha \xi^\alpha = \xi^{\overline{\alpha}} \qquad a.s. \end{aligned}$$

Then the processes Y and Y^α satisfy:

$$Y_t = \operatorname{ess\,inf}_\alpha Y_t^\alpha = Y_t^{\overline{\alpha}}, \quad \forall t \in [0, T], \ \mathbb{P}\, a.s. \tag{4.1}$$

Remark The same result holds if the generators only satisfy: for each $\epsilon > 0$, there exists a control α^ϵ such that

$$f(t, Y_t, Z_t) = \operatorname{ess\,inf} f^\alpha(t, Y_t, Z_t) \geq f^{\alpha^\varepsilon}(t, Y_t, Z_t) - \varepsilon, \qquad dt \otimes d\mathbb{P} a.s.$$

Proof (Y, Z) and (Y^α, Z^α) are solutions of two BSDEs whose generators and terminal conditions satisfy the assumptions of the comparison theorem 3.5. Hence, for any α, $Y_t \leq Y_t^\alpha$ and consequently $Y_t \leq \operatorname{ess\,inf} Y_T^\alpha$ $\mathbb{P}\, a.s.$ for any time t.

We now prove the equality using the uniqueness theorem for BSDEs and the existence of a parameter $\overline{\alpha}$ such that $f(t, Y_t, Z_t) = f^{\overline{\alpha}}(t, Y_t, Z_t)$, and $\xi = \xi^{\overline{\alpha}}$. Hence (Y, Z) and $(Y^{\overline{\alpha}}, Z^{\overline{\alpha}})$ are both solutions of the same BSDE associated with $f^{\overline{\alpha}}$; therefore, they are the same. So,

$$\operatorname{ess\,inf} Y_t^\alpha \geq Y_t = Y_t^{\overline{\alpha}} \geq \operatorname{ess\,inf} Y_t^\alpha \quad \forall t \in [0, T], \ \mathbb{P}\, a.s.$$

□

Similar results were extended to minimax problems by Hamadene & Lepeltier (1994) in connection with stochastic differential games. These techniques are also useful for solving optimization problems associated with recursive utilities (Quenez 1993, El Karoui, Peng & Quenez 1994).

4.1.2 Stochastic control problems

A lot of stochastic control problems (Krylov 1980, El Karoui 1981, Elliott 1982, Davis 1985) are specified in the following manner: the laws of the controlled process belong to a family of equivalent measures whose densities are associated with

$$dH_t^u = H_t^u[d(t, u_t)dt + n(t, u_t)^* dW_t]\,; \ H_0^u = 1\,, \tag{4.2}$$

where $(d(t, u_t), t \in [0, T])$ and $(n(t, u_t), t \in [0, T])$ are predictable processes uniformly bounded by δ_t and ν_t respectively. A feasible control $(u_t, 0 \leq t \leq T)$ is a predictable process valued in a (Polish) space U. The set of feasible controls is denoted by $\mathcal{U}$. The problem is to minimize over all feasible control processes u, the objective function

$$J(u) = \mathbb{E}\left[\int_0^T H_t^u \, k(t, u_t)dt + H_T^u \xi^u\right] , \tag{4.3}$$

where ξ^u is the terminal condition (square integrable) and $k(t, u_t)$ is the running cost associated with the control u. The processes $k(t, u_t)$ are supposed to be uniformly dominated by a square-integrable process k_t. Here, for simplicity, we suppose that δ, ν and k are bounded.

The controller acts on the law of processes by change of equivalent probability measures with controlled Radon-Nikodym derivatives L^u and by a controlled discount factor D^u with bounded rate $d(t, u_t)$, such that,

$$dL_t^u = L_t^u \, n(t, u_t)^* dW_t, \quad dD_t^u = D_t^u \, d(t, u_t)dt, \quad \text{with } H_t^u = D_t^u L_t^u.$$

Let us denote by $\mathbb{Q}^u$ the probability measure with density L_T^u on $\mathcal{F}_T$. Then the objective function can be written

$$J(u) = \mathbb{E}_{\mathbb{Q}^u}\left[\int_0^T D_t^u \, k(t, u_t)dt + D_T^u \xi^u\right] . \tag{4.4}$$

Notice that, by Proposition 3.4, $J(u) = Y_0^u$ where (Y^u, Z^u) is the solution of the linear BSDE associated with terminal condition ξ and standard generator f^u,

$$f^u(t, y, z) = k(t, u_t) + d(t, u_t).y + n(t, u_t)^* z \, .$$

The process H^u corresponds to the adjoint process associated with (Y^u, Z^u) and

$$Y_t^u = \mathbb{E}\left[\int_t^T H_{t,s}^u \, k(s, u_s)dt + H_{t,T}^u \xi^u \mid \mathcal{F}_t\right] .$$

The previous results lead to the verification theorem which is a sufficient condition for a process to be the value function.

Proposition 4.2 (Verification theorem) *The parameter (f, ξ) defined by*

$$f(t, y, z) = \text{essinf}\{f^u(t, y, z) \mid u \in \mathcal{U}\} \, \xi = \text{essinf}\{\xi^u \mid u \in \mathcal{U}\}$$

is a standard parameter with concave generator. Let (Y, Z) be the solution of the BSDE associated with terminal condition ξ. Then Y is the value function Y^* of the control problem that is, for each $t \in [0, T]$,

$$Y_t = Y_t^* = \text{ess inf}\{Y_t^u \mid u \in \mathcal{U}\} \, .$$

Proof In order to use the previous proposition, we have to show that f is a standard generator, and that the minimum is achieved by a feasible control. In fact, we have essentially to overcome some technical measurability questions about f, and about the 'optimal' control (see El Karoui, Peng & Quenez (1997) for details). □

Recall that the main tool in stochastic control is *the principle of dynamic progamming.* However, in the context of BSDEs it is just the flow property (3.3). Let us denote by $(Y_t^u(T,\xi^u),\ 0 \le t \le T)$ the solution of the BSDE associated with generator f^u and terminal conditions (T,ξ^u).

Proposition 4.3 *The value function Y satisfies the dynamic programming principle: for any time t and any stopping time S with $t \le S \le T$,*

$$Y_t(T,\xi) = \text{ess}\inf_{u\in\mathcal{U}} Y_t^u(S, Y_S(T,\xi^u)) \ a.s.,$$

which can also be written

$$Y_t(T,\xi) = \text{ess}\inf_{u\in\mathcal{U}} \mathbb{E}\left[\int_t^S H_{t,s}^u\, k(s,u_s)dt + H_{t,S}^u Y_S(T,\xi^u) \mid \mathcal{F}_t\right] \quad a.s.$$

Actually, the optimization problem is to find an (0)–optimal control u^0 which achieves the minimum for the problem $\inf\{Y_0^u \mid u \in \mathcal{U}\}$, i.e. $Y_0^* = Y_0^{u^0}$. The comparison theorem gives a *criterion* for finding 0-optimal controls.

Corollary 4.4 (Optimality criterion) *A control $(u_s^0,\ 0 \le s \le T)$ is 0-optimal if and only if*

$$\xi = \xi^{u^0}\ , \quad f(s,Y_s,Z_s) = f^{u^0}(s,Y_s,Z_s) \ \ d\mathbb{P}\otimes ds \ \ a.s.$$

In this case, u^0 is also optimal for the problem starting at time t, that is, $Y_t^* = Y_t^{u^0}$ for any time t .

Proof It is an easy consequence of the second part of the comparison theorem 3.5. □

Remark The verification theorem shows that the solution (Y,Z) of the BSDE satisfies the condition that for each feasible control u,

$$H_t^u Y_t + \int_0^t H_s^u\, k(s,u_s)ds$$

is a uniformly integrable submartingale with increasing process given by $\int_0^t H_s^u K_s^u\, ds$ where

$$K_t^u = -f(t,Y_t,Z_t) + f^u(t,Y_t,Z_t) \ge 0.$$

Furthermore, the optimality criterion shows that u^0 is 0-optimal if and only if $K_t^{u^0} = 0$, in other words if and only if $H_t^{u^0} Y_t + \int_0^t H_s^{u^0}\, k(s,{u^0}_s)ds$ is a martingale. Consequently, the previous results correspond to the classical properties of the value function.

Notice that in this example the generator f is concave. We will see in the next section that, conversely, a concave BSDE is always associated with a control problem.

4.1.3 Concave BSDE as an Infimum

Here we fix some notation and recall a few properties of convex analysis (whose proofs are, for example, in Ekeland & Temam (1976) and Ekeland & Turnbull (1979)) in order to show that a concave generator is an infimum of linear generators. Let $f(t, y, z)$ be a standard generator of a BSDE, concave with respect to y, z and let $F(t, \beta, \gamma)$ be the polar process associated with f:

$$F(t, \omega, \beta, \gamma) = \sup_{(y,z) \in \mathbb{R} \times \mathbb{R}^n} [f(t, \omega, y, z) - \beta\, y - \gamma^* z]. \tag{4.5}$$

The *effective domain of* F is by definition

$$\mathcal{D}_F = \{(\omega, t, \beta, \gamma) \in \Omega \times [0, T] \times \mathbb{R} \times \mathbb{R}^n |\ F(t, \omega, \beta, \gamma) < +\infty\}.$$

Since f is concave, f is continuous with respect to (y, z), and (f, F) satisfies the conjugacy relation

$$f(t, \omega, y, z) = \inf\{F(t, \omega, \beta, \gamma) + \beta y + \gamma^* z \mid (\beta, \gamma) \in \mathcal{D}_F^{(\omega,t)}\}.$$

For every (t, ω, y, z) the infimum is achieved in this relation by a pair (β, γ) which depends on (t, ω).

We want to associate with the polar process F a wide enough family of linear *standard* generators $f^{\beta,\gamma}$ such that the assumptions of Proposition 4.1 hold. Let us denote

$$f^{\beta,\gamma}(t, y, z) = F(t, \beta_t, \gamma_t) + \beta_t\, y + \gamma_t^*\, z,$$

where (β_t, γ_t) are predictable and bounded[5] processes, called *control parameters*. Recall that by the conjugacy relation, f is also the infimum of $f^{\beta,\gamma}$. To ensure that $f^{\beta,\gamma}$ is a standard generator, it is sufficient to suppose that (β, γ) belongs to $\mathcal{A}$ defined by

$$\mathcal{A} = \{(\beta, \gamma) \in \mathcal{P}, \mid \mathbb{E} \int_0^T F(t, \beta_t, \gamma_t)^2\, dt < +\infty\}.$$

$\mathcal{A}$ is said to be the set of *admissible control parameters.* Let (Y, Z) be the unique solution of the BSDE with concave standard generator f. In order to apply Proposition 4.1, it remains to show the following lemma.

Lemma *There exists an optimal control* $(\overline{\beta}, \overline{\gamma}) \in \mathcal{A}$, *such that*

$$f(t, Y_t, Z_t) = f^{\overline{\beta},\overline{\gamma}}(t, Y_t, Z_t) \qquad dt \otimes d\mathbb{P}\ a.s.$$

Proof Recall that for each (t, ω, y, z), the infimum in the conjugacy relation is achieved, since f is concave uniformly Lipschitz. Also, by a measurable

[5] By the Lipschitz constant of f

selection theorem and since $f(t, Y_t, Z_t), Y_t$ and Z_t are predictable processes, there exists a pair of predictable (bounded) processes $(\overline{\beta}, \overline{\gamma})$ such that

$$f(t, Y_t, Z_t) = f^{\overline{\beta}, \overline{\gamma}}(t, Y_t, Z_t) \qquad dt \otimes d\mathbb{P} \; a.s.$$

Since by assumption $f(t, Y_t, Z_t)$, Z_t and Y_t belong to $\mathbb{H}_T^{2,1}$ and $\overline{\beta}, \overline{\gamma}$ are bounded, $F(t, \overline{\beta}_t, \overline{\gamma}_t)$ belongs also to $\mathbb{H}_T^{2,1}$. Hence, the pair $(\overline{\beta}, \overline{\gamma})$ which achieves the infimum in the conjugacy relation belongs to $\mathcal{A}$. □.

For each control process $(\beta, \gamma) \in \mathcal{A}$, introduce the dual 'controlled objective' processes $(Y^{\beta,\gamma}, Z^{\beta,\gamma})$ as the unique solution of the LBSDE with data $(\xi, f^{\beta,\gamma})$. Thus, Proposition 4.1 gives directly the following result.

Proposition 4.5 *Let f be a concave standard generator and $f^{\beta,\gamma}$ the associated linear standard generators satisfying*

$$f(t, y, z) = \operatorname{ess\,inf}\{f^{\beta,\gamma}(t, y, z) \mid (\beta, \gamma) \in \mathcal{A}\}.$$

Then

$$Y_t = \operatorname{ess\,inf}\{Y_t^{\beta,\gamma} \mid (\beta, \gamma) \in \mathcal{A}\}.$$

Let us interpret the above result as associated with a control problem, with control set $\mathcal{A}$. From Proposition 3.4, the LBSDEs solution $Y_t^{\beta,\gamma}$ can be written using the adjoint process $(\Gamma_{t,s}^{\beta,\gamma}, t \leq s \leq T)$, which is the unique solution of the forward linear SDE

$$d\Gamma_s = \Gamma_s[\beta_s ds + \gamma_s dW_s]; \quad \Gamma_t = 1, \tag{4.6}$$

in the following manner,

$$Y_t^{\beta,\gamma} = \mathbb{E}\left[\int_t^T \Gamma_{t,s}^{\beta,\gamma} \, F(s, \beta_s, \gamma_s)\, ds + \Gamma_{t,T}^{\beta,\gamma} \, \xi | \mathcal{F}_t\right].$$

Here $Y^{\beta,\gamma}$ is called the *controlled objective function* of a control problem where the running cost function is the function $F(t, \beta, \gamma)$ and where the terminal cost is the random variable ξ.

4.2 Application to recursive utility

We come back to the example of recursive utility presented in Section 2. In an economic or financial context, the generator $f(t, c_t, Y_t, Z_t)$ represents the instantaneous utility (at time t) of consumption rate ($c_t \geq 0$). In general, we suppose that the consumption process c belongs to $\mathbb{H}_T^{2,1}$ and that f satisfies the standard assumptions of the generators of a BSDE with an additional condition such as

$$|f(t, c, 0, 0)| \leq k_1 + k_2|c| \quad dt \otimes \mathbb{P} \; a.s.$$

or some other growth condition which ensures that $f(t, c_t, 0, 0) \in \mathbb{H}^2$.

4.2.1 Classical Properties

In this section, we show that under natural conditions, the classical properties of utilities (see Section 2.4) are satisfied by recursive ones; actually, it is a direct consequence of the comparison theorem.

Proposition 4.6 *Let ξ^1 and ξ^2 be two terminal square integrable rewards. Let c^1 and c^2 be two consumption processes which belong to $\mathbb{H}_T^{2,1}$. Let Y^{c^1,ξ^1} and Y^{c^2,ξ^2} be the recursive utilities associated with $(\xi^1, f(t,c^1,.))$ and $(\xi^2, f(t,c^2,.))$. Then the following properties are true:*

• *(Time consistency) If for some time t, $Y_t^{c^1,\xi^1} \geq Y_t^{c^2,\xi^2}$ and $c_s^1 = c_s^2$, $0 \leq s \leq t$ $\mathbb{P}$ a.s., then $Y_s^{c^1,\xi^1} \geq Y_s^{c^2,\xi^2}$, $0 \leq s \leq t$, $\mathbb{P}$ a.s.*

• *(Monotonicity with respect to the consumption) Suppose $\xi^1 \geq \xi^2$.*
If the generator f is non-decreasing with respect to c, and if $c_t^1 \geq c_t^2$, $0 \leq t \leq T$, $\mathbb{P}$ a.s., then $Y_t^{c^1,\xi^1} \geq Y_t^{c^2,\xi^2}, 0 \leq t \leq T$, $\mathbb{P}$ a.s.

• *(Concavity) Suppose $\lambda\xi^1 + (1-\lambda\xi^2 = \xi$. If the generator f is concave with respect to c, y, z, then for each $\lambda \in [0,1]$, $t \in [0,T]$*
$\lambda Y_t^{c^1,\xi^1} + (1-\lambda)Y_t^{c^2,\xi^2} \leq Y_t^{\lambda c^1+(1-\lambda)c^2,\xi}$, $\mathbb{P}$ a.s.

Proof We show how to prove the first point. Since $Y_t^{c^1,\xi^1} \geq Y_t^{c^2,\xi^2}$ $\mathbb{P}$ *a.s.* and since $Y_s^{c^1,\xi}$ and $Y_s^{c^2,\xi}$ have the same generator on $[0,t]$, the result follows from the comparison theorem applied between time 0 and time t. The other properties are also direct consequences of the comparison theorem. □

4.2.2 Variational Formulation of the Recursive Utility

As we have just seen above, a natural assumption for a recursive utility is the concavity of the generator f with respect to (c, y, z). Consequently the recursive utility can be written as the value function of a control problem.

Fix ξ to be a square integrable terminal reward and c a square integrable consumption process. Let $Y^{c,\xi}$ be the associated recursive utility (that is the solution of the BSDE associated with the generator $f(t, c_t, ., .)$ and the terminal value ξ. For a consumption rate c_t, let $F(t, c_t, ., .)$ be the polar function of $f(t, c_t, ., .)$, i.e.,

$$F(t, c_t, \beta, \gamma) = \sup_{(y,z)\in\mathbb{R}\times\mathbb{R}^n} [f(t, c_t, y, z) - \beta.y - \gamma.z] .$$

Let $\mathcal{A}(c)$ be the set of feasible processes (β, γ) such that $\mathbb{E}\int_0^T F(t, c_t, \beta_t, \gamma_t)^2\, dt < +\infty$. Then by the results stated in Subsection 4.1, the recursive utility can be written as

$$Y_t^{c,\xi} = \operatorname{ess} \inf_{(\beta,\gamma)\in\mathcal{A}} \mathbb{E}\left[\int_t^T \Gamma_{t,s}^{\beta,\gamma}\, F(s, c_s, \beta_s, \gamma_s)\, ds + \Gamma_{t,T}^{\beta,\gamma}\xi \mid \mathcal{F}_t\right] .$$

Hence, the recursive utility $Y^{c,\xi}$ can be defined through a *felicity function* F firstly introduced by Geoffard (1995) (in the deterministic case). The felicity function $F(t, c_t, \beta_t, \gamma_t)$ at some current time t, expressed in terms of the current time unit, is a function of current consumption (c), current rate $(-\beta)$, and risk premium $(-\gamma)$. This function can be thought as an ex-post felicity when the agent knows the current interest rate and the risk-premium.

Notice that the adjoint processes $\Gamma^{\beta,\gamma}$ can be interpreted as a deflator (Duffie & Skiadias 1991). Also, the process $Y^{\beta,\gamma}$ can be interpreted as an ex-post utility, when the deflator is given by $\Gamma^{\beta,\gamma}$. Hence, the utility is equal to the minimum of ex-post utilities over all price deflators. Ex-ante, the optimal deflator is the one that minimizes the agent's ex-post utility.

Concerning the wealth process associated with some portfolio consumption strategy, we have the same kind of interpretation, as we shall see in the next section.

4.3 Nonlinear arbitrage pricing theory

A general setting of the wealth equation (which extends the examples of Section 2.3) is

$$-\,dX_t = b(t, X_t, \sigma_t^* \pi_t)dt - \pi_t^* \sigma_t dW_t\,. \tag{4.7}$$

Here b is a real process satisfying the standard hypotheses of a generator. The classical case (Section 2.2) corresponds to a linear functional

$$b(t, x, z) = -r_t x - z^* \theta_t\,,$$

where the risk premium vector θ, and r the spot rate are bounded processes.

Let $\xi \geq 0$ be a square integrable contingent claim. As in the classical case, the price of the contingent claim is the wealth process X associated with an admissible strategy π which finances ξ, i.e., the square integrable solution of the equation (4.7) with terminal value ξ.

The comparison theorem gives a sufficient condition for the price to be positive, which is

$$b(t, 0, 0) \geq 0 \quad dt \otimes d\mathbb{P} a.s.,$$

i.e. such that (X, π) is the square integrable solution of equation (4.7) with terminal value ξ.

4.3.1 Variational Formulation of the Price

As we have seen in Section 2.3, the generator b associated with the price process (or the wealth process) is always convex with respect to (x, π). It follows that the price process is the value function of a control problem. Indeed, let $B(t, \beta, \gamma)$ be the polar process associated with b,

$$B(t, \beta, \gamma) = \inf_{(x,\pi) \in \mathbb{R} \times \mathbb{R}^n} [b(t, x, \pi) + \beta.x + \gamma.\pi]\,. \tag{4.8}$$

The *effective domain of* B is by definition $\mathcal{D}_B^t = \{(\beta,\gamma)|\ B(t,\beta,\gamma) > -\infty\}$. Since b is convex, (b, B) satisfies the conjugacy relation

$$b(t,x,\sigma_t^*\pi) = \sup\{B(t,\beta,\gamma) - \beta.x - \gamma\sigma_t^*\pi \mid (\beta,\gamma) \in \mathcal{D}_B^t\}.$$

By the results on concave BSDEs, we have

Proposition 4.7 *Let (X,π) be the hedging strategy for ξ, where $(X,\sigma^*\pi)$ is the unique solution of BSDE (4.7). Then X can be written as the maximum of ex-post prices over all feasible deflators*

$$X_t = \text{esssup}\{X_t^{\beta,\gamma} \mid (\beta,\gamma) \in \mathcal{A}\},$$

where $\mathcal{A}$ is the set of (β,γ) feasible bounded controls defined by

$$\mathcal{A} = \{(\beta,\gamma) \in \mathcal{P} \mid \mathbb{E}\int_0^T B(t,\beta_t,\gamma_t)^2\, dt < +\infty\}$$

and where, for each control process $(\beta,\gamma) \in \mathcal{A}$, the ex-post strategy $(X^{\beta,\gamma}, \sigma_t^*\pi^{\beta,\gamma})$ is the unique solution of the LBSDE

$$\begin{cases} -dX_t^{\beta,\gamma} &= (B(t,\beta_t,\gamma_t) - \beta_t.X_t^{\beta,\gamma} - \gamma_t^*\sigma_t^*\pi^{\beta,\gamma})\,dt - (\pi_t^{\beta,\gamma})^*\sigma_t dW_t \\ X_T^{\beta,\gamma} &= \xi. \end{cases} \tag{4.9}$$

The ex-post strategy $(X^{\beta,\gamma},\pi^{\beta,\gamma})$ is a classical strategy in a fictitious market, with bounded interest rate process given by β and bounded risk premium given by γ. So, the price of the contingent claim ξ is the standard price in an optimal fictitious market, associated with $\overline{\beta},\overline{\gamma}$ which achieves the supremum in the conjugacy relation

$$b(t,X_t,\sigma_t^*\pi_t) = B(t,\overline{\beta}_t,\overline{\gamma}_t) - \overline{\beta}_t X_t - \overline{\gamma}_t^*\sigma_t^*\pi_t \qquad dt\otimes d\mathbb{P}\ a.s. \tag{4.10}$$

The only difference with the non-constrained case is that the optimal fictitious market associated depends on the claim to be priced.

Example We come back to the example in Section 2.3.3 (hedging claims with higher interest rate for borrowing) and solved by Cvitanic & Karatzas (1993) under slightly different assumptions. Here we supposed the matrix $(\sigma^*)^{-1}$ to be a bounded process. The hedging strategy (wealth, portfolio) (X,π) satisfies

$$dX_t = r_t X_t\,dt + \pi_t^*\sigma_t\theta_t\,dt + \pi_t^*\sigma_t dW_t - (R_t - r_t)(X_t - \sum_{i=1}^n \pi_t^i)^-\,dt; \quad X_T = \xi. \tag{4.11}$$

Like the other coefficients, the process R_t ($\geq r_t$) is supposed to be bounded. The generator b of this LBSDE is given by the convex process:

$$b(t, x, \sigma_t^* \pi) = -r_t\, x \; - \pi^* \sigma_t \theta_t \; + (R_t - r_t)(x - \pi_t^* \mathbf{1})^- \,.$$

The polar process $B(t, \beta, \gamma)$ associated with b is given by,

$$B(t, \beta, \gamma) = \begin{cases} 0 & \text{if } \gamma = \sigma_t \theta_t + (r_t - \beta)\mathbf{1} \text{ and } r_t \leq \beta \leq R_t, \\ -\infty & \text{otherwise}. \end{cases} \tag{4.12}$$

By Proposition 4.7, it follows that the unique solution $(X, \sigma^* \pi)$ of the BSDE (4.11) satisfies

$$X_t = \operatorname{ess\,sup}\{X_t^\beta \mid r_t \leq \beta_t \leq R_t\}$$

where

$$-dX_t^\beta = -\beta_t X_t^\beta - [\sigma_t \theta_t + (r_t - \beta_t)\mathbf{1}]^* \pi_t^\beta \, dt - (\pi_t^\beta)^* \sigma_t dW_t \; ; \; X_T = \xi .$$

Remark The problems with constraints on the wealth (El Karoui, Kapoudjan, Pardoux, Peng & Quenez (1997) and El Karoui, Pardoux & Quenez, this volume) or on the portfolio (El Karoui & Quenez 1991, and Cvitanic & Karatzas 1993) can be formulated formally in the same way but with a generator which can be infinite, with non-bounded effective domain. For example, the case of an incomplete market (Section 2.3.2), and, more generally, the case of the portfolio process π_t being constrained to take values in a convex set K (Cvitanic & Karatzas 1993) corresponds formally to an upper-price $(X_t, 0 \leq t \leq T)$ solution of BSDE (4.7) with generator

$$b(t, x, \pi) = -r_t x - \pi^* \sigma_t \theta_t \; + \; \mathbf{1}_K(\pi) \,,$$

where $\mathbf{1}_K(\pi)$ is the indicator function of K in the sense of convex analysis, namely equal to 0 if $\pi \in K$ and equal to ∞ otherwise. The example of an incomplete market corresponds to $K = \{\pi \in \mathbb{R}^n \,/ \pi_k = 0 \,, j \leq k \leq n.\}$

The variational formulation of the price remains almost the same as the one described in Section 4.3.1. However, as in the example of an incomplete market, the effective domain is not bounded; moreover, the supremum is not attained and X does not solve a classical BSDE.

References

Antonelli, F. (1993) 'Backward-forward stochastic differential equations', *Annals of Appl. Prob.* **3**, 777–793.

Bardhan, I.(1993) 'Stochastic income in incomplete markets: Hedging and optimal policies', *Working paper.*

Benes, V.E. (1971) 'Existence of optimal stochastic control law', *SIAM J. Control* **9**, 446–472.

Bergman, Y. (1991) 'Option pricing with divergent borrowing and lending rates', *Working Paper*, Dept of Economics, Brown University.

Bismut, J.M. (1973) 'Conjugate convex functions in optimal stochastic control,' *J. Math. Anal. Appl.* **44**, 384–404.

Bismut, J.M. (1978) 'Contrôle des systèmes linéaires quadratiques : applications de l'intégrale stochastique', *Sémin. Prob. XII*, Lect. Notes in Math. **649**, 180–264, Springer.

Buckdahn, R. (1993) 'Backward stochastic differential equations driven by a martingale', *Preprint*.

Buckdahn, R., & Pardoux, E. (1994) 'BSDEs with jumps and associated integral-stochastic differential equations', *Preprint*.

Cinlar, E., Jacod, J., Protter, P. & Sharpe, M.J. (1980) 'Semimartingale and Markov Processes', *Z. f. W.* **54**, 161–219.

Cvitanic, J. & Karatzas, I. (1992) 'Hedging contingent claims with constrained portfolios', *Annals of Appl. Prob.* **3**, 652–681.

Cvitanic, J. & Ma, J. (1994) 'Hedging options for a large investor and forward-backward SDEs', to appear in *Annals of Appl. Prob.*

Davis, M.H.A. (1973) 'On the existence of optimal policies in stochastic control', *SIAM J. of Control and Optimization* **11**, 587–594.

Delbaen, F. & Schachermayer, W. (1994) 'A general version of the fundamental theorem of asset pricing', *Math. Anal.* **123**.

Duffie, D. (1988) *Security Markets: Stochastic Models*, Academic Press.

Duffie, D. (1992) *Dynamic Asset Pricing Theory*, Princeton University Press.

Duffie, D. & Epstein, L. (1992) 'Stochastic differential utility', *Econometrica* **60**, 353–394.

Duffie, D. & Huang, M. (1994) 'Swap rates and credit quality', *Working paper*, Graduate School of Business, Stanford University.

Duffie, D., Ma, J. & Yong, J. (1994) 'Black's consol rate conjecture', *Working paper*, Graduate School of Business, Stanford University.

Duffie, D., & Singleton, K. (1994) 'Econometric modelling of term structures of defaultable bonds', *Working paper*, Graduate School of Business, Stanford University.

Duffie, D. & Skiadias, C. (1991) 'Continuous-time security pricing: a utility gradient approach', *Working paper*, Graduate School of Business, Stanford University.

Ekeland, I. & Turnbull, T. (1979) *Infinite Dimensional Optimization and Convexity*, Chicago Lectures in Math.

El Karoui, N. (1981) 'Les aspects probabilistes du contrôle stochastique', *Lectures Notes in Math.* **816**, Springer Verlag.

El Karoui, N., Kapoudjian, C., Pardoux, E., Peng, S. & Quenez, M.C. (1995) 'Reflected solutions of backward SDEs, and related obstacle problems for PDEs', *Annals of Prob.*, to appear.

El.Karoui, N., Peng, S. & Quenez, M.C. (1994) 'Optimization of utility functions', *Working Paper*, Paris VI University.

El Karoui, N., Peng, S. & Quenez, M.C. (1997) 'Backward stochastic differential equations in finance', *Mathematical Finance*, to appear.

El Karoui, N. & Quenez, M.C. (1991) 'Dynamic programming and pricing of contingent claims in incomplete market', *Siam J. Control and Optim.* **33**, 1.

Elliott, R.J. (1982) *Stochastic Calculus and Applications*, Springer Verlag.

Epstein, L. & Zin, S. (1989) 'Substitution, risk aversion and the temporal behavior of consumption and asset returns: a theorical framework', *Econometrica* **57**, 837–969.

Fleming, W.F. & Soner, M. (1993) *Controlled Markov Processes and Viscosity Solutions*, Springer Verlag.

Föllmer, H., & Schweitzer, M. (1990) 'Hedging of contingent claims under incomplete information', in *Applied Stochastic Analysis*, M.H.A. Davis and R.J. Elliot (eds.), Gordon and Breach.

Föllmer, H., & Sondermann, D. (1986) 'Hedging of non redundant contingent claims', *Contributions to Mathematical Economics. In Honor of Gérard Debreu*, Hildenbrand and Mas-Colell (eds.), North-Holland.

Geoffard, P.Y. (1995) 'Discounting and optimizing, utility maximization: a Lagrange variational formulation as a minmax problem', *Journal of Economic Theory*.

Hamadene, S. & Lepeltier, J.P. (1993) 'Zero-sum stochastic differential games and backward equations', *Systems and Control Letters* **24**, 259–263.

Harrison, M. & Kreps, D. (1979) 'Martingales and arbitrage in multiperiod securities markets', *Journal of Economic Theory* **20**, 381–408.

Harrison, M. & Pliska, S.P. (1981) 'Martingales and stochastic integrals in the theory of continuous trading', *Stochastic Processes and their Applications* **11**, 215–260.

Harrison, M. & Pliska, S.P. (1983) 'A stochastic calculus model of continuous trading: complete markets', *Stochastic Processes and their Applications* **15**, 313–316.

He H. & Pearson, N.D. (1991) 'Consumption and portfolio policies with incomplete markets and short-sale constraints: the infinite dimensional case', *Journal of Economic Theory* **54**, 259–304.

Jouini, E. & Kallal, H. (1992) 'Arbitrage and equilibrium in securities markets with shortsale constraints', *Working Paper*, Univ. Paris I.

Karatzas,I. , & Shreve, S.E. (1987) *Brownian Motion and Stochastic Calculus*, Springer Verlag.

Karatzas, I. & Shreve, S.E. (1996) *Methods of Mathematical Finance*, to appear.

Korn, R. (1992) 'Option pricing in a model with a higher interest rate for borrowing than for lending', Working paper.

Kramkov, D.O. (1994) 'Optional decomposition of supermartingales and hedging contingent claims in incomplete security markets', *submitted to Probability Theory and Related Fields.*

Krylov, N. (1980) *Controlled Diffusion Processes*, Springer Verlag.

Ma, J., Protter, P. & Yong, J. (1994) 'Solving forward-backward stochastic differential equations explicitly - a four step scheme', *Probability Theory and Related Fields* **98**, 339–359.

Merton, R. (1971) 'Optimum consumption and portfolio rules in a continuous time model', *Journal of Economic Theory* **3**, 373–413.

Pardoux, E., & Peng, S. (1990) 'Adapted solution of a backward stochastic differential equation', *Systems and Control Letters* **14**, 55–61.

Pardoux, E. & Peng, S. (1992) 'Backward stochastic differential equations and quasilinear parabolic partial differential equations', *Lecture Notes in CIS* **176**, 200–217, Springer.

Peng, S. (1991) 'Probabilistic interpretation for systems of quasilinear parabolic partial differential equations', *Stochastics* **37**, 61–74.

Peng, S. (1992a) 'A generalized dynamic programming principle and the Hamilton-Jacobi-Bellman equation', *Stochastics* **38**, 119–134.

Peng, S. (1992b) 'A nonlinear Feynman–Kac formula and applications', *Proceedings of Symposium of System Sciences and Control theory*, Chen & Yong (eds.), 173–184, World Scientific.

Peng, S. (1993) 'Backward stochastic differential equation and its application in optimal control', *Appl. Math. Optim.* **27**, 125–144.

Quenez, M.C. (1993) *Méthodes de Contrôle Stochastique en Finance*, Thèse de doctorat de l'Université Pierre et Marie Curie.

Schweizer, M. (1992) 'Mean-variance hedging for general claims', *Annals of Appl. Prob.* **2**, 171–179.

Svensson, L.E.O. & Werner, J. (1990) 'Portfolio choice and asset pricing with non-traded assets', *Research Paper* **2005**, Graduate School of Business, Stanford University.

Uzawa, H. (1968) 'Time preference, the consumption-function, and optimal asset holdings', in *Value, Capital, and Growth: Papers in Honor of Sir John Hicks*, J.N. Wolfe (ed.), Edinburgh University, Edinburgh.

Reflected Backward SDEs and American Options

N. El Karoui, E. Pardoux and M.C. Quenez

1 Introduction

We have seen in El Karoui & Quenez (1997) that the pricing of European contingent claims, even in imperfect markets, can be formulated in terms of backward stochastic differential equations. However, the case of American options has not be considered. In this article, we will see that the price of an American option corresponds to the solution of a new type of backward equation called reflected BSDEs. The solution of such an equation is forced to stay above a given stochastic process, called the obstacle. An increasing process is introduced which pushes the solution upwards, so that it may remain above the obstacle. Recall that by definition the price of an American option is constrained to be greater than the payoff of the option (which corresponds to the obstacle). Furthermore, in a perfect market, it is well known (see Bensoussan (1984), Karatzas & Shreve (1995) and Karatzas (1988)) that such options cannot be perfectly hedged by a portfolio; in this case, the price process corresponds to the minimal 'superhedging' strategy for the option, that is a strategy with a so-called 'tracking' error which is an increasing process. We will see that this property can be generalized to imperfect markets.

In this article, we will refer often to the previous one by El Karoui & Quenez (1997) concerning the different results on BSDEs and also the notation. The problem is formulated in detail in Section 2. We show that solutions of reflected BSDEs satisfy quite similar properties as in the classical case. First, we prove a comparison theorem for non-reflected BSDEs, from which we deduce the uniqueness of the solution of a reflected BSDE. We give also some a priori estimates on the spread of the solutions of two RBSDEs, which are very similar to the classical case. Then, we show the existence of a solution. The first step is to show the existence of the solution of a reflected equation whose generator is a given process. For classical BSDEs, this case is simply solved by using the representation theorem. In the case of a reflected BSDE, the solution is given by the value function of an optimal stopping time control problem (see Subsection 3.2). The second step is to use the a priori estimates so that a fixed point theorem can be applied. A second method to show the existence is based on the approximation by penalized classical BSDEs. This point of view is very classical in analysis for solving some PDEs.

The different properties concerning reflected BSDEs and optimization problems are studied in Section 3. First, we show that the solution can be associated with a classical deterministic Skohorod problem. From this, it is easy to show that the increasing process of the RBSDE can be expressed as an infimum. Furthermore, we state that the solution of the RBSDE corresponds to the value of an optimal stopping time problem (which depends on the solution itself). Finally, we study the case of reflected BSDE with concave or convex coefficients. Recall that in the case of a classical concave BSDE, the solution corresponds to the value function of an optimization problem (see El Karoui & Quenez (1997)). Similarly, the solution of a concave reflected BSDE is shown to be the value function of a mixed optimal stopping/optimal stochastic control problem. Furthermore, we have a separation theorem concerning the optimal stopping time and the optimal control. In Section 4, we study reflected BSDEs in a Markovian framework. Recall that it has been noticed in Pardoux & Peng (1992) that solutions of BSDEs are naturally connected with viscosity solutions of possibly degenerate parabolic PDEs. In this article, we show that, provided the problem is formulated within a Markovian framework, the solution of the reflected BSDE provides a probabilistic representation for the unique viscosity solution of an obstacle problem for a nonlinear parabolic partial differential equation.

Finally, we give some applications of reflected BSDEs to finance. We show that the classical results concerning the pricing of American options in a perfect market still hold in constrained cases. Recall that under some constraints such as a higher interest rate for borrowing, the strategies weath-portfolio satisfy nonlinear convex stochastic differential equations. The price of a European option corresponds to the solution of a classical convex BSDE (see El Karoui & Quenez 1997). In the case of an American option, using the results of Section 3 on concave or convex reflected BSDEs, we show that the price process is a solution of a convex reflected BSDE; furthermore, it is equal to the so-called 'upper price' defined as the smallest price which allows us to construct a 'superhedging' strategy for the option.

2 Reflected BSDE, Comparison Theorem, Existence and Uniqueness

Put $\mathcal{S}^2 = \left\{\varphi \text{ predictable}; \ \mathbb{E}\left(\sup_{0\le t\le T} |\varphi_t|^2\right) < +\infty\right\}$. The processes belonging to $\mathcal{S}^2$ will be called bounded in $\mathbb{L}^2$.

The problem can be formulated as follows:

We are given three objects:

- a terminal value $\xi \in \mathbb{L}^2$;
- a standard generator f;

- an 'obstacle' $\{S_t,\ 0 \leq t \leq T\}$, which is a continuous adapted real valued process bounded in $\mathbb{L}^2$.
  We shall always assume that $S_T \leq \xi$ a.s.
  Each triple (ξ, f, S) satisfying the above assumptions is called a set of standard data.

Let us now introduce our reflected BSDE. The solution of our RBSDE is a triple $\{(Y_t,\ Z_t,\ K_t),\ 0 \leq t \leq T\}$ of $\mathcal{F}_t$-progressively measurable processes taking values in $\mathbb{R}$, $\mathbb{R}^n$ and $\mathbb{R}_+$ respectively, and satisfying:

(i) $Z \in \mathbb{H}^2$, $Y \in \mathcal{S}^2$ and $K_T \in \mathbb{L}^2$.

(ii) $Y_t = \xi + \int_t^T f(s, Y_s, Z_s)\, ds + K_T - K_t - \int_t^T (Z_s\, dB_s),\ 0 \leq t \leq T$;

(iii) $Y_t \geq S_t,\ 0 \leq t \leq T$;

(iv) $\{K_t\}$ is continuous and increasing, $K_0 = 0$, and $\int_0^T (Y_t - S_t)\, dK_t = 0$.

Note that from (ii) and (iv) it follows that $\{Y_t\}$ is continuous.

Intuitively, $d\,K_t/dt$ represents the amount of 'push upwards' that we add to $-dY_t/dt$, so that the constraint (i) is satisfied. Condition (iv) says that the push is minimal, in the sense that we push only when the constraint is saturated, i.e. when $Y_t = S_t$. We will see that this minimality property can be derived from a comparison theorem.

Indeed, as in the classical case, we have a comparison theorem very similar to the previous one, from which the uniqueness of the solution of the RBSDE follows. The proof of the comparison theorem for RBSDEs differs from the proof for classical BSDEs because the difference of the solutions of two RBSDEs is no longer a solution of a RBSDE. However, we will see that the properties (iv) of the increasing process K allow us to derive the desired result.

Theorem 2.1 *Let (ξ, f, S) and (ξ', f', S') be two sets of standard data, and suppose in addition that*

- $\xi \leq \xi'$ *a.s.;*
- $f(t, y, z) \leq f'(t, y, z)\ dP \times dt$ *a.e.,* $\forall\, (y, z) \in \mathbb{R} \times \mathbb{R}^d$*;*
- $S_t \leq S'_t,\ 0 \leq t \leq T,$ *a.s.*

Let (Y, Z, K) be a solution of the RBSDE with data (ξ, f, S), and (Y', Z', K') a solution of the RBSDE with data (ξ', f', S'). Then

$$Y_t \leq Y'_t,\ 0 \leq t \leq T, a.s.$$

Proof Applying Itô's formula to $|(Y_t - Y'_t)^+|^2$, and taking the expectation, we have:

$$\begin{aligned} E|(Y_t - Y'_t)^+|^2 &+ E\int_t^T \mathbf{1}_{\{Y_s > Y'_s\}} |Z_s - Z'_s|^2 \, ds \\ &\leq 2E\int_t^T (Y_s - Y'_s)^+ [f(s, Y_s, Z_s) - f'(s, Y'_s, Z'_s)] \, ds \\ &\quad + 2E\int_t^T (Y_s - Y'_s)^+ (dK_s - dK'_s). \end{aligned}$$

Since[1] on $\{Y_t > Y'_t\}$, $Y_t > S'_t \geq S_t$,

$$\begin{aligned} \int_t^T (Y_s - Y'_s)^+ (dK_s - dK'_s) &= -\int_t^T (Y_s - Y'_s)^+ \, dK'_s. \\ &\leq 0 \end{aligned}$$

Now, since f is Lipschitz with constant C, it follows that

$$\begin{aligned} &E|(Y_t - Y'_t)^+|^2 + E\int_t^T \mathbf{1}_{\{Y_s > Y'_s\}} |Z_s - Z'_s|^2 \, ds \\ &\leq 2E\int_t^T (Y_s - Y'_s)^+ [f(s, Y_s, Z_s) - f(s, Y'_s, Z'_s)] \, ds \\ &\leq 2CE\int_t^T (Y_s - Y'_s)^+ (|Y_s - Y'_s| + |Z_s - Z'_s|) \, ds \\ &\leq E\int_t^T \mathbf{1}_{\{Y_s > Y'_s\}} |Z_s - Z'_s|^2 \, ds + \overline{C}E\int_t^T |(Y_s - Y'_s)^+|^2 \, ds \end{aligned}$$

(we have used the fact that $2Cyz \leq C^2y^2 + z^2$). Hence

$$E|(Y_t - Y'_t)^+|^2 \leq \overline{C}E\int_t^T |(Y_s - Y'_s)^+|^2 \, ds,$$

and from Gronwall's Lemma, $(Y_t - Y'_t)^+ = 0$, $0 \leq t \leq T$. □

From the proof of the comparison theorem, we derive the minimality property of the increasing process K. More precisely,

Corollary 2.2 *Let (Y, Z, K) be a solution of the RBSDE with standard data (ξ, f, S) and let (Y', Z', K') be a triple of processes satisfying (i), (ii), (iii) (associated with the same data (ξ, f, S)) but with K' a general continuous increasing process with $K'_0 = 0$ but not necessarily $\int_0^T (Y'_t - S_t) \, dK'_t = 0$. Then*

$$Y_t \leq Y'_t, \; 0 \leq t \leq T, \; a.s.$$

Also, we deduce immediately the following uniqueness result from the comparison theorem with $\xi' = \xi$, $f' = f$ and $S' = S$:

[1]Notice that the property $\int_0^T (Y'_t - S'_t) \, dK'_t = 0$ is not used in the proof. Hence, the result still holds if (Y', Z', K') satisfies only (i), (ii), (iii) with K' a general continuous increasing process with $K'_0 = 0$.

Corollary 2.3 *Let (ξ, f, S) be a set of standard data. There exists at most one progressively measurable triple $\{(Y_t,\ Z_t,\ K_t),\ 0 \le t \le T\}$ of the RBSDE associated with data (ξ, f, S).*

It remains to show the existence of a solution of the RBSDE. As in the case of a classical BSDE, we derive some a priori estimates concerning the spread between the solutions of two different RBSDEs very similar to those obtained in the classical case (see the proposition on 'a priori estimates' in El Karoui & Quenez (1997)); then, once we have proved the existence for a generator $f(t)$ which does not depend on y, z, we derive (using the a priori estimates) the existence of the solution of a RBSDE by a fixed point theorem. More precisely,

Theorem 2.4

- *(A priori estimates) Let $((\xi^i, f^i, S^i);\ i = 1, 2)$ be two sets of standard data of an RBSDE with the same obstacle $S^1 = S^2$, and $((Y^i, Z^i, K^i);\ i = 1, 2)$ the solutions of the associated RBSDE.*

  Then, the a priori estimates still hold.

- *(Existence) Let (ξ, f, S) be a set of standard data. Then, the RBSDE defined by (i), (ii), (iii) and (iv) has a unique solution $\{(Y_t, Z_t, K_t)\,;\, 0 \le t \le T\}$.*

Sketch of the Proof The a priori estimates can be derived by using the same arguments as in the classical case; indeed, as in the proof of the comparison theorem, the supplementary terms concerning the increasing processes K^1 and K^2 which appear in the inequalities do not create any difficulty because they are negative (by property (iv)).

Uniqueness has already been shown. Notice that instead of saying that a triple $\{(Y_t, Z_t, K_t)\,;\, 0 \le t \le T\}$ of $\mathbb{R} \times \mathbb{R}^d \times \mathbb{R}_+$–valued progressively measurable processes is a solution of our RBSDE, we could say that a pair $\{(Y_t, Z_t)\,;\, 0 \le t \le T\}$ of $\mathbb{R} \times \mathbb{R}^d$–valued progressively measurable processes satisfying (i) and (iii) is a solution of our RBSDE, meaning that, if $\{K_t,\ 0 \le t \le T\}$ is defined by (ii), then the pair (Y, K) satisfies moreover (iv). In that sense, it follows from Corollary 2.3 that there exists at most one pair $\{(Y_t,\ Z_t)\,;\, 0 \le t \le T\}$ of progressively measurable processes which solves the RBSDE.

It remains now to prove existence. In the case where f does not depend on (y, z), the existence of the solution of the RBSDE will be shown in the next section (see Proposition 3.4) by using a result of the optimal stopping problem.

Suppose now that the generator $f(t, y, z)$ is general. As in the proof of the existence theorem for classical BSDEs, we use a fixed point theorem, for the mapping ϕ defined on $\mathbb{H}^2_\beta \times \mathbb{H}^2_\beta$ which maps (y_t, z_t) into the pair (Y_t, Z_t), where (Y, Z, K) is the solution of the RBSDE associated with the generator $f(t, y_t, z_t)$ (which does not depend on Y, Z). In other words, (Y_t, Z_t, K_t) is the solution of the RBSDE associated with data $(\xi, f(t, y_t, z_t), S)$. It remains to show that for a good choice of β, ϕ is a contraction. Let (y^1, z^1), (y^2, z^2) be two elements of $\mathbb{H}^2_\beta \times \mathbb{H}^2_\beta$ and let (Y^1, Z^1, K^1), (Y^2, Z^2, K^2) be the associated solutions. Then, by the a priori estimates, we find that for β sufficiently large, there exists a positive contant $\epsilon < 1$ such that the following inequality holds

$$\| Y^1 - Y^2 \|^2_\beta + \| Z^1 - Z^2 \|^2_\beta < \epsilon(\| y^1 - y^2 \|^2_\beta + \| z^1 - z^2 \|^2_\beta),$$

for any elements (y^1, z^1), (y^2, z^2) of $\mathbb{H}^2_\beta \times \mathbb{H}^2_\beta$ and their associated solutions (Y^1, Z^1, K^1), (Y^2, Z^2, K^2). Hence, ϕ is a contraction and there exists a fixed point (Y, Z) which is the unique solution of the RBSDE. The associated increasing process K is then defined by condition (ii). □

To show existence and uniqueness, another point of view is to approximate the solution of the RBSDE by classical BSDEs which are obtained by penalizing the solution if it is smaller than the obstacle. These techniques are often used in analysis to solve some PDEs. More precisely,

Proposition 2.5 *For each $n \in \mathbb{N}$, let $\{(Y^n_t, Z^n_t)\,;\, 0 \leq T \leq T\}$ denote the unique pair of square integrable $\mathcal{F}_t$-progressively measurable processes satisfying*

$$Y^n_t = \xi + \int_t^T f(s, Y^n_s, Z^n_s)\, ds + n \int_t^T (Y^n_s - S_s)^-\, ds - \int_t^T (Z^n_s, dB_s). \quad (2.1)$$

Then, the increasing sequence Y^n_t converges a.s. and in $\mathcal{S}^2$ to Y_t, the sequence Z^n converges to Z in $\mathbb{H}^2$ and K^n converges to K in $\mathcal{S}^2$ as n tends to infinity, where K^n is defined by

$$K^n_t = n \int_0^t (Y^n_s - S_s)^-\, ds,\ 0 \leq t \leq T.$$

The proof is based on the estimations K^n, and the difficulties come from the fact that the almost sure convergence of the increasing sequence Y^n_t is not sufficient and it is necessary to prove the uniform convergence of the different processes. For details, one is referred to El Karoui, Kapoudjian, Pardoux, Peng & Quenez (1995).

3 Reflected BSDEs, the Skohorod Problem and Optimal Stopping Time/Control Problems

In this section, we study the different control problems which are associated with a RBSDE. We first show that the solution can be associated with a classical deterministic Skohorod problem. From this, it is easy to deduce that the increasing process can be written as a supremum.

3.1 RBSDEs and the Skohorod problem

Notice that in a deterministic framework, the formulation of the RBSDE corresponds to the Skohorod problem. Consequently, we will be able to apply some well known properties of the Skohorod problem. Recall the Skohorod lemma (see for example El Karoui & Chayelat-Maurel (1978) and Revuz & Yor (1994) p. 229).

Proposition 3.1 *(Skohorod lemma)*
Let x be a real-valued continuous function on $[0,\infty[$ such that $x_0 \geq 0$. There exists a unique pair (y,k) of functions on $[0,\infty[$ such that

(a) $y = x + k$

(b) y *is positive,*

(c) $\{k_t\}$ *is continuous and increasing,* $k_0 = 0$, *and* $\int_0^\infty y_t \, dk_t = 0$.

The pair (y,k) is said to be the solution of the Skohorod problem. The function k is moreover given by

$$k_t = \sup_{s \leq t} x_s^- \,.$$

Now, our problem involves a Skohorod problem and consequently, the increasing process can be written as a supremum. More precisely,

Proposition 3.2 *Let $\{(Y_t, Z_t, K_t),\ 0 \leq t \leq T\}$ be a solution of the above RBSDE satisfying conditions (ii) to (iv). Then for each $t \in [0,T]$,*

$$K_T - K_t = \sup_{t \leq u \leq T} \left(\xi + \int_u^T f(s, Y_s, Z_s)\, ds - \int_u^T (Z_s \, dB_s) - S_u \right)^- . \quad (3.1)$$

Proof Notice that $(Y_{T-t}(\omega) - S_{T-t}(\omega), K_{T-t}(\omega) - K_T(\omega), 0 \leq t \leq T)$ is the solution of a Skohorod problem. Applying the Skohorod lemma with

$$\begin{aligned} x_t &= \left(\xi + \int_{T-t}^T f(s, Y_s, Z_s)\, ds - \int_{T-t}^T (Z_s \, , dB_s) - S_{T-t} \right)(\omega), \\ k_t &= (K_T - K_{T-t})(\omega) \\ y_t &= (Y_{T-t} - S_{T-t})(\omega), \end{aligned}$$

we derive the desired result. □

It is not at all clear from (3.1) that $\{K_t\}$ will be $\mathcal{F}_t$–adapted. The adaptedness of (Y, K) will come from the adjustment of the process Z. In orther words, Z is the process which has the effect of making (Y, K) adapted.

In the next section we show that the solution of the RBSDE can be associated with a stopping time optimal problem.

3.2 RBSDEs and stopping time optimal problem

In the following proposition, we show that the square-integrable solution Y_t of the RBSDE corresponds to the value of a stopping time optimal problem.

Proposition 3.3 *Let $\{(Y_t, Z_t, K_t),\ 0 \le t \le T\}$ be a solution of the above RBSDE satisfying conditions (i) to (iv); then for each $t \in [0,T]$,*

$$Y_t = \operatorname*{ess\,sup}_{v \in \mathcal{T}_t} E\left[\int_t^v f(s, Y_s, Z_s)\, ds + S_v \mathbf{1}_{\{v<T\}} + \xi \mathbf{1}_{\{v=T\}} | \mathcal{F}_t\right] \tag{3.2}$$

where $\mathcal{T}$ is the set of all stopping times dominated by T, and

$$\mathcal{T}_t = \{v \in \mathcal{T}\,;\, t \le v \le T\}.$$

Proof Let $v \in \mathcal{T}_t$. From (i) and (iv), we may take the conditional expectation in (ii) written between times t and v, hence

$$\begin{aligned} Y_t &= E\left[\int_t^v f(s, Y_s, Z_s)\, ds + Y_v + K_v - K_t | \mathcal{F}_t\right] \\ &\ge E\left[\int_t^v f(s, Y_s, Z_s)\, ds + S_v \mathbf{1}_{\{v<T\}} + \xi \mathbf{1}_{\{v=T\}} | \mathcal{F}_t\right]. \end{aligned}$$

We now choose an optimal element of $\mathcal{T}_t$ in order to get the reversed inequality. Let

$$D_t = \inf\{t \le u \le T\,;\, Y_u = S_u\},$$

with the convention that $D_t = T$ if $Y_u > S_u$, $t \le u \le T$. Now the condition $\int_0^T (Y_t - S_t)\, dK_t = 0$, and the continuity of K implies that

$$K_{D_t} - K_t = 0.$$

It follows that

$$Y_t = E\left[\int_t^{D_t} f(s, Y_s, Z_s)\, ds + S_{D_t} \mathbf{1}_{\{D_t<T\}} + \xi \mathbf{1}_{\{D_t=T\}} | \mathcal{F}_t\right].$$

Hence the result follows. □

It is clear from the Proposition 3.3 above that in the case where f is a given stochastic process, the solution $\{Y_t\,;\, 0 \le t \le T\}$ of the RBSDE is the value function of an optimal stopping time problem (which does not depend on (Y, Z)). From this fact, this value function seems to be a natural candidate to be the solution of the RBSDE in this case. We will show in the next section that it is indeed the solution.

3.3 Existence in the case of a given generator

By the results of optimal stopping theory, we derive the following existence result:

Proposition 3.4 *Suppose that $f(t)$ is a given predictable stochastic process in $\mathbb{H}^2$. Then, the RBSDE associated with (ξ, f, S) has a unique solution.*

Proof From Proposition 3.3, it is natural to introduce the process $\{Y_t;\, 0 \leq t \leq T\}$ defined by

$$Y_t = \operatorname*{ess\,sup}_{v \in \mathcal{T}_t} E\left[\int_t^v f(s)\, ds + S_v \mathbf{1}_{\{v<T\}} + \xi \mathbf{1}_{\{v=T\}} | \mathcal{F}_t\right],\ 0 \leq t \leq T.$$

It is clear that $Y_t \geq S_t$. Also, optimal stopping theory (cf. El Karoui 1981, Karatzas & Shreve 1995) shows that there exists a continuous version of Y_t which can be written as the solution of the RBSDE (see El Karoui, Kapoudjian, Pardoux, Peng & Quenez 1995). □

In the next section, we study the case of an RBSDE associated with a linear generator and show that, as in the case of a generator defined as a given process, the solution $\{Y_t\,;\, 0 \leq t \leq T\}$ of the RBSDE is the value function of an optimal stopping time problem (which does not depend on the solution (Y, Z)).

3.4 Linear RBSDE

In the linear case, that is when $f(t, y, z)$ is a linear function of (y, z), Proposition 3.3 shows that, as for a classical linear BSDE, the solution can be written using the adjoint process.

Proposition 3.5 *Let (β_t, γ_t) be a bounded $(\mathbb{R}, \mathbb{R}^n)$-valued predictable vector process, and φ_t be an element of $\mathbb{H}^2_T(\mathbb{R})$. Let f be the standard generator defined by*

$$f(t, y, z) = \varphi_t + \beta_t y + \gamma_t^* z.$$

Let $(\Gamma_{t,s}; t \leq s \leq T)$ be the adjoint process satisfying the linear SDE:

$$d\Gamma_{t,s} = \Gamma_{t,s}(\beta_s\, ds + \gamma_s^*\, dB_s)\ ; \quad \Gamma_{t,t} = 1.$$

Then the unique solution $\{Y_t,\, Z_t\,, K_t\,;\, 0 \leq t \leq T\}$ of the RBSDE with generator f satisfies, for each $0 \leq t \leq T$,

$$Y_t = \operatorname*{ess\,sup}_{v \in \mathcal{T}_t} E\left[\Gamma_{t,v}\xi \mathbf{1}_{\{v=T\}} + \Gamma_{t,v} S_v \mathbf{1}_{\{v<T\}} + \int_t^v \Gamma_{t,s}\varphi_s\, ds / \mathcal{F}_t\right],$$

which can also be written

$$Y_t = \operatorname*{ess\,sup}_{v \in \mathcal{T}_t} X_t(v, \tilde{S}_v)$$

where $(X_t(v, \tilde{S}_v), 0 \leq t \leq v)$ *is the solution associated with the BSDE associated with generator* f*, terminal time* v *and terminal condition* $\tilde{S}_v$ *where*

$$\tilde{S}_t = \xi \mathbf{1}_{\{t=T\}} + S_t \mathbf{1}_{\{t<T\}}.$$

Furthermore, the stopping time $D_t = \inf\{t \leq s \leq T\,;\, Y_s = S_s\}$ *is optimal.*

Proof Let us denote the process $\Gamma_{0,s}$ by Γ_s. It follows from Itô's formula that

$$Y_t\Gamma_t = \xi\Gamma_T + \int_t^T \Gamma_s\varphi_s\,ds + \int_t^T \Gamma_s\,dK_s - \int_t^T \Gamma_s(Z_s + Y_s\gamma_s)^*\,dB_s.$$

Let $(Y'_t,\, Z'_t,\, K'_t) = (Y_t\Gamma_t,\, \Gamma_t(Z_t + Y_t\gamma_t),\, \int_0^t \Gamma_s\,dK_s)$, $0 \leq t \leq T$. This triplet solves the RBSDE with final condition $\xi\Gamma_T$, coefficient $\{\Gamma_t\varphi_t,\, 0 \leq t \leq T\}$ and obstacle $\Gamma_t S_t$, without Condition (i) in Section 2.

Also we only have that

$$E[(\xi\Gamma_T)^{2-\varepsilon} + \int_0^T (\varphi_t\Gamma_t)^{2-\varepsilon}\,dt] < \infty$$

for each $\varepsilon > 0$, and not for $\varepsilon = 0$; the argument leading to Proposition 3.3 is still valid here, hence

$$\Gamma_t Y_t = \operatorname*{ess\,sup}_{v\in\mathcal{T}_t} E\left[\Gamma_T\xi\mathbf{1}_{\{v=T\}} + \Gamma_v S_v \mathbf{1}_{\{v<T\}} + \int_t^v \Gamma_{t,s}\varphi_s\,ds/\mathcal{F}_t\right],$$

and D_t is optimal. Consider now the solution $(X_t(v, \tilde{S}_v), 0 \leq t \leq v)$ as the solution associated with the BSDE associated with generator f, terminal time v and terminal condition $\tilde{S}_v = \xi\mathbf{1}_{\{v=T\}} + S_v\mathbf{1}_{\{v<T\}}$. Then, by classical results on linear BSDEs, it follows that

$$X_t(v, \tilde{S}_v) = E\left[\Gamma_{t,v}\tilde{S}_v + \int_t^v \Gamma_{t,s}\varphi_s\,ds/\mathcal{F}_t\right],$$

and hence, Y_t can also be written

$$Y_t = \operatorname*{ess\,sup}_{v\in\mathcal{T}_t} X_t(v, \tilde{S}_v).$$

Furthermore, the fact that D_t is optimal follows from the argument given in the proof of Proposition 3.3. □

In the next section, we will see that in the case where f is a concave (or convex) function of (y, z), we have that $\{Y_t\,,\, 0 \leq t \leq T\}$ is the value function of a mixture of an optimal stopping time problem and a 'classical' optimal stochastic control problem.

3.5 Concave RBSDEs and the optimal stopping time/control problem

We now suppose that $f(t, y, z)$ is a concave function of (y, z). We use the same notation as El Karoui & Quenez (1997, Section 3) on concave BSDEs. Recall that the conjugate function of f is denoted by $F(t, \beta, \gamma)$ and that $\mathcal{A}$ denotes the set of bounded progressively measurable $\mathbb{R} \times \mathbb{R}^d$–valued processes $\{(\beta_t, \gamma_t) \,;\, 0 \leq t \leq T\}$ which are such that

$$E \int_0^T F(t, \beta_t,\, \gamma_t)^2 \, dt < \infty.$$

To each $(\beta, \gamma) \in \mathcal{A}$, we associate the unique solution $\{(Y_t^{\beta,\gamma},\, Z_t^{\beta,\gamma},\, K_t^{\beta,\gamma}) \,;\, 0 \leq t \leq T\}$ of the linear RBSDE with generator $f^{\beta,\gamma}(t, y, z) = F(t, \beta_t, \gamma_t) + \beta_t y + \gamma_t^* \, z$. We shall denote by $\{(Y_t, Z_t, K_t) \,;\, 0 \leq t \leq T\}$ the unique solution of the RBSDE with generator $f(t, y, z)$.

Recall that by the conjugacy relation,

$$f(t, Y_t, Z_t) = \operatorname*{ess\,inf}_{(\beta,\gamma)\in\mathcal{A}} f^{\beta,\gamma}(t, Y_t, Z_t) \quad dt \times dP \text{ a.e.}$$

and hence, by the comparison theorem (for RBSDEs) 2.1,

$$Y_t \leq Y_t^{\beta,\gamma}, \; \forall\, (\beta, \gamma) \in \mathcal{A}.$$

Furthermore, as in the case of a classical concave BSDE (see El Karoui & Quenez 1997), by a selection measurable theorem, there exists $(\overline{\beta},\, \overline{\gamma}) \in \mathcal{A}$ such that

$$f(t, Y_t, Z_t) = f^{\overline{\beta},\overline{\gamma}}(t, Y_t, Z_t) \quad dt \times dP \text{ a.e.}, \tag{3.3}$$

hence,

$$(Y_t, Z_t, K_t) = (Y_t^{\overline{\beta},\overline{\gamma}}, Z_t^{\overline{\beta},\overline{\gamma}},\, K_t^{\overline{\beta},\overline{\gamma}})\, , 0 \leq t \leq T, \quad a.s.$$

It follows that

$$Y_t = Y_t^{\overline{\beta},\overline{\gamma}} \geq \inf_{(\beta,\gamma)\in\mathcal{A}} Y_t^{\beta,\gamma},$$

Consequently, as in the case of a concave classical BSDE, we have

Proposition 3.6 *For each t, Y_t can be written as the value function of an optimal problem given by*

$$Y_t = \operatorname*{ess\,inf}_{(\beta,\gamma)\in\mathcal{A}} Y_t^{\beta,\gamma}.$$

Furthermore, the control $(\overline{\beta}, \overline{\gamma})$ is optimal, that is

$$Y_t = Y_t^{\overline{\beta},\overline{\gamma}}.$$

Now, by the previous results on linear RBSDEs (see Proposition 3.5), we know that for each $(\beta, \gamma) \in \mathcal{A}$, the process $Y_t^{\beta,\gamma}$ can be written

$$Y_t^{\beta,\gamma} = \operatorname*{ess\,sup}_{v \in \mathcal{T}_t} X_t^{\beta,\gamma}(v, \tilde{S}_v) \tag{3.4}$$

where $X_t^{\beta,\gamma}(v, \tilde{S}_v)$ is the solution of the BSDE associated with generator $f^{\beta,\gamma}$, terminal time v and terminal condition $\tilde{S}_v$. Recall that

$$X_t^{\beta,\gamma}(v, \tilde{S}_v) = E[\Gamma_{t,v}^{\beta,\gamma} \tilde{S}_v + \int_t^v \Gamma_{t,s}^{\beta,\gamma} F(s, \beta_s, \gamma_s)\, ds / \mathcal{F}_t].$$

By Proposition 3.6 and equation (3.4), we deduce that Y_t can be written as the value function of an optimal-stopping time/control problem, that is

Theorem 3.7 *For each t, Y_t is the value function of a minimax problem, that is*

$$Y_t = \operatorname*{ess\,inf}_{(\beta,\gamma) \in \mathcal{A}} \operatorname*{ess\,sup}_{v \in \mathcal{T}_t} X_t^{\beta,\gamma}(v, \tilde{S}_v).$$

Moreover, ess inf *and* ess sup *can be interchanged, that is Y_t is also the value function of a maxmin problem given by*

$$Y_t = \operatorname*{ess\,sup}_{v \in \mathcal{T}_t} \operatorname*{ess\,inf}_{(\beta,\gamma) \in \mathcal{A}} X_t^{\beta,\gamma}(v, \tilde{S}_v).$$

and the triple $(\overline{\beta}, \overline{\gamma}, D_t)$, where $D_t = \inf\{t \le s \le T\,;\, Y_s = S_s\}$, is optimal.

Remark Notice that the optimal control $(\overline{\beta}, \overline{\gamma})$ does not depend on D_t and the definition of D_t does not depend on $(\overline{\beta}, \overline{\gamma})$. The optimal control $(\overline{\beta}, \overline{\gamma}, D_t)$ satisfies a so-called 'separation' principle.

Proof The first equality follows directly from Proposition 3.6 and equation (3.4). We have to prove ess inf and ess sup can be interchanged. We certainly have

$$\begin{aligned} Y_t &= \operatorname*{ess\,inf}_{(\beta,\gamma) \in \mathcal{A}} \operatorname*{ess\,sup}_{v \in \mathcal{T}_t} X_t^{\beta,\gamma}(v, \tilde{S}_v) \\ &\ge \operatorname*{ess\,sup}_{v \in \mathcal{T}_t} \operatorname*{ess\,inf}_{(\beta,\gamma) \in \mathcal{A}} X_t^{\beta,\gamma}(v, \tilde{S}_v). \end{aligned}$$

On the other hand, by Proposition 3.5, the stopping time $D_t^{\beta,\gamma}$, given by $D_t^{\beta,\gamma} = \inf\{t \le s \le T\,;\, Y_s^{\beta,\gamma} = S_s\}$, is optimal for the optimization problem

$$\sup_{v \in \mathcal{T}_t} X_t^{\beta,\gamma}(v, \tilde{S}_v),$$

and consequently,

$$\begin{aligned} Y_t &= \operatorname*{ess\,inf}_{(\beta,\gamma) \in \mathcal{A}} X_t^{\beta,\gamma}(D_t^{\beta,\gamma}, \tilde{S}_{D_t^{\beta,\gamma}}) \\ &\le \operatorname*{ess\,sup}_{v \in \mathcal{T}_t} \operatorname*{ess\,inf}_{(\beta,\gamma) \in \mathcal{A}} X_t^{\beta,\gamma}(v, \tilde{S}_v). \end{aligned}$$

The desired result follows.

Furthermore, by noticing that $D_t = D_t^{\overline{\beta},\overline{\gamma}}$, it follows that

$$Y_t = Y_t^{\overline{\beta},\overline{\gamma}} = X_t^{\overline{\beta},\overline{\gamma}}(D_t, \tilde{S}_{D_t})$$

and hence, the triple $(\overline{\beta}, \overline{\gamma}, D_t)$ is optimal for the problem

$$Y_t = \operatorname*{ess\,inf}_{(\beta,\gamma)\in\mathcal{A}} \operatorname*{ess\,sup}_{v\in\mathcal{T}_t} X_t^{\beta,\gamma}(v, \tilde{S}_v). \qquad \square$$

Note that one has a similar representation of Y_t in the case where f is a convex function of (y, z), with $\operatorname{ess\,inf}_{(\beta,\gamma)} \operatorname{ess\,sup}_v[\cdot]$ replaced by $\operatorname{ess\,sup}_{(\beta,\gamma)} \operatorname{ess\,sup}_v[\cdot]$.

4 Relation between a RBSDE and an Obstacle Problem for a Nonlinear Parabolic PDE

In this section, we will show that the reflected BSDE studied in the previous sections allows us to give a probabilistic representation of solutions of some obstacle problems for PDEs. For that purpose, we will put the RBSDE in a Markovian framework. The framework is the same as in the case of Markovian BSDEs and the assumptions on the state process and the coefficients b, σ, f and g are the same. Furthermore, the obstacle S_t is supposed to satisfy

$$S_s = h(s, S_s^{t,x})$$

where $h : [0, T] \times \mathbb{R}^d \to \mathbb{R}$ satisfies

$$h(t, x) \leq K(1 + |x|^p), \; t \in [0, T], \; x \in \mathbb{R}^d. \tag{4.1}$$

We assume moreover that $h(T, x) \leq g(x)$, $x \in \mathbb{R}^d$.

We will denote by $(Y_s^{t,x}, Z_s^{t,x}, K_s^{t,x}\, t \leq s \leq T)$ the solution of the RBSDE associated with obstacle S_s, generator $f(s, S_s^{t,x}, y, z)$ and terminal condition $g(S_T^{t,x})$. As in the case of a classical BSDE, we know that $Y_t^{t,x}$ is a deterministic function of (t, x) denoted by $u(t, x)$. We have the following theorem:

Theorem 4.1 *Suppose that the coefficients f, b, σ, g, h are jointly continuous in t and x. Then, the function $u(t, x)$ is the unique viscosity solution of the obstacle problem:*

$$\begin{cases} \min\Big(u(t,x) - h(t,x), \, -\dfrac{\partial u}{\partial t}(t,x) - L_t u(t,x) - f(t,x,u(t,x), \\ \qquad\qquad (\nabla u \sigma)(t,x))\Big) = 0, \quad (t,x) \in (0,T) \times \mathbb{R}^d \\ u(T,x) = g(x), \; x \in \mathbb{R}^d; \end{cases} \tag{4.2}$$

where L_t is the infinitesimal operator associated with the Itô process $S^{t,x}$.

Sketch of the proof We are going to use the approximation of the RBSDE in Proposition 2.5 by penalization.

For each $(t,x) \in [0,T] \times \mathbb{R}^d$, $n \in \mathbb{N}^*$, let $\{({}^nY_s^{t,x}, {}^nZ_s^{t,x}),\ t \le s \le T\}$ denote the solution of the BSDE:

$$\begin{aligned} {}^nY_s^{t,x} &= g(S_T^{t,x}) + \int_s^T f(r, S_r^{t,x}, {}^nY_r^{t,x}, {}^nZ_r^{t,x})\, dr \\ &\quad + n\int_s^T ({}^nY_r^{t,x} - h(r, S_r^{t,x}))^- - \int_s^T {}^nZ_r^{t,x}\, dB_r,\ t \le s \le T. \end{aligned}$$

By the previous results on classical BSDEs in the Markovian case, we know that

$$u_n(t,x) = {}^nY_t^{t,x},\ 0 \le t \le T,\ x \in \mathbb{R}^d,$$

is the viscosity solution of the parabolic PDE:

$$\begin{cases} \dfrac{\partial u_n}{\partial t}(t,x) + L_t u_n(t,x) + f_n(t,x,u_n(t,x), (\nabla u_n \sigma)(t,x)) = 0,\ 0 \le t \le T; \\ u(T,x) = g(x); \end{cases}$$

where $f_n(t,x,r,p\sigma(t,x)) = f(t,x,r,p\sigma(t,x)) + n(r - h(t,x))^-$, and $x \in \mathbb{R}^d$.

But from the results on the approximation of the RBSDE (Proposition 2.5) by penalization, for each $0 \le t \le T$, $x \in \mathbb{R}^d$,

$$u_n(t,x) \uparrow u(t,x) \text{ as } n \to \infty.$$

Since u_n and u are continuous, it follows from Dini's theorem that the above convergence is uniform on compact sets. Using this property and classical techniques of the theory of viscosity solutions, it is possible to show that $u(t,x)$ is a viscosity solution of the obstacle problem. Concerning the proof of the uniqueness of the viscosity solution of the obstacle problem, one is referred to El Karoui, Kapoudjian, Pardoux, Peng & Quenez (1995). □

In the next section, we are concerned with the application of RBSDEs to the pricing of American options.

5 Application of RBSDEs to Finance

We have seen in the study of classical concave and convex BSDEs and the 'application to European option pricing in the constrained case' (see El Karoui & Quenez 1997, Section 3) that in some constraint cases, the strategy wealth-portfolio (X_t, π_t) is a pair of adapted processes in $\mathbb{H}^2(\mathbb{R}) \times \mathbb{H}^2(\mathbb{R}^n)$ which satisfies the following BSDE

$$-\,dX_t = b(t, X_t, \pi_t)dt - \pi_t^* \sigma_t dW_t\,. \tag{5.1}$$

Here, b is an $\mathbb{R}$-valued function defined on $[0,T] \times \Omega \times \mathbb{R} \times \mathbb{R}^n$ satisfying the standard hypotheses of a generator and convex with respect to x, π. Furthermore, we will suppose that the volatility matrix σ_t of the n risky assets is invertible and that $(\sigma_t)^{-1}$ is uniformly bounded. The classical case is a special case corresponding to a linear functional

$$b(t, x, \pi) = -r_t x - \pi^* \sigma_t \theta_t ,$$

where θ is the risk premium vector. We can suppose, without loss of generality, that $\sigma_t = \mathrm{Id}$.

We are concerned with the problem of pricing an American contingent claim at each time t which consists of the selection of a stopping time $v \in \mathcal{T}_t$ and a payoff S_v on exercise if $v < T$ and ξ if $v = T$. Here, $(S_t, 0 \leq t \leq T)$ satisfies the assumptions of the 'obstacle'. Put

$$\tilde{S}_s = \xi \mathbf{1}_{\{s=T\}} + S_s \mathbf{1}_{\{v<T\}} \; 0 \leq s \leq T.$$

The pricing method for pricing the American contingent claim is based on the same arguments as in the non constrained case (see Karatzas (1989) or Karatzas & Shreve (1995)). Fix $t \in [0, T]$. Suppose for the moment that the choice of $v \in \mathcal{T}_t$ has been made; then, there exists a unique strategy $(X_s(v, \tilde{S}_v), \pi_s(v, \tilde{S}_v)) \in \mathbb{H}^2 \times \mathbb{H}^2$, denoted also by (X_s^v, π_s^v), which replicates $\tilde{S}_v$, i.e. is the solution of the classical BSDE associated with terminal time v, terminal condition $\tilde{S}_v$ and generator b:

$$-dX_s^v = b(s, X_s^v, \pi_s^v) ds - (\pi_s^v)^* dW_s, \; 0 \leq s \leq v \qquad X_v^v = \tilde{S}_v . \tag{5.2}$$

Then, the price of the American contingent claim $(\tilde{S}_s, 0 \leq s \leq T)$ at time t is given by the right-continuous adapted process X satisfying at each time t

$$X_t = \operatorname*{ess\,sup}_{v \in \mathcal{T}_t} X_t(v, \tilde{S}_v).$$

Recall that b is convex and let $B(t, \beta, \gamma)$ be the polar process associated with b. By the results of Section 3 on concave or convex classical BSDEs (see El Karoui & Quenez 1997), we know that X_t^v can be written as the maximum of ex-post prices over all feasible deflators

$$X_t(v, \tilde{S}_v) = \operatorname{ess\,sup}\{X_t^{\beta,\gamma}(v, \tilde{S}_v) \;\mid\; (\beta, \gamma) \in \mathcal{A}\} ,$$

where $\mathcal{A}$ is the set of (β, γ) *feasible control parameters*, defined by

$$\mathcal{A} = \{(\beta, \gamma) \in \mathcal{P} \mid \mathbb{E} \int_0^T B(t, \beta_t, \gamma_t)^2 \, dt < +\infty \}$$

and where the ex-post price $(X^{\beta,\gamma}(v, \tilde{S}_v), \pi^{\beta,\gamma}(v, \tilde{S}_v))$ is, for each control process $(\beta, \gamma) \in \mathcal{A}$, the unique solution of the LBSDE associated with terminal time v, terminal condition $\tilde{S}_v$ and a linear generator; more precisely,

$$\begin{cases} -dX_t^{\beta,\gamma} &= (B(t, \beta_t, \gamma_t) - \beta_t . X_t^{\beta,\gamma} - \gamma_t . \pi_t^{\beta,\gamma}) \, dt - (\pi_t^{\beta,\gamma})^* dW_t \\ X_v^{\beta,\gamma} &= \tilde{S}_v . \end{cases} \tag{5.3}$$

Consequently, X_t is the value of the optimization problem

$$X_t = \operatorname*{ess\,sup}_{v \in \mathcal{T}_t} \operatorname*{ess\,inf}_{(\beta,\gamma) \in \mathcal{A}} X_t^{\beta,\gamma}(v, \tilde{S}_v).$$

Then, applying the previous results on RBSDEs, it follows that the price $(X_t, 0 \leq t \leq T)$ corresponds to the solution of the RBSDE associated with terminal condition ξ, generator b and obstacle S, that is

Proposition 5.1 *There exists* $(\pi_t) \in \mathbb{H}^2$ *and* (K_t) *an increasing adapted continuous process with* $K_0 = 0$ *such that*

$$-\,dX_t = b(t, X_t, \pi_t)dt + dK_t - \pi_t^* dW_t \quad ; \quad X_T = \xi, \tag{5.4}$$

with $X_t \geq S_t$, $0 \leq t \leq T$, *and* $\int_0^T (X_t - S_t)\, dK_t = 0$.

Furthermore, the stopping time $D_t = \inf\{t \leq s \leq T\,;\, X_s = S_s\}$ *is optimal that is*

$$X_t = \operatorname*{ess\,sup}_{v \in \mathcal{T}_t} X_t(v, \tilde{S}_v) = X_t(D_t, \tilde{S}_{D_t}).$$

The process K_t may be interpreted as a cumulative consumption process. Such a triple (Y_t, π_t, K_t) satisfying equation (5.4) with $Y_t \geq S_t$, $0 \leq t \leq T$ (but not necessarily $\int_0^T (Y_t - S_t)\, dK_t = 0$) is called a superhedging strategy for the American option (S, ξ). Consequently, by the corollary of the comparison theorem 2.2 for RBSDEs, the price X_t is equal to the so-called 'upper price' defined as the smallest of the superhedging strategies for (S, ξ).

References

Barles, G. (1994) *Solutions de viscosité des équations de Hamilton–Jacobi du premier ordre et Applications*, Math. Appl **17**, Springer.

Barles, G. & Burdeau, J. (1994) 'The Dirichlet problem for semilinear second order degenerate elliptic equations and applications to stochastic exit time problems', preprint.

Bensoussan, A. (1984) 'On the theory of option pricing', *Acta Applicandae Mathematicae* **2**, 139–158.

Bensoussan, A. & Lions, J.L. (1978) *Applications des Inéquations Varitionnelles en Contrôle Stochastique*, Dunod.

Crandall, M., Ishii, H. & Lions, P.L. (1992) 'User's guide to the viscosity solutions of second order partial differential equations', *Bull. A.M.S.* **27**, 1–67.

Darling, R. (1994) 'Constructing gamma–martingales with prescribed limits using backward SDEs', preprint.

Davis, M. & Karatzas, I. (1994) 'A deterministic approach to optimal stopping,' in *Probability, Statistics and Optimization*, F.P. Kelly (ed.), 455–466, Wiley.

Duffie, D. & Epstein, L. (1992) 'Stochastic differential utility', *Econometrica* **60**, 353–394.

El Karoui, N. (1981) 'Les aspects probabilistes du contrôle stochastique', in *Ecole d'Eté de Saint Flour*, LNM **876**, Springer.

El Karoui, N. & Chaleyat–Maurel, M. (1978) 'Un problème de réflexion et ses applications au temps local et aux EDS sur $\mathbb{R}$, cas continu', in *Temps locaux*, Astérisque **2–53**, SMF.

El Karoui, N. & Jeanblanc–Picqué, M. (1993) 'Optimization of consumption with labor income', preprint.

El Karoui, N., Peng, S. & Quenez, M.C. (1994) 'Backward stochastic differential equations in finance', Preprint Université Paris VI, **260**.

El Karoui, N. & Quenez, M.C. (1997) 'Imperfect markets and backward stochastic differential equations', this volume.

El Karoui, N., Kapoudjian, C., Pardoux, E., Peng, S. & Quenez, M.C. (1995) 'Reflected solutions of backward SDEs, and related obstacle problems for PDEs', preprint.

Hamadene, S. & Lepeltier, J.P. (1995) 'Zero-sum stochastic differential games and backward equations,' *Systems & Control Letters* **24**, 259–263.

Jacka, S. (1993) 'Local times, optimal stopping and semimartingales', *Annals of Probability* **21**, 329–339.

Karatzas, I. (1988) 'On the pricing of American options', *Appl. Math. Optimization* **17**, 37–6.

Karatzas, I. (1989) 'Optimization problems in the theory of continuous trading', *SIAM. J. Control. Optim.* **27**, 1221-1259.

Karatzas, I. & Shreve, S. (1996) *Methods of Mathematical Finance*, book in preparation.

Pardoux, E. & Peng, S. (1990) 'Adapted solution of a backward stochastic differential equation', *Systems & Control Letters* **14**, 55-61.

Pardoux, E. & Peng, S. (1992) 'Backward SDEs and quasilinear PDEs', in *Stochastic Partial Differential Equations and their Applications*, B.L. Rozovskii, R.B. Sowers (eds.), LNCIS **176**, Springer.

Pardoux, E. & Peng, S. (1995) 'Some backward SDEs with non Lipschitz coefficients', Proc. Conf. Metz, to appear.

Revuz, D. & Yor, M. (1994) *Continuous Martingales and Brownian Motion*, Springer.

Numerical Methods for Backward Stochastic Differential Equations

D. Chevance

1 Introduction

Backward stochastic differential equations (BSDEs) were introduced by Pardoux & Peng (1990) to give a probabilistic representation for the solutions of certain nonlinear partial differential equations, thus generalizing the Feynman-Kac formula.

This sort of equation has also found many applications in finance, notably in contingent claim valuation when there are constraints on the hedging portfolios (see El Karoui & Quenez 1995, El Karoui, Peng & Quenez 1994, Cvitanic & Karatzas 1992) and in the definition of stochastic differential utility (see Duffie & Epstein 1992, El Karoui, Peng & Quenez 1994). A financial application of forward-backward SDEs can be found in Duffie, Ma & Yong (1993).

However little research has yet been performed on numerical methods for BSDEs. Here we give a review of three different contributions in that field.

In Section 2, we present a random time discretization scheme introduced by V. Bally to approximate BSDEs. The advantage of Bally's scheme is that one can get a convergence result with virtually no other regularity assumption than the ones needed for the existence of a solution to the equation. However that scheme is not fully numerical and its actual implementation would require further approximations.

In Section 3, we give an account of a four step algorithm developed by J. Ma and P. Protter to solve a class of more general equations called forward-backward SDEs. It is based on solving the associated PDE by a deterministic-type method and also makes use of the Euler scheme for stochastic differential equations. The convergence rate turns out to be as good as for the solution of a simple SDE.

Finally, in Section 4 we describe a numerical method examined in Chevance (1996) for solving a BSDE associated with a forward SDE. It involves the discretization of the BSDE both in time and in space. This leads to an algorithm which can be implemented in practice, but there one needs much stronger regularity assumptions than in Bally's scheme.

2 A Random Time Discretization Scheme

2.1 Hypotheses

For $T > 0$ fixed, let $(W_t)_{0 \le t \le T}$ be a standard d-dimensional Wiener process defined on a probability space $(\Omega, \mathcal{F}, \mathbf{P})$ and $(\mathcal{F}_t)_{0 \le t \le T}$ denote its natural filtration. In the sequel, $M^2_{\mathcal{F}}(0, T; \mathbb{R})$ (resp. $M^2_{\mathcal{F}}(0, T; \mathbb{R}^m)$ and $M^2_{\mathcal{F}}(0, T; \mathbb{R}^{d \times m})$) will denote the set of $\mathbb{R}$-valued (resp. $\mathbb{R}^m$-valued and $\mathbb{R}^{d \times m}$-valued) processes $(z_t)_{0 \le t \le T}$ which are progressively measurable w.r.t. $(\mathcal{F}_t)$ and satisfy $\mathbf{E} \int_0^T |z_t|^2 \, dt < \infty$.

Bally (1995) proposed a random time discretization scheme for a backward stochastic differential equation of the type:

$$p_t = X + \int_t^T f(\omega, s, p_s, q_s) \, ds - \int_t^T q_s^* \, dW_s \quad (0 \le t \le T) \tag{2.1}$$

with the following hypotheses:

- X is a square integrable, $\mathcal{F}_T$-measurable, m-dimensional random vector.
- $f : \Omega \times [0; T] \times \mathbb{R}^m \times \mathbb{R}^{d \times m} \to \mathbb{R}^m$ is measurable w.r.t. $\mathcal{P} \otimes \mathcal{B}(\mathbb{R}^m) \otimes \mathcal{B}\left(\mathbb{R}^{d \times m}\right)$, where $\mathcal{P}$ denotes the σ-algebra of progressively measurable subsets of $\Omega \times [0; T]$.
- $\mathbf{E} \displaystyle\int_0^T f(t, 0, 0)^2 \, dt < \infty$
- $|f(\omega, t, x, y) - f(\omega, t, x, y)| \le L\left(|x - x'| + |y - y'|\right)$,
  $\forall \omega \in \Omega, \forall t \in [0; T], \forall x, x' \in \mathbb{R}^m, \forall y, y' \in \mathbb{R}^{d \times m} \qquad (L > 0)$.

Under these hypotheses, equation (2.1) has a unique solution pair $(p_t, q_t)_{0 \le t \le T}$ in $M^2_{\mathcal{F}}(0, T; \mathbb{R}^m) \times M^2_{\mathcal{F}}(0, T; \mathbb{R}^{d \times m})$; see Pardoux & Peng (1990).

For numerical applications it is legitimate to suppose that the parameters of the BSDE are given in a discrete form. That's why we are going to replace X and f by simple functionals of the Brownian motion.

We start from a time net:

$$\pi : 0 = s_0 < s_1 < \ldots < s_{n^*} = T.$$

For $s_k \le t < s_{k+1}$, we set

$$\Delta(t) = \left(W_{s_1}, W_{s_2} - W_{s_1}, \ldots, W_{s_k} - W_{s_{k-1}}, W_t - W_{s_k}\right)$$

with $\Delta(T) = \left(W_{s_1}, \ldots, W_T - W_{s_{n^*-1}}\right)$, and make the following assumptions:

- $X = G(\Delta(T))$ where $G : \mathbb{R}^{n^* \times p} \to \mathbb{R}^m$ is a Lipschitz continuous function.
- $f(\omega, t, x, y) = F(\Delta(s_k), t, x, y)$ for $s_k \le t < s_{k+1}$, with F continuous in t.
- $|F(u_0, \ldots, u_k, t, 0, 0)| \le M \qquad (M > 0)$.
- $|F(u_0, \ldots, u_k, t, x, y) - F(u'_0, \ldots, u'_k, t, x', y')|$
  $\le L\left(\sum_{i=0}^{k} |u_i - u'_i| + |x - x'| + |y - y'|\right) \qquad (L > 0)$.

2.2 The discretization scheme

Let us now introduce the discretizing scheme for the BSDE (2.1).

Let the discretization instants form a sequence:

$$\Pi : 0 = T_n < T_{n-1} < \ldots < T_0 = T.$$

One can assume that this new set of instants is included in the set $\{s_0, \ldots, s_{n^*}\}$ and, for $0 \le i \le n$, denote by $k(i)$ the index such that $T_i = s_{k(i)}$.

Let $\lambda > 0$ be fixed and let us consider the discretized equation:

$$\overline{p}_t = X + \sum_{t \le T_k < T} \frac{1}{\lambda} f\left(T_k, \overline{p}_{T_k}, \overline{q}_{T_k}\right) - \int_t^T \overline{q}_s^* \, dW_s \quad (0 \le t \le T) \tag{2.2}$$

where $(\overline{p}_t, \overline{q}_t)_{0 \le t \le T}$ belongs to $M^2_{\mathcal{F}}(0, T; \mathbb{R}) \times M^2_{\mathcal{F}}(0, T; \mathbb{R}^d)$.

To construct a solution of that equation, we define two operators P_r and $Q_r = (Q^1_r, \ldots, Q^d_r)$ for $s_k \le t < s_{k+1}$ by:

$$P_q H(u_0, \ldots, u_k, t) = \int_{\mathbb{R}^{(q-k) \times d}} H\left(u_o, \ldots, u_{k-1}, u_k + v_k, v_{k+1}, \ldots, v_{q-1}\right) \times g\left(v_k, \ldots, v_{q-1}, t\right) dv_k \ldots dv_{q-1}$$

and, for $1 \le j \le d$:

$$Q^j_q H(u_0, \ldots, u_k, t) = \int_{\mathbb{R}^{(q-k) \times d}} H\left(u_o, \ldots, u_{k-1}, u_k + v_k, v_{k+1}, \ldots, v_{q-1}\right) \times \frac{u^j_k}{s_{k+1} - t} g\left(v_k, \ldots, v_{q-1}, t\right) dv_k \ldots dv_{q-1}$$

where

$$g((v_k, \ldots, v_{q-1}, t) = (2\pi(s_{k+1} - t))^{-d/2} \exp\left(-\frac{|v_i|^2}{2(s_{k+1} - t)}\right)$$
$$\times \prod_{i=k+1}^{q-1} (2\pi(s_{i+1} - s_i))^{-d/2} \exp\left(-\frac{|v_i|^2}{2(s_{i+1} - s_i)}\right).$$

Let us set $H_0 = G$ and define the processes $\underline{p}_t$, $\overline{p}_t$, $\overline{q}_t$ and the applications H_i $(0 \le i \le n-1)$ by the following recurrence:

$$\overline{p}_T = H_0(\Delta(T)).$$

For $T_{i+1} \le t < T_i$:

$$\underline{p}_t = P_{k(i)} H_i(\Delta(t), t)$$

and

$$\overline{q}_t^j = Q^j_{k(i)} H_i(\Delta(t), t) \qquad (1 \le j \le d).$$

For $T_{i+1} < t < T_i$, $\overline{p}_t = \underline{p}_t$ and

$$H_{i+1}(\Delta(T_{i+1})) = \underline{p}_{T_{i+1}} + \frac{1}{\lambda} F\left(\Delta(T_{i+1}), T_{i+1}, \underline{p}_{T_i}, \overline{q}_{T_i}\right)$$

$$\overline{p}_{T_{i+1}} = H_{i+1}(\Delta(T_{i+1})).$$

Proposition 2.1 *$(\overline{p}_t, \overline{q}_t)_{0 \le t \le T}$ is the only right-continuous solution of (2.2).*

Sketch of the proof Note that, for $s_k \le t < s_{k+1}$:

$$D^j_t H_i(\Delta(T_i)) = \frac{\partial}{\partial u^j_k} H_i(\Delta(T_i))$$

where D^j denotes the Malliavin derivative with respect to W^j and $\partial/\partial u^j_k$ the partial derivative w.r.t. the last variable of H_i.

Then, with the help of an integration by parts, the Clark–Ocone formula leads to:

$$H_i(\Delta(T_i)) = P_{k(i)} H_i(\Delta(t), t) + \sum_{j=1}^{d} \int_t^{T_i} Q^j_{k(i)} H_i(\Delta(t), t)\, dW^j(s).$$

Hence it is easy to show by recurrence that $(\overline{p}_t, \overline{q}_t)$ is a solution of (2.2).

The uniqueness of a right-continuous solution is obtained by remarking that one has uniqueness of q_{T_i} in L^2, for every i, by the right-continuity of (q_t). □

Now let us define the sequence $T_0, \ldots, T_n$ so that the solution of (2.2) converges to the solution of the BSDE (2.1).

On a distinct probability space $(\widetilde{\Omega}, \widetilde{\mathcal{F}}, \widetilde{\mathbf{P}})$, let N be a Poisson process of intensity λ.

Let τ_k, $k \geq 1$ be the jump times of N, with $\tau_0 = 0$.

For every $\widetilde{\omega}$ in $\widetilde{\Omega}$, we set $n = \min\{k \geq 1; \tau_k > T\}$ and for $0 \leq k \leq n$, $T_k = (T - \tau_k) \vee 0$.

We shall now work on the product space $\left(\overline{\Omega}, \overline{\mathcal{F}}, \overline{P}\right) = \left(\widetilde{\Omega} \times \Omega, \widetilde{\mathcal{F}} \otimes \mathcal{F}, \widetilde{P} \otimes P\right)$ and shall write $\overline{\mathbf{E}}Z$ for the expectation of an integrable random variable on $\overline{\Omega}$.

Let $(\overline{p}_t, \overline{q}_t)_{0 \leq t \leq T}$ be defined on Ω as the only right-continuous solution of (2.2), for every $\widetilde{\omega}$ in $\widetilde{\Omega}$. They can be constructed by the procedure showed above.

Then the following theorem gives an estimation of the error made on the exact solution $(p_t, q_t)_{0 \leq t \leq T}$ of the BSDE (2.1).

Theorem 2.2 *For each $\delta > 0$, there is a constant K_δ such that:*

$$\sup_{0 \leq t \leq T} \overline{\mathbf{E}} \left|\overline{p}_t - p_t\right|^2 + \overline{\mathbf{E}} \int_0^T \left|\overline{q}_t - q_t\right|^2 dt \leq \frac{K_\delta}{\lambda^{1-\delta}},$$

and if X and f are not simple functionals of W but $(\overline{p}_t, \overline{q}_t)_{0 \leq t \leq T}$ is a solution of (2.2) where (X, f) has been replaced by a couple of simple functionals $(\overline{X}, \overline{f})$, we have:

$$\sup_{0 \leq t \leq T} \overline{\mathbf{E}} \left|\overline{p}_t - p_t\right|^2 + \overline{\mathbf{E}} \int_0^T \left|\overline{q}_t - q_t\right|^2 dt$$
$$\leq K_\delta \left[\frac{1}{\lambda^{1-\delta}} + \overline{\mathbf{E}} \left|\overline{X} - X\right|^2 + \int_0^T \mathbf{E} \left|\overline{f}(s, p_s, q_s) - f(s, p_s, q_s)\right|^2 dt\right].$$

The randomness of the time net is essential in getting an estimation of q_t. A deterministic time net cannot lead to a similar result unless stronger regularity assumptions are made on the parameters of the BSDE.

But the algorithm used to determine p_t and q_t is not truly numerical because it calculates a sequence of functions, not just a sequence of real numbers.

3 A 4-step Scheme for Forward-backward SDEs

Ma, Protter & Yong (1994) and Douglas, Ma & Protter (1995), present a numerical method to solve the following forward-backward stochastic differential equations:

$$\begin{cases} y_t = \xi + \int_0^t b(s, y_s, p_s, q_s)\, ds + \int_0^t \sigma(s, y_s, p_s, q_s)\, dW_s \\ p_t = \psi(y_T) + \int_t^T f(s, y_s, p_s, q_s)\, ds - \int_t^T g(s, y_s, p_s, q_s)\, dW_s. \end{cases} \tag{3.1}$$

Here $(W_t)_{0\le t\le T}$ is still a standard d-dimensional Wiener process on a probability space $(\Omega, \mathcal{F}, \mathbf{P})$ with its natural filtration $(\mathcal{F}_t)$, and (y, p, q) takes values in $\mathbb{R}^n \times \mathbb{R}^m \times \mathbb{R}^{d\times m}$.

To make things simpler, from now on we will assume that $n = m = d = 1$, $\sigma(s, y, p, q) \equiv \sigma(s, y, p)$ and $g(s, y, p, q) \equiv q$.

For α in $]0; 1[$, let us define the Hölder space $C^{1+(\alpha/2),2+\alpha}$ as the space of all functions $\varphi(t, x)$ on $[0; T] \times \mathbb{R}$ which are differentiable in t and twice differentiable in x with φ_t and φ_{xx} being $\alpha/2$- and α-Hölder continuous in (t, x).

The space $C^{1+(\alpha/2),2+\alpha}$ is endowed with the norm defined by:

$$\|\varphi\|_{1,2,\alpha} = \sup_{[0;T]\times\mathbb{R}} |\varphi(t,x)| + \sup_{[0;T]\times\mathbb{R}} |\varphi_t(t,x)| + \sup_{[0;T]\times\mathbb{R}} |\varphi_x(t,x)| + \sup_{[0;T]\times\mathbb{R}} |\varphi_{xx}(t,x)| + \sup_{(t,x)\neq(t',x')} \frac{|\varphi_t(t,x)-\varphi_t(t',x')| + |\varphi_{xx}(t,x)-\varphi_{xx}(t',x')|}{\left(|x-x'|^2+|t-t'|\right)^{\alpha/2}}.$$

We make the following hypotheses:

(A1) The functions b, f and σ are continuously differentiable in t and twice continuously differentiable in y, p, q. For $\varphi = b, f, \sigma$, there exists a constant α in $]0; 1[$ such that for every (p, q), $\varphi(\cdot,\cdot,p,q) \in C^{1+(\alpha/2),2+\alpha}$ and a constant $L > 0$ such that $\|\varphi(\cdot,\cdot,p,q)\|_{1,2,\alpha} \le L$ for all (p, q) in $\mathbb{R}^2$.

(A2) There are two constants $0 < \mu \le C$ such that: $\mu \le \sigma(t, y, p) \le C$ for all (t, y, p) in $[0; T] \times \mathbb{R}^2$.

(A3) There exists a constant β in $]0; 1[$ such that $\psi \in C^{4+\beta}$.

Under the assumptions (A1)–(A3), the FBSDE (3.1) has a unique $(\mathcal{F}_t)$-adapted, square-integrable solution $(y_t, p_t, q_t)_{0\le t\le T}$ (see Douglas, Ma & Protter 1995).

The proposed algorithm for solving equation (3.1) consists in the following four step scheme:

Step 1. Define a function $h : [0; T] \times \mathbb{R}^3 \to \mathbb{R}$ by:

$$h(t, y, p, q) = -q\sigma(t, y, p) \qquad \forall (t, y, p, q).$$

Step 2. Using the function h above, solve the following quasilinear parabolic equation for $\theta(t, x)$:

$$\begin{cases} \theta_t + \frac{1}{2}\sigma(t,x,\theta)^2\theta_{xx} + b\left(t,x,\theta,h(t,x,\theta,\theta_x)\right) \\ \qquad\qquad + f(t,x,\theta,h(t,x,\theta,\theta_x)) = 0, \qquad (t,x)\in[0;T]\times\mathbb{R} \\ \theta(T,x) = \psi(x), \qquad x\in\mathbb{R}. \end{cases} \tag{3.2}$$

Step 3. Using θ and h, solve the following forward SDE:

$$y_t = \xi + \int_0^t \widetilde{b}(s, y_s)\, ds + \int_0^t \widetilde{\sigma}(s, y_s)\, dW_s \tag{3.3}$$

with

$$\begin{cases} \widetilde{b}(s,x) = b\left(t, x, \theta(t,x), h\left(t, x, \theta(t,x), \theta_x(t,x)\right)\right) \\ \widetilde{\sigma}(t,x) = \sigma\left(t, x, \theta(t,x)\right) \end{cases}.$$

Step 4. Set

$$\begin{cases} p_t = \theta(t, y_t) \\ q_t = h\left(t, y_t, \theta(t, y_t), \theta_x(t, y_t)\right). \end{cases}$$

Ma, Protter & Yong (1994) prove:

Theorem 3.1 *Under assumptions (A1)–(A3), the four step scheme is always applicable and yields the only adapted, square-integrable solution of (3.1).*

Moreover the unique classical solution θ to (3.2) is in $C^{1+(\alpha/2),2+\alpha}$, and all the partial derivatives of θ up to the second order in t and fourth order in x are bounded.

Let us now specify the numerical methods employed to perform Steps 2 and 3.

To solve the PDE (3.2), Ma and Protter use a method introduced earlier by Douglas & Russel (1982), which combines the finite difference method and the method of characteristics.

The finite difference grid is built up with a time mesh size T/n and a spatial mesh size $C(T/n)$, where C is a constant determined from b, σ and f. One uses the same number of time steps n to discretize the forward SDE (3.3) with the Euler scheme.

Thus we define $(\overline{y}_t^n)_{0 \le t \le T}$ by:

$$\overline{y}_t^n = \xi + \int_0^t \widetilde{b}\left(\eta(s), \overline{y}_{\eta(s)}^n\right) ds + \int_0^t \widetilde{\sigma}\left(\eta(s), \overline{y}_{\eta(s)}^n\right) dW_s$$

with $\eta(s) = \sup\left\{k\frac{T}{n}; k \in \mathbf{N}, k\frac{T}{n} \le s\right\}$.

When the triple $(\overline{y}_t^n, \overline{p}_t^n, \overline{q}_t^n)$ has been constructed in accordance with the above algorithm, the error with respect to the exact solution of (3.1) is bounded as follows:

Theorem 3.2

$$\mathbf{E}\left[\sup_{0 \le t \le T} |\overline{y}_t^n - y_t| + \sup_{0 \le t \le T} |\overline{p}_t^n - p_t| + \sup_{0 \le t \le T} |\overline{q}_t^n - q_t|\right] = \mathcal{O}\left(\frac{1}{\sqrt{n}}\right)$$

and if φ is a C^2, uniformly Lipschitz function, then:

$$|\mathbf{E}\varphi\left(\overline{y}_T^n, \overline{q}_T^n\right) - \mathbf{E}\varphi\left(\overline{y}_T^n, \overline{q}_T^n\right)| = \mathcal{O}\left(\frac{1}{n}\right).$$

Notice that both inequalities exhibit the same convergence rate as in the case of a simple forward SDE discretized with the Euler scheme.

4 Time and Space Discretization of BSDEs

4.1 Preliminaries

In this section, we propose a comprehensive, ready to implement, numerical method for solving BSDEs in the Markovian case, that is when the BSDE is associated with a forward SDE. We consider the most basic context for those equations.

For $T > 0$ fixed, let $(W_t)_{0 \le t \le T}$ be a standard one-dimensional Wiener process defined on a probability space $(\Omega, \mathcal{F}, \mathbf{P})$. Let $(\mathcal{F}_t)_{0 \le t \le T}$ denote its natural filtration.

We are given a real number ξ and the functions b, σ, ψ and f satisfying the assumptions:

(H1) $b : [0;T] \times \mathbb{R} \to \mathbb{R}$ measurable, $\sigma : [0;T] \times \mathbb{R} \to \mathbb{R}$ measurable
$b^2(t,x) + \sigma^2(t,x) \le L_1(1+x^2), \quad \forall t \in [0;T], \forall x \in \mathbb{R}$
$|b(t,x) - b(t,x')| + |\sigma(t,x) - \sigma(t,x')| \le L_1|x-x'|, \quad \forall t \in [0;T], \forall x, x' \in \mathbb{R}$
$(L_1 > 0)$

(H2) $\psi : \mathbb{R} \to \mathbb{R}$ measurable
$|\psi(x)| \le C_1\left(1 + |x|^{p_1}\right), \quad \forall x \in \mathbb{R} \qquad (C_1 > 0,\ p_1 \in \mathbb{N})$

(H3) $f : [0;T] \times \mathbb{R} \times \mathbb{R} \to \mathbb{R}$ measurable
$|f(t,x,0)| \le C_2\left(1 + |x|^{p_2}\right), \quad \forall t \in [0;T], \forall x \in \mathbb{R} \qquad (C_2 > 0,\ p_2 \in \mathbb{N})$
$|f(t,x,z) - f(t,x,z')| \le L_2|z-z'|, \quad \forall t \in [0;T], \forall x, z, z' \in \mathbb{R} \quad (L_2 > 0).$

Let $(y_t)_{0 \le t \le T}$ be the unique solution in $M^2_{\mathcal{F}}(0,T)$ of the stochastic differential equation:

$$y_t = \xi + \int_0^t b(s, y_s)\, ds + \int_0^t \sigma(s, y_s)\, dW_s \quad (0 \le t \le T). \tag{4.1}$$

Let us consider the *backward stochastic differential equation* (BSDE):

$$p_t = \psi(y_T) + \int_t^T f(s, y_s, p_s)\, ds - \int_t^T q_s\, dW_s \quad (0 \le t \le T). \tag{4.2}$$

Under assumptions (H1)–(H3), equation (2.1) has a unique solution pair $(p_t, q_t)_{0 \le t \le T}$ in $M^2_{\mathcal{F}}(0,T;\mathbb{R}) \times M^2_{\mathcal{F}}(0,T;\mathbb{R})$, and there exist two measurable deterministic functions $u(t,x)$ and $v(t,x)$ such that $p_t = u(t, y_t)$ and $q_t = v(t, y_t)$ for all t.

The process $(p_t)_{0\leq t\leq T}$ may be seen also as the unique solution in $M^2_{\mathcal{F}}(0,T)$ of the equation:

$$p_t = \mathbf{E}\left[\psi(y_T) + \int_t^T f(s, y_s, p_s)\,ds \,\middle|\, \mathcal{F}_t\right] \quad (0 \leq t \leq T); \tag{4.3}$$

p_0 is equal to a constant almost surely, and our aim is to compute that constant numerically.

4.2 Time discretization

We are going to write an Euler scheme which approaches the BSDE (4.3). For that purpose we need the following assumptions:

(H4) b, σ are of class $C^{2,4}$, with all derivatives bounded.

(H5) ψ is of class C^4, with $\psi^{(4)}$ growing at most like a polynomial function at infinity.

(H6) f is of class $C^{1,4,4}$, with $\partial f/\partial z$ bounded and all the derivatives of f growing at most like a polynomial function at infinity, uniformly in t and z.

Let Q be a probability law on $\mathbb{R}$, satisfying:

(H7) $\int_{\mathbb{R}} x\,dQ(x) = 0,\ \int_{\mathbb{R}} x^2\,dQ(x) = 1,\ \int_{\mathbb{R}} x^3\,dQ(x) = 0,\ \int_{\mathbb{R}} |x|^5\,dQ(x) < \infty.$

Let $n \geq 1$ be fixed. Let us set $h = T/n$ and $t_j = jh$ for $0 \leq j \leq n$. Let $(U_1^n, U_2^n, \ldots, U_n^n)$ be an n-sized sample of the law Q. Then we set $\overline{\mathcal{F}}_j^n = \sigma\{U_k^n; 1 \leq k \leq j\}$ for $1 \leq j \leq n$ and $\overline{\mathcal{F}}_0^n = \{\emptyset, \Omega\}$.

Let the process $(\overline{y}_j)_{0\leq j\leq n}$ be defined by the recurrence:

$$\begin{cases} \overline{y}_0 = \xi \\ \overline{y}_j = \overline{y}_{j-1} + hb(t_{j-1}, \overline{y}_{j-1}) + \sqrt{h}\sigma(t_{j-1}, \overline{y}_{j-1})U_j^n \quad (1 \leq j \leq n) \end{cases} \tag{4.4}$$

and the process $(\overline{p}_j)_{0\leq j\leq n}$ by:

$$\begin{cases} \overline{p}_n = \psi(\overline{y}_n) \\ \overline{p}_j = \mathbf{E}\left[\overline{p}_{j+1} + hf\left(t_{j+1}, \overline{y}_{j+1}, \overline{p}_{j+1}\right) \middle| \overline{\mathcal{F}}_j^n\right] \quad (0 \leq j \leq n-1) \end{cases} . \tag{4.5}$$

Theorem 4.1 *Under assumptions (H4)–(H7), there exists a constant K independent of n, such that:*

$$|\overline{p}_0 - p_0| \leq Kh = K\frac{T}{n}.$$

The random variables $(\overline{p}_j)_{0\leq j\leq n}$ are determined if we know the sequence of functions $\left(\overline{\varphi}_j\right)_{0\leq j\leq n}$ defined by the recurrence:

$$\begin{cases} \overline{\varphi}_n = \psi \\ \overline{\varphi}_j(x) = \int_{\mathbb{R}} A^h_{j+1}\overline{\varphi}_{j+1}\left(x + hb(t_j,x) + \sqrt{h}\sigma(t_j,x)u\right) dQ(u), \\ \qquad\qquad\qquad \forall x \in \mathbb{R}, \qquad 0 \leq j \leq n-1, \end{cases} \tag{4.6}$$

where we set $A^h_j g(x) = g(x) + hf\left(t_j, x, g(x)\right)$ for all functions g, for $x \in \mathbb{R}$ and $1 \leq j \leq n$.

Actually the following proposition is easy to prove:

Proposition 4.2 *For all j we have:*

$$\overline{p}_j = \overline{\varphi}_j(\overline{y}_j) \ p.s.$$

In particular $\overline{p}_0 = \overline{\varphi}_0(\xi)$ a.s.

If we want to use relation (4.6) in order to compute $\overline{\varphi}_0(\xi)$, it will involve the computation of at least 2^j different values of $\overline{\varphi}_j$ for every j, by choosing Q such that $Q(1) = Q(-1) = 1/2$.

So, to obtain an algorithm that can be used in practice, we must introduce a new approximation in order to reduce the number of possible states at every instant.

4.3 Space discretization

The idea is to choose a finite set of discretization points in $\mathbb{R}$ for each j, and replace every value of $\overline{y}_j$ by the closest discretization point.

A finite, non-empty subset D of $\mathbb{R}$ being given, we define the projection π_D from $\mathbb{R}$ over D by:

$$\begin{cases} \pi_D(x) \in D \\ |x - \pi_D(x)| \leq |x - z|, \ \ \forall z \in D \\ \forall z \in D, \ \ |x - \pi_D(x)| = |x - z| \Rightarrow \pi_D(x) \leq z \end{cases} . \tag{4.7}$$

For $N \geq 1$, let $S = \left(Y^1, Y^2, \ldots, Y^N\right)$ be an N-sized sample of the law of $(\overline{y}_j)_{0\leq j\leq n}$. S represents a series of simulations of the process $(\overline{y}_j)_{0\leq j\leq n}$. For $0 \leq j \leq n$, let $S_j = \left\{Y^1_j, Y^2_j, \ldots, Y^N_j\right\}$. The elements of S_j will be the discretization points at instant j. So let us set:

$$\begin{cases} \hat{y}_0 = \xi \\ \hat{y}_j = \pi_{S_j}\left(\hat{y}_{j-1} + hb(t_{j-1}, \hat{y}_{j-1}) + \sqrt{h}\sigma(t_{j-1}, \hat{y}_{j-1})U^n_j\right) \ \ (1 \leq j \leq n). \end{cases} \tag{4.8}$$

We now have to compute the value at time 0 of the process $(\hat{p}_j)_{0\le j\le n}$ defined as follows:

$$\begin{cases} \hat{p}_n = \psi(\hat{y}_n) \\ \hat{p}_j = \mathbf{E}_S\left[\hat{p}_{j+1} + hf\left(t_{j+1}, \hat{y}_{j+1}, \hat{p}_{j+1}\right) \middle| \overline{\mathcal{F}}_j^n\right] \quad (0 \le j \le n-1) \end{cases} \tag{4.9}$$

where the symbol $\mathbf{E}_S$ means that we take the conditional expectation on a probability space where S is considered fixed in $\mathbb{R}^{Nn}$. This is equivalent to considering the sequence of functions $(\hat{\varphi}_j)_{0\le j\le n}$ defined by the recurrence:

$$\begin{cases} \hat{\varphi}_n = \psi \\ \hat{\varphi}_j(x) = \displaystyle\int_{\mathbb{R}} A_{j+1}^h \hat{\varphi}_{j+1} \circ \pi_{S_{j+1}}\left(x + hb(t_j, x) + \sqrt{h}\sigma(t_j, x)u\right) dQ(u), \\ \qquad\qquad\qquad\qquad \forall x \in \mathbb{R} \qquad 0 \le j \le n-1 \end{cases} \tag{4.10}$$

where $A_j^h g(x) = g(x) + hf(t_j, x, g(x))$ for all function g, for $x \in \mathbb{R}$ and $1 \le j \le n$.

This equivalence is because we still have:

Proposition 4.3 *For all j:*

$$\hat{p}_j = \hat{\varphi}_j(\hat{y}_j) \ a.s.$$

In particular $\hat{p}_0 = \hat{\varphi}_0(\xi)$ a.s.

As π_j takes its values in S_j, we immediately see that in order to compute $\hat{\varphi}_0(\xi)$, it is enough to evaluate the matrix $\left(\hat{\varphi}\left(Y_j^i\right); 1 \le i \le N, 0 \le j \le n\right)$.

In dimension 1, the projection π_j is easy to program if we begin by arranging the values Y_j^i in increasing order. Consequently, the total number of operations needed to get the value of $\hat{\varphi}_0(\xi)$ is of order $nN \log N$.

In order to establish a result of convergence for this algorithm, let us replace respectively assumptions (H5), (H6) and (H7) by the following assumptions:

(H8) ψ is of class C^4, with ψ' such that:

$$\int_{-\infty}^{+\infty} |\psi'(x)|\, dx < \infty$$

and $\psi^{(4)}$ growing at most like a polynomial function at infinity.

(H9) f is of class $C^{1,4,4}$, with $\dfrac{\partial f}{\partial x}$ such that:

$$\sup_{t\in[0;T]} \int_{-\infty}^{+\infty} \sup_{z\in\mathbb{R}} \left|\frac{\partial f}{\partial x}(t, x, z)\right| dx < \infty$$

In addition, $\partial f/\partial z$ is bounded over $[0;T] \times \mathbb{R}^2$ and all the derivatives of f grow at most like a polynomial function at infinity, uniformly in t and z.

(H10) Q is a probability law on $\mathbb{R}$ with compact support, such that:

$$\int_{\mathbb{R}} x\, dQ(x) = 0,\ \int_{\mathbb{R}} x^2\, dQ(x) = 1,\ \int_{\mathbb{R}} x^3\, dQ(x) = 0,\ \int_{\mathbb{R}} |x|^5\, dQ(x) < \infty.$$

Then:

Theorem 4.4 *Under assumptions (H4) and (H8)–(H10), for every $p \geq 1$, there exists a constant C_p independent of n and N, such that:*

$$\|\hat{p}_0 - p_0\|_p \leq C_p \left(h + \frac{n}{N} \right).$$

In particular, if n and N are linked by $N = kn^2$, with k constant, the theorem says that:

$$\|\hat{p}_0 - p_0\|_p \leq C_p \left(1 + \frac{1}{kT} \right) h.$$

The proofs of Theorems 4.1 and 4.4 will be given in Chevance (1996).

References

Bally, V. (1995) 'An approximation scheme for BSDEs and applications to control and nonlinear PDEs', prépublication **95-15** du Laboratoire de Statistique et Processus de l'Université du Maine.

Chevance, D. (1996) PhD thesis, Université de Provence.

Cvitanic, J. & Karatzas, I. (1992) 'Hedging contingent claims with constrained portfolios', *Annals of Applied Probability* **3**, 652–681.

Douglas Jr., J., Ma, J. & Protter, P. (1995) 'Numerical methods for forward-backward stochastic differential equations', to appear in *Annals of Applied Probability.*

Douglas Jr., J. & Russel, T.F. (1982) 'Numerical methods for convection-dominated diffusion problems based on combining the method of characteristics with finite element or finite difference procedures', *SIAM J. Numer. Anal.* **19**, 871–885.

Duffie, D. & Epstein, L. (1992) 'Stochastic differential utility', *Econometrica* **60**, 353–394.

Duffie, D., Ma, J. & Yong, J. (1993) 'Black's consol rate conjecture', *Annals of Applied Probability* **5**, 356–382.

El Karoui, N., Peng, S. & Quenez, M.-C. (1994) 'Backward stochastic differential equations in finance', prépublication **260** du Laboratoire de Probabilités de l'Université Paris VI.

El Karoui, N. & Quenez, M.-C. (1995) 'Dynamic programming and pricing of contingent claims in incomplete market', *SIAM Journal of Control and Optimization* **33**, 29–66.

Ma, J., Protter, P. & Yong, J. (1994) 'Solving forward-backward stochastic differential equations explicitely - a four step scheme', *Probab. Theory Relat. Fields* **98**, 339–359.

Pardoux, E. & Peng, S. (1990) 'Adapted solution of a backward stochastic differential equation', *Systems and Control Letters* **14**, 55–61.

Viscosity Solutions and Numerical Schemes for Investment/Consumption Models with Transaction Costs

Agnès Tourin and Thaleia Zariphopoulou

1 Introduction

In this article we examine a general investment and consumption decision problem for a single agent. The investor consumes at a nonnegative rate and he distributes his current wealth between two assets. One asset is a *bond*, i.e. a riskless security with instantaneous rate of return r. The other asset is a *stock*, whose price is driven by a Wiener process.

When the investor makes a transaction, he pays transaction fees which are assumed to be proportional to the amount transacted. More specifically, let x_t and y_t be the investor's holdings in the riskless and the risky security prior to a transaction at time t. If the investor increases (or decreases) the amount invested in the risky asset to $y_t + h_t$ (or $y_t - h_t$), the holding of the riskless asset decreases (increases) to $x_t - h_t - \lambda h_t$ (or $x_t + h_t - \mu h_t$). The numbers λ and μ are assumed to be nonnegative and one of them must always be positive. The control objective is to maximize, in an infinite horizon, the expected discounted utility which comes only from consumption. Due to the presence of the transaction fees, this is a singular control problem.

Our goals are to derive the Hamilton–Jacobi–Bellman (HJB) equation that the value function solves and to characterize the latter as its unique weak solution, to come up with numerical schemes which converge to the value function as well as the optimal investment and consumption rules and to perform actual numerical computations and compare the results to the ones obtained in closed form by Davis & Norman.

We continue with the description of the model. The price P_t^0 of the bond is given by

$$\begin{cases} dP_t^0 = rP_t^0 dt \qquad (t > 0), \\ P_0^0 = p_0, \end{cases} \tag{1.1}$$

where $r > 0$ is the *interest rate*. The price P_t of the stock satisfies

$$\begin{cases} dP_t = bP_t dt + \sigma P_t dW_t \qquad (t > 0), \\ P_0 = p, \end{cases} \tag{1.2}$$

where b is the *mean rate of return*, σ is the *dispersion coefficient* and the process W_t, which represents the source of uncertainty in the market, is a standard Brownian motion defined on the underlying probability space (Ω, F, P). As usual, F_t is the augmentation under P of $F_t^w = \sigma(W_s : 0 < s \leq t)$ for $t > 0$. The market coefficients r, b and σ are assumed to be constant with $\sigma \neq 0$ and $b > r > 0$.

The amounts x_t and y_t, invested at time t in bonds and stock, respectively, are the *state variables* and they evolve according to the equations

$$\begin{cases} dx_t = (rx_t - C_t)dt - (1+\lambda)dM_t + (1-\mu)dN_t, \\ dy_t = by_t + \sigma y_t dw_t + dM_t - dN_t, \\ x_0 = x, y_0 = y, \end{cases} \tag{1.3}$$

where (x, y) is the *endowment* of the investor. For simplicity, we assume here that all financial charges are paid from the holdings in the bond.

The *control processes* are the *consumption rate* $C_.$ and the processes $M_.$ and $N_.$ which represent, respectively, the *cumulative purchases* and *sales of stock*. We say that the controls $(C_., M_., N_.)$ are admissible if:

(i) C_t is F_t-measurable, $C_t \geq 0$ a.s. and $E\int_0^t e^{-rs}C_s ds < +\infty$ a.s., $\forall t \geq 0$.

(ii) M_t, N_t are F_t-measurable, right continuous and nondecreasing processes.

(iii) If x_t, y_t are the state trajectories given by (1.3), when the controls M_t, N_t are used, then, for all $t \geq 0$,

$$\begin{cases} x_t + (1+\lambda)y_t \geq 0 \quad \text{a.s. if} \quad y_t \leq 0, \\ x_t + (1-\mu)y_t \geq 0 \quad \text{a.s. if} \quad y_t \geq 0. \end{cases} \tag{1.4}$$

We denote by $\mathcal{A}(x, y)$ the set of admissible policies.

The total *expected discounted utility* J from consumption, is given by

$$J(x, y, C, M, N) = E\int_0^{+\infty} e^{-\beta t}U(C_t)dt \tag{1.5}$$

with $(C, M, N) \in \mathcal{A}(x, y)$ and $(x, y) \in \overline{\Omega}$ where

$$\Omega = \{(x, y) \in \mathcal{R} \times \mathcal{R} : x + (1+\lambda)y > 0 \text{ if } y < 0 \text{ and } x + (1-\mu)y > 0 \text{ if } y > 0\}.$$

The utility function $U : [0,+\infty) \longrightarrow [0,+\infty)$ is assumed to have the following properties:

$$\begin{cases} \text{(i)} & U \in C^2((0,\infty)) \text{ and is strictly increasing, nonnegative and concave} \\ & \text{in } [0,+\infty). \\ \text{(ii)} & \text{There exists } K > 0 \text{ and } \gamma \in (0,1) \text{ such that, for all } c \geq 0, \\ & U(c) \leq K(1+c)^\gamma. \\ \text{(iii)} & \lim_{c \to 0} U'(c) = +\infty \text{ and } \lim_{c \to +\infty} U'(c) = 0. \end{cases}$$

The *discount factor* $\beta > 0$ weights consumption now versus consumption later. Note that the controls $M_.$ and $N_.$ are acting implicitly through the state constraints given by (1.4).

The *value function* u is given by

$$u(x,y) = \sup_{(C,M,N) \in A(x,y)} J(x,y,C,M,N). \tag{1.6}$$

To guarantee that the value function is well defined when U is unbounded, we assume that

$$\beta > r\gamma + \Big(\gamma(b-r)/\sigma^2(1-\gamma)\Big). \tag{1.7}$$

This condition implies that the value function which corresponds to $\lambda = \mu = 0$ and $U(c) = K(1+c)^\gamma$, and thereby all value functions for $0 < \lambda, \mu < 1$, is finite (see Karatzas *et al.* (1987) or Zariphopoulou (1994)).

Our goal is first to derive the Hamilton-Jacobi-Bellman equation associated with the above singular stochastic control problem and to characterize u as its unique weak solution. It turns out that the Bellman equation here is a Variational Inequality with gradient constraints.

Due to the nature of our goals, in the Introduction we only state our result regarding the characterization of the value function. The rest of our results are far more complicated to state here. We hence choose to present them in the main body of the article.

Theorem 1.1 *The value function u is the unique constrained viscosity solution of*

$$\begin{aligned} \min\Big[\beta u - \frac{1}{2}\sigma^2 y^2 u_{yy} - byu_y - rxu_x - \max_{c \geq 0}[-cu_x + U(c)], \\ (1+\lambda)u_x - u_y, -(1-\mu)u_x + u_y\Big] = 0 \quad \text{in } \Omega, \end{aligned} \tag{1.8}$$

in the class of concave increasing and uniformly continuous functions.

The fact that the value function turns out to be the *unique* viscosity solution of (1.8) plays a very crucial role for the convergence of the numerical schemes proposed in Section 5.

We continue with a short discussion about the history of the model. Transaction costs are an essential feature of some economic theories and are often incorporated in the two-asset portfolio selection model. Constantinides (1979) and (1986) assumes that the transaction costs deplete only the riskless asset and that the stock price is a logarithmic Brownian motion. In the continuous time framework, Taksar, Klass & Assaf (1986) assume that the investor does not consume but maximizes the long term expected rate of growth of wealth. In the same framework, but under more general assumptions, Fleming, Grossman, Vila & Zariphopoulou (1989) study the finite horizon problem, the average cost per unit time problem and an asymptotic growth problem.

Davis & Norman (1990) relax the assumption that the transaction costs are charged only to the nonrisky asset. They consider a particular class of utility functions of the form $U(c) = c^p (0 < p < 1)$ and they get an explicit form for the value function. They also prove that the optimal strategy confines the investor's portfolio to a certain wedge-shaped region in the portfolio plane. Their results are presented in detail in Section 5. In a paper which appeared after a preliminary version of this article was circulated, Shreve & Soner (1994) examine the above class of utility functions and they relax some assumptions on the market parameters in order for the value function to be finite. Moreover, they prove that the value function is a smooth solution of the HJB equation and that the boundary of the aforementioned wedge-shaped region is also smooth.

Finally, there are several directions in which the two-asset problem with transaction costs can be extended. Firstly, more than one risky asset can be allowed. Although this extension is straightfoward, the computational requirements are enormous. Secondly, fixed transaction costs can be introduced. Some single-period models with fixed transaction costs are discussed by Leland (1985), Brennan (1975) and Goldsmith (1976). Kandel & Ross (1983) introduce quasi-fixed transaction costs and portfolio management fees. In a different direction, a model with proportional fees when the rate of return of the risky asset is a continuous time Markov chain is examined by Zariphopoulou (1994). See also Whalley & Wilmott (1996).

The article is organized as follows: Section 2 is about some basic properties of the value function. In Section 3, we study the solutions of the Variational Inequality (1.8) and we characterize the value function as its unique solution. Section 4 reviews the results of Davis & Norman for the HARA utility functions. In Section 5, we present the numerical algorithms and we study the behavior of the transaction regions. Finally, in Section 6, we summarize the main conclusions of the article.

2 Basic Properties of the Value Function

In this section we derive some basic properties of the value function.

Proposition 2.1 *The value function u is jointly concave in x and y, strictly increasing in x and increasing in y.*

Sketch of the proof The joint concavity of the value function comes from the concavity of the utility function and the linearity of the dynamics. Indeed, if (C_1, L_1, M_1) and (C_2, L_2, M_2) are optimal policies for the points (x_1, y_1) and (x_2, y_2), then $(\lambda C_1 + (1-\lambda)C_2, \lambda M_1 + (1-\lambda)M_2, \lambda N_1 + (1-\lambda)N_2)$ is admissible for $(\lambda x_1 + (1-\lambda)x_2, \lambda y_1 + (1-\lambda)y_2)$.

The monotonicity of u follows from the same monotonicity of the utility function and the linear dynamics of the state trajectories. The strict monotonicity in x comes from the fact that u is also concave (for a detailed proof of a similar question, see Proposition 2.1 in Zariphopoulou (1994)).

Proposition 2.2 *The value function u is uniformly continuous on $\overline{\Omega}$.*

Proof Since u is concave, it is obviously continuous in Ω.

We next show that u is continuous on the boundary. The continuity at the point $(0,0)$ follows as in Proposition 2.2 in Zariphopoulou (1994).

We now show that

$$\lim_{(x_n,y_n)\to(x_0,y_0)} u(x_n, y_n) = u(x_0, y_0)$$

where

$$(x_0, y_0) \in l_1 = \{(x,y) \in \mathcal{R}^+ \times \mathcal{R}^- : x + (1+\lambda)y = 0\}$$

or

$$(x_0, y_0) \in l_2 = \{(x,y) \in \mathcal{R}^- \times \mathcal{R}^+ : x + (1-\mu)y = 0\}.$$

We only examine the case $(x_0, y_0) \in l_1$ since the other is treated similarly. To this end, consider a point $(x_0, y_0) \in l_1$ and a sequence

$$(x_n, y_n) \in l_1^+ = \{(x,y) \in \mathcal{R}^+ \times \mathcal{R}^- : x + (1+\lambda)y > 0\}$$

such that

$$\lim_{n\to+\infty} (x_n, y_n) = (x_0, y_0).$$

Since u is locally Lipschitz, by concavity, it suffices to show that

$$\lim_{n\to+\infty} |u(x_0, y_n) - u(x_0, y_0)| = 0.$$

Finally, since u is increasing, we only need to show that

$$u(x_0, y_n) \le u(x_0, y_0) + \epsilon$$

for any $\epsilon > 0$ and n sufficiently large.

Let (C^n, M^n, N^n) be an ϵ-optimal policy at (x_0, y_n). Then

$$u(x_0, y_n) \le E \int_0^{+\infty} e^{-\beta t} U(C_t^n) dt + \epsilon\,.$$

Moreover the control $(C^n, \overline{M^n}, N^n)$, where

$$d\overline{M_t^n} = dM_t^n + \frac{(1-\mu)}{(\lambda+\mu)}(y_n - y_0)\delta_0(t)$$

is admissible for

$$(\overline{x_n}, \overline{y_n}) = \left(x_0 + \frac{(1-\mu)(1+\lambda)}{(\lambda+\mu)}(y_n - y_0), y_0 - \frac{(1-\mu)}{\lambda+\mu}(y_n - y_0) \right) \in l_1.$$

Therefore,

$$E \int_0^{+\infty} e^{-\beta t} U(C_t^n) dt \le u(\overline{x_n}, \overline{y_n}) + \epsilon\,.$$

Combining the last two inequalities and using the fact that u is continuous on $l_1 - \{0, 0\}$ we conclude.

Finally, since u is uniformly continuous on compact subsets of $\overline{\Omega}$, we remark that its uniform continuity on $\overline{\Omega}$ follows from the fact that, by concavity, u is Lipschitz continuous in $[x, +\infty) \times [y, +\infty)$ with Lipschitz constant $1/|(x, y)|$ for every $(x, y) \in \Omega$. □

We conclude this section by stating (for a proof see, for example, Lions (1983)) a fundamental property of the value function known as the *Dynamic Programming Principle.*

Proposition 2.3 *If θ is a stopping time (i.e. a nonnegative, F-measurable random variable), then*

$$u(x, y) = \sup_{\mathcal{A}(x,y)} E\left\{ \int_0^{\theta} e^{-\beta t} U(C_t) dt + e^{-\beta\theta} u(x_\theta, y_\theta) \right\}. \tag{2.1}$$

3 Viscosity Solutions

In this section we characterize the value function as the *unique constrained viscosity solution* of the (HJB) equation (1.8). The characterization of u as a constrained solution is natural because of the presence of state constraints given by (1.4).

The notion of viscosity solutions was introduced by Crandall & Lions (1984) for first-order and by Lions (1983) for second-order equations. For a general overview of the theory we refer to the 'User's Guide' by Crandall, Ishii & Lions (1992) and to Fleming & Soner (1993).

Next, we recall the notion of constrained viscosity solutions, which was introduced by Soner (1986) and Capuzzo-Dolcetta & Lions (1995) for first-order equations and by Lions (1983) for second-order equations (see also Ishii & Lions (1990)). To this end, we consider a nonlinear second-order partial differential equation of the form

$$F(X, v, Dv, D^2v) = 0 \text{ in } \Omega, \tag{3.1}$$

where Dv and D^2v stand respectively for the gradient vector and the second derivative matrix of v; F is continuous in all its arguments and degenerate elliptic, meaning that

$$F(X, p, q, A + B) \leq F(X, p, q, A) \text{ if } B \geq 0. \tag{3.2}$$

Definition 3.1 *A continuous function $u : \mathbf{R} \to \mathbf{R}$ is a constrained viscosity solution of (3.1) if*

(i) *u is a* viscosity subsolution *of* (3.1) *on $\overline{\Omega}$, that is for any $\phi \in C^2(\overline{\Omega})$ and any local maximum point $X_0 \in \overline{\Omega}$ of $u - \phi$*

$$F(X_0, u(X_0), D\phi(X_0), D^2\phi(X_0)) \leq 0$$

and

(ii) *u is a* viscosity supersolution *of* (3.1) *in Ω, that is for any $\phi \in C^2(\overline{\Omega})$ and any local minimum point $X_0 \in \Omega$ of $u - \phi$*

$$F(X_0, u(X_0), D\phi(X_0), D^2\phi(X_0)) \geq 0.$$

Theorem 3.2 *The value function u is a constrained viscosity solution of* (1.8) *on $\overline{\Omega}$.*

Proof (i) We first show that v is a viscosity subsolution of (1.8) on $\overline{\Omega}$. Let $\phi \in C^2(\overline{\Omega})$ and $X_0 = (x_0, y_0) \in \overline{\Omega}$ be a maximum of $u - \phi$; without loss of generality we may assume that

$$v(X_0) = \phi(X_0) \quad \text{and} \quad u \leq \phi \text{ on } \overline{\Omega}. \tag{3.3}$$

We need to show that

$$\begin{aligned} \min\Big[\beta\phi(X_0) - \frac{1}{2}\sigma^2 y_0^2 \phi_{yy}(X_0) - by_0\phi_y(X_0) - rx_0\phi_x(X_0) \\ - \max_{c \geq 0}(-c\phi_x(X_0) + U(c)),\ (1+\lambda)\phi_x(X_0) - \phi_y(X_0), \\ -(1-\mu)\phi_x(X_0) + \phi_y(X_0)\Big] \leq 0. \end{aligned} \tag{3.4}$$

We argue by contradiction and we assume that

$$(1+\lambda)\phi_x(X_0) - \phi_y(X_0) > 0, \tag{3.5}$$

$$-(1-\mu)\phi_x(X_0) + \phi_y(X_0) > 0, \tag{3.6}$$

and

$$\beta\phi(X_0) - \frac{1}{2}\sigma^2 y_0^2 \phi_{yy}(X_0) - by_0\phi_y(X_0) - rx_0\phi_x(X_0) - \max_{c\geq 0}(-c\phi_x(X_0) + U(c)) > \theta \tag{3.7}$$

for some $\theta > 0$.

From the fact that ϕ is smooth, the above inequalities become

$$(1+\lambda)\phi_x(X) - \phi_y(X) > 0, \tag{3.8}$$

$$-(1-\mu)\phi_x(X) + \phi_y(X) > 0, \tag{3.9}$$

and

$$\beta\phi(X) - \frac{1}{2}\sigma^2 y^2 \phi_{yy}(X) - by\phi_y(X) - rx\phi_x(X) - \max_{c\geq 0}(-c\phi_x(X) + U(c)) > \theta \tag{3.10}$$

for some $\theta > 0$ where $X = (x, y) \in \mathcal{B}(X_0)$ a neighborhood of X_0.

We now consider the optimal trajectory $X_0^*(t) = (x_0^*(t), y_0^*(t))$ where $X_0^*(0) = (x_0, y_0)$ with optimal policies (C_t^*, M_t^*, N_t^*) being used. (The existence of optimal policies was shown in Zhu (1986) (Th. (iv) 2)). We will need the following lemma which shows that X_0^* has no jumps a.s. at $t = 0^+$.

Lemma 3.3 *Assume that inequality* (3.5) *(resp.* (3.6)*) holds and let A be the event that the optimal trajectory $X_0^*(t)$ has a jump at least of size ϵ at $t = 0^+$ along the direction $(-(1+\lambda), 1)$ (resp. $((1-\mu), -1)$). If*

$$(x_0 - (1+\lambda)\epsilon, y_0 + \epsilon) \in \mathcal{B}(X) \quad (\text{resp. } (x_0 + (1-\mu)\epsilon, y_0 - \epsilon) \in \mathcal{B}(X_0))$$

then $P(A) = 0$.

Since the proof is similar to the one of Lemma 1 in Davis, Panas & Zariphopoulou (1993), it is not presented here.

We now continue the proof of the theorem. We define the random time τ to be $\tau(\omega) = \inf\{t \geq 0 \colon X_0^*(t) \notin \mathcal{B}(X_0)\}$. Notice that by the preceding lemma, $\tau(\omega) > 0$ a.s. Combining (3.8), (3.9) and (3.10) we get

$$E\int_0^\tau \theta e^{-\beta s}ds < E\int_0^\tau e^{-\beta s}[\beta\phi(X_s^*) - \tfrac{1}{2}\sigma^2(y_s^*)^2\phi_{yy}(X_s^*)$$

$$-by_s^*\phi_y(X_s^*) - rx_s^*\phi_x(X_s^*) - \{-C_s^*\phi_x(X_s^*) + U(C_s^*)\}]ds$$

$$+E\int_0^{\tau(\omega)} e^{-\beta s}[(1+\lambda)\phi_x(X_s^*) - \phi_y(X_s^*)]dM_s^*$$

$$+E\int_0^{\tau(\omega)} e^{-\beta s}[-(1-\mu)\phi_x(X_s^*) - \phi_y(X_s^*)]dN_s^*$$

$$= E(I_1(\tau)) - E\int_0^\tau e^{-\beta s}U(C_s^*)ds + E(I_2(\tau)) + E(I_3(\tau)). \tag{3.11}$$

Applying Itô's formula to $e^{-\beta\tau}\phi(X_0^*(\tau))$ gives

$$E\left\{e^{-\beta\tau}\phi(X_0^*(\tau))\right\} = \phi(X_0) - [E(I_1(\tau)) + E(I_2(\tau)) + E(I_3(\tau))]. \tag{3.12}$$

Combining (3.3), (3.11) and (3.12) we get

$$E\{u(X_0^*(\tau))\} \le u(X_0) - \left[E\int_0^\tau e^{-\beta s}U(C_s^*)ds + \theta\frac{1 - Ee^{-\beta\tau}}{\beta}\right], \tag{3.13}$$

which violates the Dynamic Programming Principle, together with the optimality of (C_t^*, M_t^*, N_t^*). Therefore, at least one of the argument of the minimum operator in (3.4) must be nonpositive and hence the value function is a viscosity subsolution of (1.8).

(ii) In the second part of the proof, we show that u is a viscosity supersolution of (1.8) in Ω; for this we must show that, for all smooth functions $\varphi(X)$, such that $u - \varphi$ has a local minimum at $X_0 \in \Omega$, the following holds:

$$\min[\beta\varphi(X_0) - \frac{1}{2}\sigma^2 y_0^2\varphi_{yy}(X_0) - by_0\varphi_y(X_0) - rx_0\varphi_x(X_0)$$

$$- \max_{c\ge 0}(-c\varphi_x(X_0) + U(c)), (1+\lambda)\varphi_x(X_0) - \varphi_y(X_0),$$

$$- (1-\mu)\varphi_x(X_0) + \varphi_y(X_0)] \ge 0,$$

where, without loss of generality, $u(X_0) = \varphi(X_0)$ and $u \ge \varphi$ on $\overline{\Omega}$. In this case, we prove that each argument of the above minimum operator is nonnegative.

Consider the trading strategy $L(t) = L_0 > 0$ and $M(t) = 0$ for $t \ge 0$. By the Dynamic Programming Principle,

$$u(x_0, y_0) \ge u(x_0 - (1+\lambda)L_0, y_0).$$

This inequality holds for φ as well, and, by taking the left-hand side to the right-hand side, dividing by L_0 and sending $L_0 \to 0$, we get

$$(1+\lambda)\varphi_x(X_0) - \varphi_y(X_0) \geq 0.$$

Similarly, by using the trading strategy $L(t) = 0$ and $M(t) = M_0 > 0$, for $t \geq 0$, we obtain

$$-(1-\mu)\varphi_x(X_0) + \varphi_y(X_0) \geq 0.$$

Finally consider the case where the investor does not trade but consumes at a constant rate $C_t = C$ for $0 < t \leq \tau$ where $\tau = n \wedge \tau_1 \wedge \tau_2$ with $n \in \mathbf{N}$

$$\tau_1 = \inf\{t : x_t + (1+\lambda)y_t \geq 0 \text{ a.s. if } y_t \leq 0\},$$

$$\tau_2 = \inf\{t : x_t + (1-\mu)y_t \geq 0 \text{ a.s. if } y_t \geq 0\}$$

and x_t, y_t are the state trajectories, given by (1.3), under policy $(C, 0, 0)$. The Dynamic Programming Principle yields

$$u(x_0, y_0) \leq E\left[\int_0^\tau e^{-\beta s} U(C) ds + e^{-\beta\tau} u(x_\tau, y_\tau)\right].$$

The same inequality holds for φ which, in turn, combined with Itô's rule applied to $e^{-\beta\tau}\varphi(x_\tau, y_\tau)$ gives

$$E\int_0^\tau e^{-\beta s}[-\beta u(X_s) + \frac{1}{2}\sigma^2 y_s^2 \varphi_{yy}(X_s) - by_s\varphi_y(X_s) - rx_s\varphi_x(X_s)$$

$$+ C\varphi_x(X_s) - U(C)]ds \leq 0.$$

Dividing by n, and sending $n \to +\infty$ we get

$$\beta\varphi(X_0) - \frac{1}{2}\sigma^2 y_0^2 \varphi_{yy}(X_0) - by_0\varphi_y(X_0) - rx_0\varphi_x(X_0)$$

$$- \max_{c \geq 0}(-c\varphi_x(X_0) + U(c)) \geq 0$$

(for a detailed argument, see Zariphopoulou (1994)). This completes the proof.

We conclude this section by presenting a comparison result for constrained viscosity solutions of (1.8). This result will be used later in Section 5 to obtain convergence of the numerical schemes employed for the value function and the optimal policies.

Theorem 3.4 *Let u be an upper semi-continuous viscosity subsolution of (1.8) on $\overline{\Omega}$ with sublinear growth and v be a bounded from below uniformly continuous viscosity supersolution of (1.8) in Ω. Then $u \leq v$ on $\overline{\Omega}$.*

Sketch of the proof We first construct a strictly positive supersolution of (1.8) in Ω. To this end, let $w(x, y)$ be the value function defined as in (1.6) with U replaced by some U_1 such that $U_1(c) > U(c)$ for $c > 0$, $U_1(0) = U(0) \geq 0$ and $\lambda = \mu = 0$. This value function is the solution to the classical Merton consumption-portfolio problem in the absence of transaction costs and satisfies $w(x, y) = v(z)$, where $z = x + y$ and v solves

$$\begin{cases} \beta v = -\dfrac{(b-r)^2}{2\sigma^2}\dfrac{v'^2}{v''} + rzv' + \max\limits_{c\geq 0}\{-cv' + U_1(c)\} \quad (z > 0) \\ v > 0, v' > 0 \;\text{ and }\; v'' < 0, \quad (z > 0). \end{cases} \tag{3.14}$$

We now let

$$V(x, y) = v(x + ky) + K + C_1 x + C_2 y \qquad ((x, y) \in \overline{\Omega})$$

where the constants K, C_1, C_2 and k are positive and C_1, C_2 and k satisfy

$$1 - \mu < k < 1 + \lambda, \quad (1 + \lambda)C_1 \geq C_2 \geq \frac{\beta - r}{\beta - b}(1 - \mu)C_1$$

and we claim that V is a strictly positive supersolution of (1.8).

The choice of k implies

$$x + ky > 0 \;\text{ whenever }\; (x, y) \in \overline{\Omega}$$

which combined with (3.14) yields $v > 0$ and $v' > 0$ on $\overline{\Omega}$, together with the fact that C_1 and C_2 are positive constants.

It then follows that

$$\begin{cases} (1+\lambda)V_x(x, y) - V_y(x, y) = (1 + \lambda - k)v'(x + ky) \\ \qquad\qquad + [(1 + \lambda)C_1 - C_2] \\ -(1 - \mu)V_x(x, y) + V_y(x, y) = (-1 + \mu + k)v'(x + ky) + \\ \qquad\qquad [-(1 - \mu)C_1 + C_2]. \end{cases} \tag{3.15}$$

Moreover,

$$\begin{aligned} &\beta V(x, y) - \tfrac{1}{2}\sigma^2 y^2 V_{yy}(x, y) - byV_y(x, y) - rxV_x(x, y) \\ &\quad - \max_{c\geq 0}\{-cV_x(x, y) + U(c)\} \\ &= \beta v(z) - \tfrac{1}{2}\sigma^2 (ky)^2 v''(z) - b(ky)v'(z) - rxv'(z) \\ &\quad - \max_{c\geq 0}\{-c(v'(z) + C_1) + U(c)\} + \beta K + \beta C_1 x + \beta C_2 y - bC_2 y - rC_1 x. \end{aligned}$$

Next, observe that, because $C_1 > 0$,

$$\max_{c\geq 0}\{-c(v'(z)+C_1)+U(c)\} \leq \max_{c\geq 0}\{-cv'(z)+U(c)\}.$$

Moreover, the choice of the constants C_1 and C_2, together with the fact that $b > r$, yields

$$(\beta - r)C_1 x + (\beta - b)C_2 y \geq 0 \qquad ((x,y)\in\overline{\Omega})$$

and

$$(1+\lambda)C_1 \geq C_2 \geq (1-\mu)C_2.$$

Combining the above inequalities and (3.14) we get,

$$\begin{aligned}
&\beta v(z) - \tfrac{1}{2}\sigma^2(ky)^2 v''(z) - b(ky)v'(z) - rxv'(z)\\
&\quad - \max_{c\geq 0}\{-c(v'(z)+C_1)+U(c)\} + \beta K + (\beta - r)C_1 x + (\beta - b)C_2 y\\
&\geq \beta K + \left[-\frac{(b-r)^2}{2\sigma^2}\frac{(v'(z))^2}{v''(z)} - \tfrac{1}{2}\sigma^2(ky)^2 v''(z) - (b-r)kyv'(z)\right]\\
&\quad + \left\{\max_{c\geq 0}[-cv'(z)+U_1(c)] - \max_{c\geq 0}[-cv'(z)+U(c)]\right\}\\
&= \beta K + g_1(z,y) + g_2(z,y).
\end{aligned}$$

We now observe that the term g_1 in the above sum is nonnegative, since the maximum value of the quadratic $\mathcal{D}(g_1) = \frac{1}{2}\sigma^2 q^2 v'' + (b-r)qv'$ is $-\frac{(b-r)^2}{2\sigma^2}\frac{(v')^2}{v''}$, at the point $q = ky$. Moreover

$$\begin{aligned}
g_2(z) = \quad & g_2(x+ky) = \max_{c\geq 0}\{-cv'(x+ky)+U_1(c)\}\\
& - \max_{c\geq 0}\{-cv'(x+ky)+U(c)\} > 0
\end{aligned} \tag{3.16}$$

due to the choice of U_1.

Let

$$\begin{aligned}
H(X,V,DV,D^2V) = \min\Big\{&\beta V - \tfrac{1}{2}\sigma^2 y^2 V_{yy} - byV_y - rxV_x -\\
&\max_{c\geq 0}\{-cV_x + U(c)\}, (1+\lambda)V_x - V_y, -(1-\mu)V_x + V_y\Big\}.
\end{aligned}$$

Combining (3.15) and (3.16) yields

$$\begin{aligned}
&H(X,V,DV,D^2V) \geq\\
&\qquad \min\{\beta K, (1+\lambda)C_1 - C_2, -(1-\mu)C_1 + C_2\} = \delta > 0 \text{ in } \Omega.
\end{aligned}$$

To conclude the proof of the theorem we will need the following lemma. Its proof follows along the lines of Theorem VI.5 in Ishii & Lions (1990) and therefore it is omitted.

Lemma 3.5 *Let u be an upper semi-continuous with sublinear growth viscosity subsolution of (1.8) on $\overline{\Omega}$ and v be a bounded from below, uniformly continuous, viscosity supersolution of $H(X, u, Du, D^2u) - h(X)$, where $h \geq \delta > 0$ in Ω, for some constant δ. Then $u \leq v$ on $\overline{\Omega}$.*

We now conclude the proof of the theorem. We define the function $w^\theta = \theta v + (1-\theta)V$ where $0 < \theta < 1$ and we observe that w^θ is a viscosity supersolution of $H - h = 0$ with $h(X) \equiv \delta$. (See also Davis, Panas & Zariphopoulou (1993), Th. 2, for a similar argument.) Applying the above lemma to u and w^θ we get

$$u \leq w^\theta \quad \text{on } \overline{\Omega};$$

sending θ to 1 concludes the proof.

4 The Case of HARA Utilities

Davis & Norman (1990) solve explicitly the problem in the case of Hyperbolic Absolute Risk Aversion utility functions $U(c) = \frac{1}{\gamma}c^\gamma, 0 < \gamma < 1$ and $U(c) = \log c$ for $\gamma = 0$.

They first remark that the solvability region has to be depleted into three regions: the so-called *sell* and *buy regions* (sales and purchases respectively take place instantaneously), and the *non-transaction region.*

In order to find the location of the free boundaries, they use the homothetic property of the value function and they reduce the problem to a one-dimensional one. More precisely, they set $v(x, y) = y^\gamma \Psi\left(\frac{x}{y}\right)$ where the function Ψ satisfies:

$$\begin{cases} \Psi(x) = \frac{1}{2}A(x + (1-\mu)y)^\gamma \quad (x \leq x_1), \\ \\ \beta_1\Psi(x) + \beta_2 x\Psi'(x) + \beta_3 x^2\Psi''(x) + \frac{1-\gamma}{\gamma}\Psi'(x)^{\frac{-\gamma}{1-\gamma}} = 0 \quad (x_1 \leq x \leq x_2), \\ \\ \Psi(x) = \frac{1}{\gamma}B(x + (1+\lambda)y)^\gamma \quad (x \geq x_2), \end{cases}$$

where $\beta_1 = -\frac{1}{2}\sigma^2\gamma(1-\gamma) + b\gamma - \beta$, $\beta_2 = \sigma^2(1-\gamma) + r - \beta$, $\beta_3 = \frac{1}{2}\sigma^2$ and the points x_1, x_2 and the coefficients A and B are explicitly determined.

They prove that the existence of such a function Ψ provides a sufficient condition for the optimality of a policy (C, M, N) such that the corresponding process (x_t, y_t) is a reflecting diffusion in the non-transaction region and M and N are the local times at the lower and upper boundaries respectively. They also prove the existence of such a solution Ψ.

Finally, they propose an algorithm that solves the above problem by integrating backwards a system of differential equations.

5 Numerical Schemes

This section is devoted to the construction of finite difference schemes in order to compute the unique viscosity solution of the Variational Inequality (1.8). The approach which relies on the theory of viscosity solutions and the Dynamic Programming Principle yields *monotone, stable and consistent schemes.* The convergence of such schemes was originally proved in some situations by Crandall & Lions (1984), Barles & Souganidis (1991) and Souganidis (1985). More recently, it was proved for parabolic equations, arising in problems of option pricing, by Barles, Daher & Romano in (1991) and by Davis, Panas & Zariphopoulou in (1993). Then an extension to a MUSCL filtered scheme developed by Lions & Souganidis (1995) for conservation laws and Hamilton–Jacobi equations is proposed; their scheme is filtered in the sense that it preserves an upper bound on some weak second-order finite differences. Although the convergence of such a scheme is not proved in this situation, this approach allows us to get more precise numerical results, especially near the origin.

In the first scheme constructed here, the first-order operators are approximated by a monotone finite difference scheme. As far as the second-order operator is concerned, the first-order part is approximated by a monotone explicit scheme based on the Dynamic Programming Principle whereas the second-order term is approximated by an implicit Crank–Nicholson scheme. Thus splitting into two half-iterations allows one to choose a time step of the same order as the mesh size. This method is known as the *time splitting* method or method of *fractional steps.*

We next present the numerical scheme we developed. To this end, we first write (1.8) in the concise form

$$\min\Big\{L_0(x,y,u,u_x,u_y,u_{yy}),L_1(u_x,u_y),L_2(u_x,u_y)\Big\}=0 \tag{5.1}$$

where

$$L_0(x,y,u,u_x,u_y,u_{yy})=\beta u-\tfrac{1}{2}\sigma^2y^2u_{yy}-byu_y-rxu_x-\max_{c\geq 0}\{-cu_x+U(c)\}$$

and

$$L_1(u_x,u_y)=(1+\lambda)u_x-u_y,\quad L_2(u_x,u_y)=-(1-\mu)u_x+u_y.$$

We consider the domain $\mathcal{D}\subset\mathbf{R}^2$

$$\mathcal{D}=\{(-X_{\max}+(i-1)\Delta x,-Y_{\max}+(j-1)\Delta y),1\leq i\leq M,1\leq j\leq L\}\cap\overline{\Omega}$$

where M and L denote the number of grid points on the x and y axes, $\Delta x,\Delta y>0$ are the mesh sizes and $X_{\max}$ and $Y_{\max}$ the maximal values for x

and y respectively. The value of our numerical approximation at each point $(x_i, y_j) \in \mathcal{D}$, will be denoted by V_{ij}.

We then define the first-order differences

$$D_x^+ V_{ij} = \frac{V_{i+1,j} - V_{ij}}{\Delta x}, \quad D_y^+ V_{ij} = \frac{V_{i,j+1} - V_{ij}}{\Delta y},$$

and

$$D_x^- V_{ij} = \frac{V_{ij} - V_{i-1,j}}{\Delta x}, \quad D_y^- V_{ij} = \frac{V_{ij} - V_{i,j-1}}{\Delta y}.$$

Inside the domain, the first-order operators L_1 and L_2 are approximated in a monotone way using the appropriate backward and forward finite differences

$$g_1(D_x^- V_{ij}, D_y^+ V_{ij}) = (1+\lambda) D_x^- V_{ij} - D_y^+ V_{ij} \tag{5.2}$$

$$g_2(D_x^+ V_{ij}, D_y^- V_{ij}) = -(1-\mu) D_x^+ V_{ij} - D_y^- V_{ij}. \tag{5.3}$$

Next, we consider the first-order operator L_0' obtained by eliminating the second-order term from L_0, i.e.

$$L_0'(x, y, u, u_x, u_y) = \beta u - byu_x - rxu_x - \max_{c \geq 0}\{-cu_x + U(c)\}.$$

The solution of the equation $L_0'(x, y, \overline{u}, \overline{u}_x, \overline{u}_y) = 0$ can be characterized (see, for example, Lions (1983)) as the value function of the deterministic control problem

$$\overline{u}(x, y) = \max_{c \geq 0} \left\{ \int_0^\infty e^{-\beta t} U(C_t) dt \right\} \tag{5.4}$$

where the state trajectories $\overline{x}_t$ and $\overline{y}_t$ solve

$$\begin{cases} d\overline{x}_t = (rx_t - C_t)dt - (1+\lambda)dM_t + (1-\mu)dN_t \\ \\ d\overline{y}_t = by_t dt + dM_t - dN_t \\ \\ x_0 = x, \quad y_0 = y. \end{cases}$$

We are going to construct a monotone scheme to approximate the value function $\overline{u}$. To this end, we apply the Dynamic Programming Principle to (5.4) (see for example, Alziary de Roquefort (1991), Capuzzo-Dolcetta (1983), Falcone (1985), (1987) and Rouy & Tourin (1995)), to get

$$\overline{u}(x, y) = \max_{c \geq 0} \left\{ \int_0^T e^{-\beta t} U(C_t) dt + e^{-\beta T} \overline{u}(\overline{x}_T, \overline{y}_T) \right\}.$$

We choose $T = \Delta\tau$ arbitrarily small and assume that the control remains constant in the time interval $[0, T]$. The above equality then yields

$$\sup_{c \geq 0} \left\{ U(c) + \frac{\overline{u}(x + \Delta\tau(rx - c), y + \Delta\tau by) - \overline{u}(x, y)}{\Delta\tau} e^{-\beta\Delta\tau} \right.$$
$$\left. + \overline{u}(x, y) \frac{e^{-\beta\Delta\tau} - 1}{\Delta\tau} \right\} = 0.$$

In order to approximate the operator L'_0, one has to find an explicit formulation for the following optima corresponding respectively to the cases $y_j \geq 0$ and $y_j < 0$:

$$\max\Big\{\sup_{0\leq c\leq rx_i}(U(c) - \beta V_{ij} + D_x^+ V_{ij}(rx_i - c) + D_y^+ V_{ij} by_j),$$

$$\sup_{c\geq rx_i}(U(c) - \beta V_{ij} + D_x^- V_{ij}(rx_i - c) + D_y^+ V_{ij} by_j)\Big\},$$

$$\max\Big\{\sup_{0\leq c\leq rx_i}(U(c) - \beta V_{ij} + D_x^+ V_{ij}(rx_i - c) + D_y^- V_{ij} by_j),$$

$$\sup_{c\geq rx_i}(U(c) - \beta V_{ij} + D_x^- V_{ij}(rx_i - c) + D_y^- V_{ij} by_j)\Big\}.$$

Finally, inside the domain, a numerical approximation V of the solution $\overline{u}$ of the equation $L'_0(x, y, \overline{u}, \overline{u}_x, \overline{u}_y) = 0$ will satisfy at each grid point,

$$g(D_x^- V_{ij}, D_x^+ V_{ij}, D_y^- V_{ij}, D_y^+ V_{ij}) = 0.$$

Below and only to simplify the presentation, we restrict ourselves to the case of the HARA utility functions. We note, however, that all our arguments can be easily extended to general utilities.

When U is a HARA utility function, given by $U(c) = \frac{1}{\gamma}c^\gamma$ with $\gamma \in (0, 1)$, g takes the following form at each grid point such that $y_j \geq 0$ (the extension to the case when $y_j < 0$ being straightforward):

(i) if $(D_x^+ V_{ij})^{\frac{1}{\gamma-1}} \leq rx_i$ and $(D_x^- V_{ij})^{\frac{1}{\gamma-1}} \leq rx_i$ then

$$g(D_x^- V_{ij}, D_x^+ V_{ij}, D_y^- V_{ij}, D_y^+ V_{ij}) = -\beta V_{ij} + \frac{1-\gamma}{\gamma}(D_x^+ V_{ij})^{\frac{\gamma}{\gamma-1}} + rx_i D_x^+ V_{ij} + by_j D_y^+ V_{ij},$$

(ii) if $(D_x^+ V_{ij})^{\frac{1}{\gamma-1}} \leq rx_i$ and $(D_x^- V_{ij})^{\frac{1}{\gamma-1}} > rx_i$ then

$$g(D_x^- V_{ij}, D_x^+ V_{ij}, D_y^- V_{ij}, D_y^+ V_{ij}) = -\beta V_{ij} + \frac{1-\gamma}{\gamma}(D_x^- V_{ij})^{\frac{\gamma}{\gamma-1}} + rx_i D_x^- V_{ij} + by_j D_y^+ V_{ij},$$

(iii) if $(D_x^+ V_{ij})^{\frac{1}{\gamma-1}} > rx_i$ and $(D_x^- V_{ij})^{\frac{1}{\gamma-1}} > rx_i$ then

$$g(D_x^- V_{ij}, D_x^+ V_{ij}, D_y^- V_{ij}, D_y^+ V_{ij}) = -\beta V_{ij} + \frac{1-\gamma}{\gamma}(D_x^- V_{ij})^{\frac{\gamma}{\gamma-1}} + rx_i D_x^+ V_{ij} + by_j D_y^+ V_{ij},$$

(iv) if $(D_x^+ V_{ij})^{\frac{1}{\gamma-1}} > rx_i$ and $(D_x^- V_{ij})^{\frac{1}{\gamma-1}} \leq rx_i$
then

$$g(D_x^- V_{ij}, D_x^+ V_{ij}, D_y^- V_{ij}, D_y^+ V_{ij}) = -\beta V_{ij} + \frac{1}{\gamma}(rx_i)^\gamma + by_j D_y^+ V_{ij}.$$

In addition to the above approximations one has to construct an approximation for the points located on the boundary of the domain $\mathcal{D}$.

For the rest of the section, we assume that the *non-transaction region* is strictly included in Ω. As the value function is a constrained viscosity solution of the Variational Inequality, it seems natural to impose state constraints on the boundary of Ω. Furthermore, since the value function satisfies respectively $u_y + (1+\lambda)u_x = 0$ and $u_y - (1-\mu)u_x = 0$ along these lines, we know exactly the trajectory of the state and choose the appropriate finite difference.

Next, at the point $i = 1$, $j = 1$, we impose the Dirichlet condition $V_{11} = 0$. Actually this value follows directly from the Variational Inequalities themselves, evaluated at the origin.

Finally, we impose Neumann conditions at the points located on $x = (M-1)\Delta x$ and $y = (L-1)\Delta y$. We have to assign given values to the normal derivatives but the results may vary strongly with the prescribed values, especially the location of the free boundaries. Actually, from the numerical experiments, it turns out that the error is essentially concentrated near the boundary.

Such a phenomenon has already been noticed by Barles, Daher & Romano (1991) for the heat equation and the Black–Scholes formula and by Fitzpatrick & Fleming (1991) for an Investment–Consumption model. Finally, we compute the value function and the free boundaries in a sufficiently large domain and only take into account the results inside the domain.

In the second half-iteration, we solve the monodimensional heat equation using a Crank–Nicholson scheme (see Raviart & Thomas (1983)). Such a scheme requires boundary conditions which are chosen as follows: on the boundary of Ω, we impose Dirichlet conditions whose values are provided by the first half-iteration. At the points located on $y = (L-1)\Delta y$, we have already imposed Neumann conditions. Thus the second half-iteration consists of inverting a tridiagonal matrix.

Let us recall that we have to choose the time step in order that the scheme be monotone. At each step, we choose the greatest value among those which preserve the monotonicity of the scheme. Actually, it yields a time step which is not far from being constant but may evolve a little during the convergence.

Finally, we compute the approximation using the following algorithm where i_0 denotes the first index such that there exists a point $(x_{i_0}, y_j) \in \mathcal{D}$ and for all i, j_i denotes the first index such that $(x_i, y_{j_i}) \in \mathcal{D}$:

Algorithm

1st step

- $V_{ij}^0 = x_i + y_j$
- C given

(n+1)st step

- V^n is given
- Construction of V_1:

$$V_{1,ij}^{n+1} = V_{ij}^n - \min(g_1, g_2) \times \Delta t.$$

- Construction of V on the boundary of Ω:

$$V_{ij_i}^{n+1} = V_{ij_i}^n + \Delta t \frac{(V_{i-1j_i+1}^n - V_{ij_i}^n)}{\sqrt{\Delta x^2 + \Delta y^2}} \text{ if } y_{j_i} \leq 0,$$

$$V_{ij_i}^{n+1} = V_{ij_i}^n + \Delta t \frac{(V_{i+1j_i-1}^n - V_{ij_i}^n)}{\sqrt{\Delta x^2 + \Delta y^2}} \text{ if } y_{j_i} > 0,$$
$$V_{11}^{n+1} = 0.$$

- Construction of V_2.

First half-iteration inside $\mathcal{D}$

$$V_{2,ij}^{n+1/2} = V_{ij}^n + \Delta t g(D_x^- V_{ij}^n, D_x^+ V_{ij}^n, D_y^- V_{ij}^n D_y^+ V_{ij}^n).$$

Second half-iteration inside $\mathcal{D}$

$$\frac{V_{2,ij}^{n+1} - V_{2,ij}^{n+1/2}}{\Delta t} = \frac{1}{2} y_j^2 \sigma^2 \Bigg[\frac{1}{2} \left(V_{2,ij+1}^{n+\frac{1}{2}} + V_{2,ij-1}^{n+\frac{1}{2}} - 2V_{2,ij}^{n+\frac{1}{2}} \right)$$
$$+ \frac{1}{2} \left(V_{2,ij+1}^{n+1} + V_{2,ij-1}^{n+1} - 2V_{2,ij}^{n+1} \right) \Bigg],$$

given that $V_{2,iL}^{n+1} = V_{2,iL-1}^{n+1} + C$.

- Construction of V^{n+1} from V_1^{n+1} and V_2^{n+1}:

$$V_{ij}^{n+1} = \max \left(V_{1,ij}^{n+1}, V_{2,ij}^{n+1} \right),$$

$$V_{iL}^{n+1} = V_{iL-1}^{n+1} + C \quad i_0 < i \leq M,$$

$$V_{Mj}^{n+1} = V_{M-1j}^{n+1} + C \qquad j_{M-1} < j \leq L.$$

- If $\sup_{ij} |V_{ij}^n - V_{ij}^{n+1}| < \epsilon$ then stop, where ϵ is a tolerance bound prescribed by the user.
- After the convergence is established, find the non-transaction region:

$$(NT) = \Big\{ (x_i, y_j) \in \mathcal{D} \ \text{ where } \ \min\{L_0, L_1, L_2\} = L_0 \Big\}.$$

The algorithm has been implemented on a HP workstation. In order to obtain satisfactory free boundaries, one has to let the algorithm converge for a few hours, the time being highly dependent on the choice of the utility function and the initial condition. To improve the efficiency of the algorithm, we constructed a second-order approximation with a TVD limiter (see for example Harten (1984), Osher (1985) for further details on total-variation-stable schemes). It is well known that such an approximation is not convergent in our case and in order to circumvent its lack of monotonicity, we applied a *filtered* scheme: at each iteration, when an upper bound on some second-order finite differences is not preserved, we replace the second-order operator by the first-order monotone one; although we are not able to justify the choice of the above criterion (and perhaps it is not the right one), this approach allows us to improve considerably the numerical results.

Numerical experiments Next, we present numerical experiments corresponding to two different classes of utility functions

(i) $U(c) = \frac{1}{\gamma} c^\gamma \quad 0 < \gamma < 1$

(ii) $U(c) = \frac{1}{\gamma 1} c^{\gamma 1} + \frac{1}{\gamma 2} c^{\gamma 2}$, with $0 < \gamma_1,\ \gamma_2 < 1$

Figures 1–3 show the non-transactional regions corresponding to the parameters $b = 0.12, \sigma = 0.4, \beta = 0.1$ and $r = 0.07$ computed with the second-order algorithm; in each case $X_{\max} = Y_{\max} = 10$ and $M = L = 401$; the first one corresponds to the utility function $U_1(c) = 2\sqrt{c}$ and the transaction costs are both equal to 0.01. In the second example the transaction costs have been set to 0.001. Figure 3 shows the computed non-transactional region corresponding to the utility $U_2(c) = 3c^{\frac{1}{3}} + \frac{3}{2}c^{\frac{2}{3}}$; both transaction costs have been set to 0.01.

6 Conclusions

In this article, we presented a class of numerical schemes for the value function and the optimal policies of an optimal investment and consumption model which was formulated as a singular stochastic control problem. Although, due to the presence of singular policies, the value function is not, in general, smooth, the convergence of the scheme is guaranteed by the uniqueness of

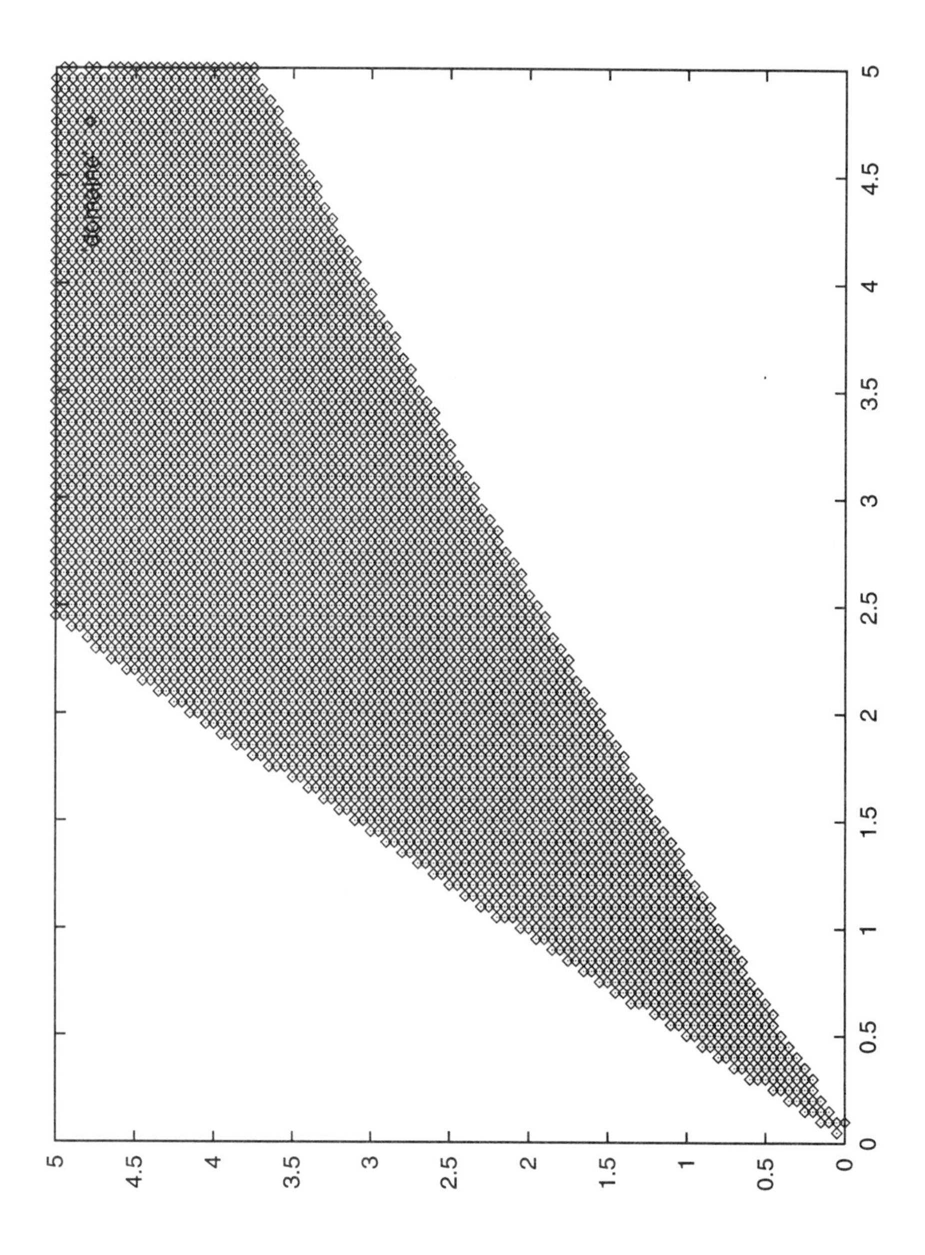

Figure 1

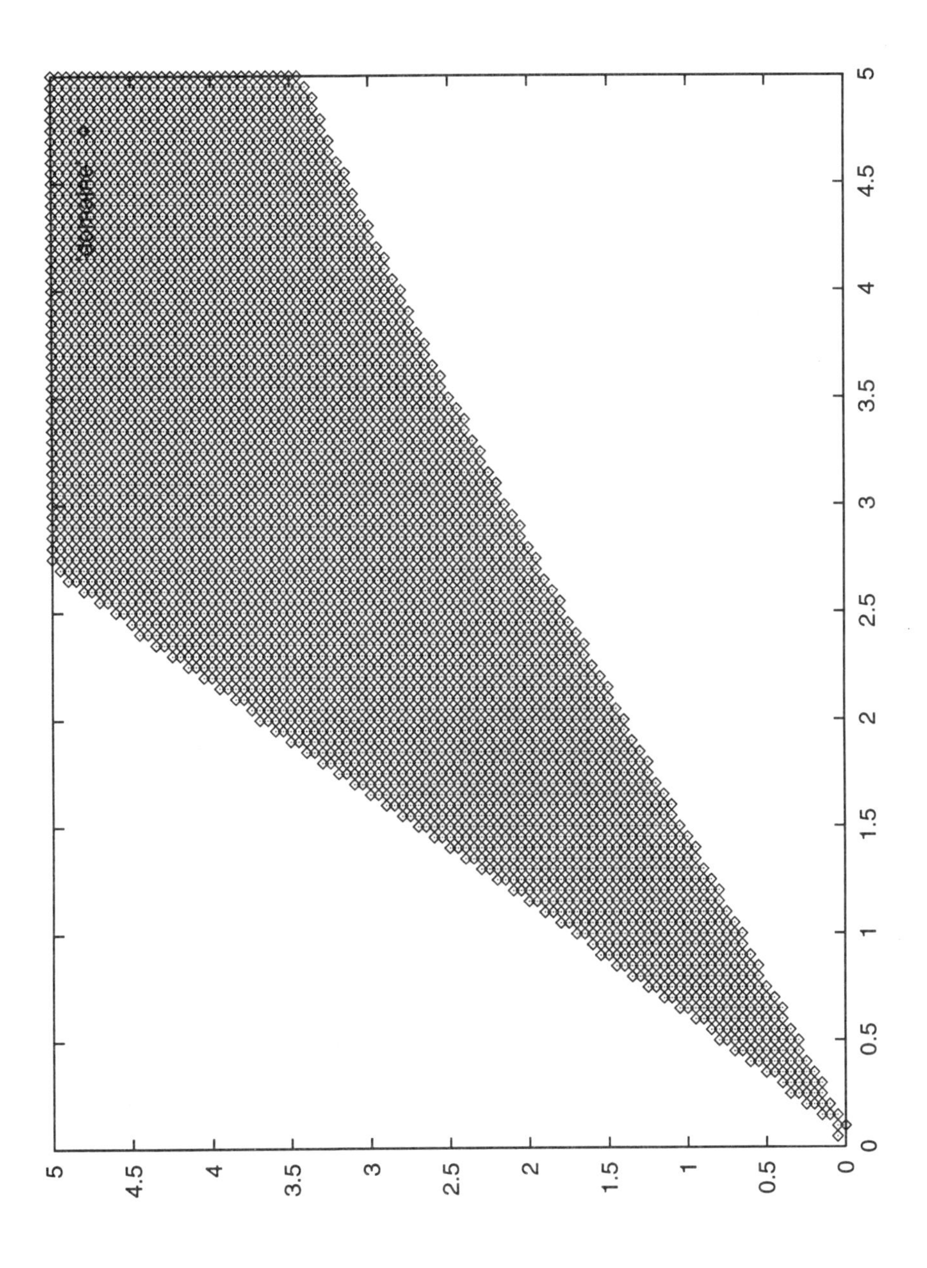

Figure 2

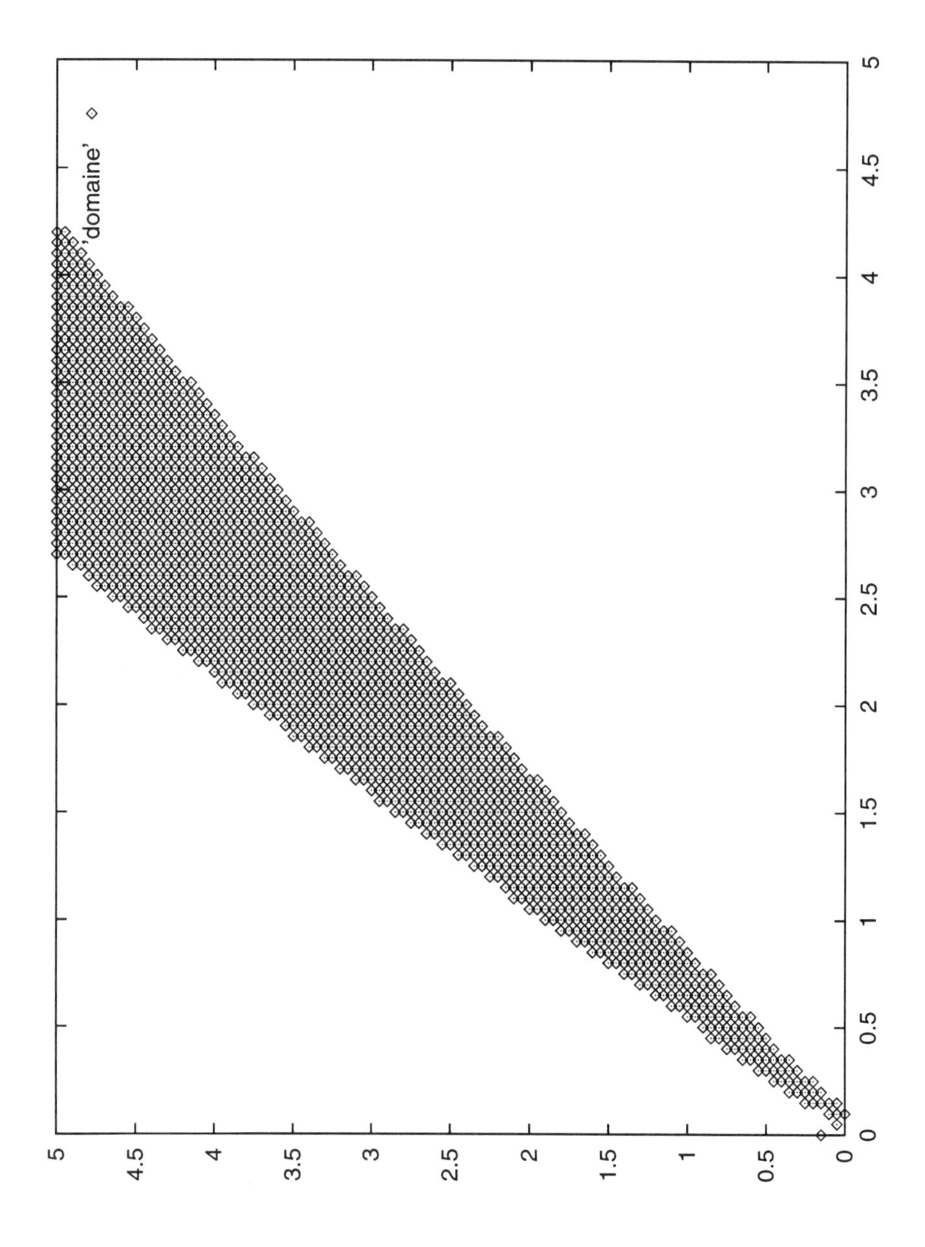

Figure 3

viscosity solutions of the Hamilton–Jacobi–Bellman equation. The numerical schemes developed here can be applied to a number of control problems with singular policies as well as problems in which some state dynamics are governed by stochastic processes and some others by deterministic ones; the latter problems give rise to degenerate HJB equations which are, in general, very hard to solve.

References

Alziary de Roquefort, B. (1991) 'Jeux différentiels et approximation numérique de fonction valeur, 2eme partie: étude numérique', *RAIRO Modél. Math. Anal. Numér.* **25**, 535–560.

Barles, G., Daher, C. & Romano, M. (1991) 'Convergence of numerical schemes for parabolic equations arising in finance theory', *Report, Caisse Autonome de Refinancement.*

Barles, G. & Souganidis, P.E. (1991) 'Convergence of approximation schemes for fully nonlinear second order equations', *J. Asym. Anal.* **4**, 271–283.

Brennan, M.J. (1975) 'The optimal number of securities in a risky asset portfolio when there are fixed costs of transacting: Theory and some empirical results', *J. Fin. and Quant. An.* **10**, 483–496.

Capuzzo-Dolcetta, I. (1983) 'On a discrete approximation of the Hamilton–Jacobi equation of dynamic programming', *App. Math. Optim.* **10**, 367–377.

Capuzzo-Dolcetta, I. & Falcone, M. (1989) 'Discrete dynamic programming and viscosity solutions of the Bellman equation', *Annales de l'Institut Henry Poincaré Analyse Non Linéaire* **6**, 161–181.

Capuzzo-Dolcetta, I. & Lions, P.L. (1995) 'Hamilton-Jacobi equations with state constraints', *Trans. Amer. Math. Soc.* **318**, 643–683.

Constantinides, G.M. (1979) 'Multiperiod consumption and investment behavior with convex transactions costs', *Management Sci* **25**, 1127–1137.

Constantinides, G.M. (1986) 'Capital market equilibruim with transaction costs', *J. Political Economy* **94**, 842–862.

Crandall, M.G., Ishii, H. & Lions, P.L. (1992) 'User's guide to viscosity solutions of second-order partial differential equations', *Bull. AMS* **27**, 1–67.

Crandall, M.G. & Lions, P.L. (1983) 'Viscosity solutions of Hamilton–Jacobi equations', *Trans. Amer. Math. Soc.* **277**, 1–42.

Crandall, M.G. & Lions, P.L. (1984) 'Two approximations of solutions of Hamilton–Jacobi equations', *Math. Comp.* **43**, 1–19.

Davis, M.H.A. & Norman, A.R. (1990) 'Portfolio selection with transaction costs', *Math. Op. Res.* **15**, 676–713.

Davis, M.H.A., Panas, V. & Zariphopoulou, T. (1993) 'European option pricing with transaction costs', *SIAM J. Control Optim.* **31**, 470–493.

Falcone, M. (1985) 'Numerical solution of deterministic continuous control problems', *Proceedings of the International Symposium on Numerical Analysis*, Madrid.

Falcone, M. (1987) 'A numerical approach to the infinite horizon problem of deterministic control theory', *Appl. Math. Optim.* **15**, 1–13.

Fitzpatrick, B.G. & Fleming, W.H. (1991) 'Numerical methods for an optimal investment/consumption model', *Math. Op. Res.* **16**, 823–841.

Fleming, W.H., Grossman, S., Vila, J.L. & Zariphopoulou, T. (1989) *Optimal portfolio rebalancing with transaction costs*, preprint.

Fleming, W.H. & Soner, H.M. (1993) *Controlled Markov Processes and Viscosity Solutions*, Springer Verlag.

Fleming, W.H. & Zariphopoulou, T. (1991) 'An optimal investment/consumption model with borrowing', *Math. Op. Res.* **16**, 802–822.

Goldsmith, D. (1976) 'Transaction costs and the theory of portfolio selection', *J. Finance* **31**, 1127-1139.

Harten, A. (1984) 'On a class of high resolution total-variation-stable finite-difference schemes', *SIAM Journal of Numerical Analysis* **21**, 1–23.

Ishii, H. & Lions, P.L. (1990) 'Viscosity solutions of fully nonlinear second-order elliptic partial differential equations', *J. Diff. Eqns.* **83**, 26–78.

Karatzas, I., Lehoczky, J., Sethi, S. & Shreve, S. (1987) 'Explicit solution of a general consumption/investment problem', *Mathematics of Operations Research* **11**, 261–294.

Kandel, S. & Ross, S.A. (1983) *Some intertemporal models of portfolio selection with transaction costs*, working paper No. 107, University of Chicago, Grad. School Bus., Center Res. Security Prices.

Leland, H.E. (1985) 'Option pricing and replication with transaction costs', *J. Finance* **40**, 1283–1301.

Levy, H. (1978) 'Equilibrium in an imperfect market: a constraint on the number of securities in the portfolio', *American Economic Review* **68**, 643–658.

Lions, P.L. (1983) 'Optimal control of diffusion processes and Hamilton–Jacobi–Bellman equations 1: The Dynamic Programming Principle and applications; 2: Viscosity solutions and uniqueness', *Comm. P.D.E.* **8**, 1101–1174; 1229–1276.

Lions, P.L. & Souganidis, P.E. (1995) 'Convergence of MUSCL and filtered schemes for scalar conservation laws and Hamilton–Jacobi equations', *Numeritsche*, to appear.

Merton, R.C. (1971) 'Optimum consumption and portfolio rules in a continuous-time model', *J. Economic Theory* **3**, 373–413.

Osher, S. (1985) 'Convergence of generalized MUSCL schemes', SIAM Journal of Numerical Analysis **29**, 947–961.

Raviart P.A. & Thomas J.M. (1983) *Introduction à l'Analyse Numérique des Equations aux Dérivées Partielles*, Masson.

Rouy, E. & Tourin, A. (1995) 'A viscosity solutions approach to shape-from-shading', *SIAM Journal of Numerical Analysis* **29**, 867–884.

Shreve, S.E. & Soner, H.M. (1994) *Optimal investment and consumption with transaction costs*, preprint.

Soner, H.M. (1986) 'Optimal control with state space constraints', *SIAM J. Control Optim.* **24**, 552–562; 1110–1122.

Souganidis, P.E. (1985) 'Convergence of approximation schemes for viscosity solutions of Hamilton–Jacobi equations', *J. Differential Equation* **57**, 1–43.

Taksar, M., Klass, M.J. & Assaf, D. (1986) *A diffusion model for optimal portfolio selection in the presence of brokerage fees*, Dept. of Statistics, Florida State University.

Tourin, A. (1992) *Thèse de Doctorat*, Université Paris IX-Dauphine (1992).

Tourin, A. & Zariphopoulou, T. (1994) 'Numerical schemes for investment models with singular transactions', *Computational Economics* **7**, 287–307.

Whalley, A.E. & Wilmott, P. (1996) 'An asymptotic analysis of an optimal hedging model for option pricing with transaction costs', *Math. Finance*, to appear.

Zariphopoulou, T. (1992) 'Investment/consumption model with transaction costs and markov-chains parameters', *SIAM J. Control Opt.* **30**, 613–636.

Zariphopoulou, T. (1994) 'Investment and consumption models with constraints', *SIAM J. Control Optim.* **32**, 59–84.

Zhu, H. (1986) *Characterization of Variational Inequalities in Singular Control*, PhD Thesis, Brown University.

Does Volatility Jump or Just Diffuse? A Statistical Approach.

Renzo G. Avesani and Pierre Bertrand

1 Introduction

1.1 From discrete time models (ARCH-GARCH) to continuous time models (SDE)

In recent years accurate analysis of financial time series has brought to the attention of researchers the existence of several anomalies. In particular, volatility estimation has always occupied a central place in the research on financial markets: this is due to the practical implications that an accurate measure of this quantity would have in terms of derivatives pricing. All the original models developed for pricing options and other interest rates derivatives such as the Black–Scholes model shared the feature of a constant volatility parameter. It is true that, for example, in the Cox, Ingersoll, & Ross (1985) term structure model, the volatility parameter is specified so that the stochastic process (X_t) describing the underlying asset shows a state dependent evolution, i.e., higher (lower) volatility for higher (lower) level of the state process. At the same time the specification chosen by Cox, Ingersoll, & Ross (1985) assumes σ constant:

$$dX_t = \beta(\theta - X_t)dt + \sigma\sqrt{X_t}\,dW(t).$$

This situation has been called into question in recent years due to two innovations which took place respectively in applied work and in theoretical finance.

On the first account we experienced an explosion of the econometric literature on ARCH-GARCH models which showed the pervasiveness of heteroscedasticity in the evolution of financial time series. On the other side, starting with Hull & White (1990), a new family of models incorporates the idea that volatility follows itself a diffusion process. These developments produced a new vein of empirical research aimed at the estimation of stochastic volatility using different approaches as in Gourieroux, Monfort & Renault (1993), Harvey, Ruiz & Shephard (1994), and Nelson (1990). At the same time growing evidence has been collected pointing to the presence of parameter instability in the estimation of the Cox–Ingersoll–Ross type of models.

Brown & Dybvig (1986) and Barone, Cuoco & Zautzik (1989) for example find in different situations that the constancy of the long term mean of the short term interest rate (θ) and of the recall parameter (β) is not supported by the data[1].

More problems pointing to the same issue of parameter instability have been evidenced more recently even inside the ARCH-GARCH approach by Hamilton & Susmel (1994). What has been questioned is the autoregressive nature of the process the variance should follow. In fact, the excessive variance persistence ARCH-GARCH models imply, reduces their forecasting performance. In particular it has been observed that big shocks have different implications for the future evolution of the variance with respect to small shocks. In the ARCH-GARCH set-up what we are supposed to observe is that small shocks are followed by small shocks and big shocks are followed by big shocks giving rise to the now well known volatility clustering effect first noticed by Mandelbrot (1963). What is troublesome with this approach is that it implies higher persistence for big shocks than for small shocks due to the autoregressive nature of the volatility process. Therefore a big shock implies a longer transition period before the process returns to its *natural* state. Financial time series instead seem to suggest a much faster adjustment path. Therefore the crucial question is: *how much persistence is there in financial volatility*?

From the analysis so far performed it appears that more work needs to be done in order to identify the nature of volatility evolution in financial time series data. This seems particularly needed since it has been only in recent times that data on financial transactions at a very high frequency became available. This type of data is particularly apt for testing the continuous time models that characterize a large part of theoretical finance. In fact so far most of the estimations of financial models were performed using daily data, which are a poor approximation for continuous time. It is our claim, supported also by Müller, Dacorogna, Davé, Olsen, Pictet, & Weizsäcker (1995) and Goodhart, Ito, & Payne (1994), that the observation of almost continuous time real financial processes will highlight a completely different set of dynamics from the ones discrete time econometric models brought to our attention.

The difficulties of this approach are just begining to unravel and they point out the necessity of linking continuous time stochastic process analysis with estimation. In order to reach this target we need to state our problem in terms of stochastic differential equations. We consider a stochastic process satisfying the following stochastic differential equation:

$$dX_t = b(t, X_t)\,dt + \sigma(t, X_t)\,dW_t. \tag{1.1}$$

[1]For example the September 1995 issue of *The Journal of Fixed Income* is almost entirely devoted to the problem of parameter instability in term structure models.

We have discretized observations of the process (X_t) at the times t_i and we want to identify an appropriate model for the diffusion coefficient $\sigma(t, X_t)$.

In this article we will develop a statistical approach and later present our work on the subject. Let us briefly mention two other approaches, most often used by the markets.

1.2 Two approaches to continuous time models

Implied Volatility

Volatility estimation is a crucial problem for every derivative instrument pricing formula. As a point of reference we use the Black–Scholes model where the price of an asset is assumed to satisfy the following stochastic differential equation

$$dS_t = S_t \left[r_0 \, dt + \sigma_0 \, dW_t \right], \qquad S(0) = S_0 \tag{1.2}$$

with constant drift r_0 and constant volatility σ_0. With this model Black & Scholes (1973) compute the value of a European call option as a function $u_{BS}(t, S_t, K, T, \sigma_0)$ which depends on time t, the value of the underlying asset S_t, the strike price K, the maturity T and the volatility σ_0. In this formula all the parameters are observable except for the volatility σ_0. Since the function $u_{BS}(t, S_t, K, T, \sigma_0)$ is a monotonic function of volatility, when the price of an option is observed in the market we can invert the Black and Scholes formula and get a unique value $\sigma_{impl}(t, S_t, K, T)$ called the *implied volatility*.

Let us stress the fact that this implied volatility is meaningful if and only if the Black–Scholes model is the appropriate one. In this case, this implied volatility should be constant and should correspond to the volatility of the underlying asset, often called the *historical volatility*. It is now well known that data observed in financial markets do not exactly comply with the hypothesis of the Black–Scholes model[2].

Many generalizations of the Black–Scholes model have been proposed. Up till now, two approaches have been used. The first one is based on implied volatility. In this case a model for the evolution of the underlying price is specified and the option prices are derived by identifying the unique volatility function with makes the theoretical price coincide with the quoted price. As an example of this approach we can consider the contributions of Derman & Kani (1994) and Dupire (1994).

We think that this approach presents a major problem since if the implied volatility does not correspond to the historical one then we are confronted with an inconsistency of the underlying model. Therefore, in our opinion the identification and estimation of the *historical* volatility has to be done beforehand.

[2]For alternative models see Hull & White (1987), Longstaff & Schwartz (1991).

Micro Ecomomic Derivation of a Stochastic Volatility

A second approach emphasizes the econonomic behavior of agents operating in the market[3]. On the other hand, Bensoussan, Crouhy & Galay (1994) derive an extension of the Black–Scholes model where stochastic volatility arises from the impact of a change in the value of the firm's asset on financial leverage. Both these approaches generate a stochastic volatility which is a given parametric function of the type $\sigma(t) = g(\theta, t, S_t)$, where the (multidimensional) parameter θ should be estimated from observed data. Statistical work is therefore needed in order to close the gap between theoretical and applied work. A good reference for parametric volatility estimation is Genon-Catalot & Jacod (1993).

Nonetheless we have to remark that since the volatility functions $g(\theta, t, x)$ proposed by the two approaches just mentioned are continuous and the parameter θ is assumed to be constant, these models do not allow for volatility jumps. Even when S_t is a discontinuous semi-martigale, the models produce volatility jumps at the same time that we observe jumps in the process S_t. Heuristic considerations and observed data suggest that volatility may jump at times when the underlying process remains continuous. Good support for our ideas is provided in Roure (1992) by the '*Gulf War effect*'. From daily observations of the French stock price index (CAC 40), Roure computes historical volatility with a simple estimator and observes that the volatility of CAC 40 index jumped from 10 % to 45% at the begining of the Gulf War and went back to 7% at its end.

1.3 Statistical approach to continuous time models

The aim of the statistical approach is to specify a class of model depending on unknown parameters and try to identify the parameter from the observed data. There are two different approaches which can be pursued in order to solve this problem:

1. **parametric estimation** when there are a finite number of unknown parameters ($\theta \in \mathbb{R}^q$),

2. **non-parametric estimation** when we estimate a functional parameter (which belongs to an infinite dimensional space).

In order to be useful, a statistical model has to be consistent; this means that the estimated parameters converge to the true ones. The speed of congergence of the different estimators is often obtained *via* the Central Limit Theorem.

A troublesome problem is the *misspecification* of the model. In fact if the true parameters belong to the class of model described by the parametrization,

[3]Platen & Schweizer (1994), for example, take into account traders' strategies.

the comparison in terms of speed of convergence is meaningful. If, on the other hand, the specification of the class of model chosen is not the true one then even a good estimator could give a very bad result. As an example consider the problem of estimating the volatitity of the original Black–Scholes model. As we know this model assumes that volatility, σ_0, is constant, while it has been observed that both historical and implied volatility seem to be time varying functions $\sigma(t)$. In this case the best estimator for constant volatility $\hat{\sigma}_0$ will converge to the mean value of the function $\sigma(t)$, i.e. $\lim_{\Delta \to 0} \hat{\sigma}_0^2 = (1/T) \int_0^T \sigma(t)^2 \, dt$ but the true volatility function would be almost always very different from the misspecified estimator $\hat{\sigma}_0$, except when $\sigma(t)$ is constant.

For these reasons, in our work, we chose a very large class of models within which to investigate the nature of the unknown volatility function.

Finally, procedures to test if a model does or does not correspond to the real data should be developed. This is called *test of model adequacy*. These tests unfortunately are not very often implemented though they should be central in any future research on volatility modelling[4].

Volatility estimation: the state of the art

Stochastic processes are observed only at discrete times. If we could observe the trajectory $(X_t, t \in [0,T])$ at each point in time (i.e. continuously), then the volatility would be the derivative of the quadratic variation almost surely (Lipster & Shiryaev, 1977).

This question was first analyzed by Dacunha-Castelle & Florens-Zmirou (1986). The parametric case, i.e., $\sigma(t,x) = \theta f(x)$, was studied by Dohnal (1987) for regular sampling times and by Genon-Catalot & Jacod (1993) in a general framework. However Fournié & Talay (1991) tried to estimate the Cox–Ingersoll–Ross model with constant coefficients on daily observations of the French short term interest rate and they concluded that at least the diffusion coefficient was time varying.

The non-parametric case for a non-time-varying diffusion coefficient, i.e., $\sigma(t,x) = \sigma(x)$, was treated by Florens-Zmirou (1993) for a one-dimensional process and by Brugière (1993) for multidimensional processes.

Florens-Zmirou (1989) gave a non-parametric estimator for a time varying volatility $\sigma(x,t) = \sigma(t)h(x)$. Genon-Catalot, Laredo & Picard (1992) introduced the non-parametric estimation by wavelet methods. A convenient presentation of this method and thresholding, a fast developing field of research in recent years, is contained in Hall & Patil (1995). Both Florens-Zmirou (1989) and Genon-Catalot *et al.* (1992) consider a time varying coefficient $\sigma(t)$ which is deterministic and a $\mathcal{C}^1$ function of time. However, for financial applications it seems more reasonable to consider a stochastic volatility $\sigma(t)$. In this case the coefficent $\sigma(t)$ would be less regular than $\mathcal{C}^1$.

[4]See Fournié (1993) for an adequacy test of the Cox–Ingersoll–Ross model for French interest rates.

Our contribution, as specified in Avesani & Bertrand (1996), assumes that volatility is best described by a mixture of a diffusion and a jump process.

Remark on the statistical estimation of the drift

It should be noted that the estimation of the drift coefficient in a continuous process raises different issues. In fact we cannot derive a consistent estimator even though we observe the entire trajectory of $(X_t, t \in [0,T])$. We only get convergence of the drift estimator when the upper bound of the interval of observation T becomes infinite, i.e. $T \to +\infty$ or when the volatility converges to 0. A good treatment of these questions is provided by Kutoyants (1984), (1994) and the references therein.

On estimating stochastic volatility with jumps

In the following sections, we will present a new method to detect volatility jumps, based on a *(non-parametric)* functional estimator.

In Section 2, we describe the model we will use and present some results on the existence and positiveness of the time varying coefficients of the Cox–Ingersoll–Ross type stochastic differential equation.

In Section 3, we compare performance of two non-parametric estimators: a wavelet estimator which is an orthonormal projection into the Haar basis and a symmetric kernel estimator which turns out to be a moving average estimator improved by a centring procedure. We would like to emphasize the dependence of the kernel estimator on the window size and the sampling interval. Surprisingly, the kernel estimator turns out to be better and more robust as soon as we have to detect a volatility jump.

In Section 4, we use the idea of varying the window size of the kernel estimator to build an estimator of the volatility jump times. We prove its consistency and present a Central Limit Theorem in order to prove that the errors of the estimator approach a normally distributed random variable. For technical reasons this Central Limit Theorem is proved under the assumption that the jump times are deterministic.

In the last section, we apply the kernel volatility estimator and the volatility jump time estimator to real financial data: the quotations of the Italian BTP Futures and the Italian one month Euro deposit rates. We also propose an economic analysis of the statistical results.

2 Description of the Model

We assume that there exists a stochastic basis $(\Omega, \mathcal{F}, \mathcal{F}_t, \mathbb{P})$ and a one-dimensional Wiener process (W_t) adapted to $(\Omega, \mathcal{F}, \mathcal{F}_t, \mathbb{P})$. We consider a stochastic process satisfying the following stochastic differential equation:

$$dX_t = b(t, X_t)\,dt + \sigma(t)\,h(X_t)\,dW_t \qquad (2.1)$$

where the function $h(x)$ is assumed to be known, the volatility coefficient $\sigma(x)$ is an unknown function of time and has to be estimated; the drift coefficient $b(t,x)$ could be unknown.

We observe one sample path of the process $(X_t, t \in [0,T])$ at discrete times t_i for $i = 0,\dots,N$. To simplify the notation, and without loss of generality[5], we suppose that the sampling intervals $t_{i+1} - t_i$ are constant and equal to Δ. We assume that Δ is small in comparison to T.

For the volatility evolution we will consider both a Hölder continuous function of time (to mimic a continuous diffusion) and a piecewise constant function (to mimic a jump process). El Karoui, Jeanblanc-Picqué & Shreve (1996) used the same characterization of stochastic volatility. As an example we can think of a Wiener process (W_t) with the standard filtration $(\mathcal{G}_t^W)$, and $\sigma(t)$ another stochastic process independent of the increments of W_t. Let $(\mathcal{F}_t)$ be the enlarged filtration; $\sigma(t)$ is $\mathcal{F}_t$-adapted and (W_t) is still a Wiener process for this filtration.

We will impose the following assumptions:

(A0) *$\sigma(t)$ is adapted to the filtration $\mathcal{F}_t$, $b(t,x)$ is a non-anticipative map, $b \in \mathcal{C}^1(\mathbb{R}^+, \mathbb{R})$ and $\exists L_T > 0$ such that $\forall t \in [0,T]$, $\mathbb{E}\sigma^4(t) \le L_T$.*

Moreover, we want to consider both a jump process and a diffusion process, respectively corresponding to the following assumptions:

(A1) *$\sigma(t) = \sum_{\rho=0}^{f} \sigma_\rho \mathbf{1}_{[t_\rho, t_{\rho+1})}(t)$ where t_ρ are the jump times. We assume there is a finite number of jumps, i.e. $f < \infty$. We denote $\delta\sigma_\rho^2 = \sigma_{\rho-1}^2 - \sigma_\rho^2$.*

(A2) *$\exists m > 0$ such that $\sigma^2(\cdot)$ is almost surely Hölder continuous of order m with a constant $K(\omega)$ and $\mathbb{E}K(\omega)^2 < \infty$.*

We need to control $\int_{t_i}^{t_{i+1}} b^4(s, X_s) ds$, thus we will use:

(B1) $\qquad \exists K_T > 0, \quad \forall t \in [0,T], \quad \mathbb{E}|b(t,X_t)|^4 \le K_T.$

In the following we work on the simplified model:

$$dX_t = b_1(t, X_t)dt + \sigma(t)\, dW_t. \tag{2.2}$$

In fact, under reasonable assumptions, the model (2.1) becomes (2.2) after the change of variable $Y_t = H(X_t)$ and $H(x) = \int_0^x h^{-1}(\xi) d\xi$.

Applications We give two examples of applications in financial models.

Example 1 Let S_t be an asset price satisfying the following stochastic differential equation

$$dS_t = S_t \left[a(t, S_t)dt + \sigma(t) dW_t \right]. \tag{2.3}$$

Then $X_t = \ln(S_t)$ satisfies (2.2) with $b(t,x) = a(t, e^{(x)}) - \frac{1}{2}\sigma^2(t)$ and $b(t,x)$ fulfils the assumption (B1).

[5] The case of irregular sampling intervals for the Centred Moving Average Estimator is treated in Avesani & Bertrand (1996).

Example 2 Let X_t be an asset price satisfying the stochastic differential equation:

$$dX_t = c(t)\,(\alpha(t) - X_t)\,dt + \sigma(t)\,\sqrt{X_t}\,dW_t \tag{2.4}$$

i.e., the Cox–Ingersoll–Ross differential equation with time varying parameters where $c(t), \alpha(t)$, and $\sigma(t)$ are $\mathcal{F}_t$ adapted. If $\mathbb{P}(X_t > 0, \forall t \in [0,T]) = 1$, then $Y_t = 2(X_t)^{1/2}$ satisfies (2.2) with $b(t,x) = 2x^{-1}\left[c(t)\alpha(t) - \frac{1}{4}\sigma^2(t)\right] - \frac{1}{2}\alpha(t)x$.

The Cox–Ingersoll–Ross model has been used to model the evolution of interest rates as a solution of a stochastic differential equation (2.4) with constant coefficients, i.e. $\sigma(t) = \sigma_0$, $c(t) = c_0$ and $\alpha(t) = \alpha_0$. The statistical work done so far in order to identify the values of the coefficients c_0, α_0, and σ_0 from real world financial data shows that these parameters are time-dependent (Brown & Dybvig 1986, Barone *et al.* 1989, Fournié & Talay 1991). It is therefore natural to consider the Cox–Ingersoll–Ross model with time-dependent stochastic coefficients.

In this situation we have to ensure that, as in the case of constant coefficients, the solution of this stochastic differential equation exists, is positive and does not vanish. In the case of constant coefficients a necessary and sufficient condition to have a strictly positive strong solution is

$$\sigma^2 \leq 2c\,\alpha \quad \text{and} \quad X(0) > 0\ a.s.$$

see for instance Ikeda & Watanabe (1989). We now present a generalization of these existence and positiveness results to the case of stochastic time varying coefficients and prove the sufficiency of the following conditions:

$$\forall t > 0, \sigma(t)^2 \leq 2c(t)\,\alpha(t) \quad \text{and} \quad X(0) > 0\ a.s. \tag{2.5}$$

This is stated in the theorem:

Theorem 2.1 *Assume that $c(t)$, $\sigma(t)$ and $\alpha(t)$ are $\mathcal{F}_t$ adapted, Condition (2.5) is satisfied and there exist $K, \lambda > 0$, such that $\forall t > 0$, $| c(t) | \leq K$ and $\lambda \leq \sigma(t) \leq K$. Then (2.4) has a unique strong solution (X_t) and $\mathbb{P}(X_t > 0, \forall t > 0) = 1$.*

Proof See Avesani & Bertrand (1996), Appendix A. □

3 Volatility Estimation

3.1 Description of the estimators

Two non-parametric estimators were proposed for a time-varying volatility: a wavelet estimator in Genon-Catalot *et al.* (1992) and a kernel estimator in

Florens-Zmirou (1989). Both studied a volatility which is a deterministic $\mathcal{C}^1$ function of time.

In the applications the sampling interval Δ is given and we have to choose the number of observations (window size denoted by A) taken into account to estimate $\sigma(t)$. In the ensuing notation, rather than arbitrarily fix A, we emphasize the dependence of the estimators on the window size and the sampling interval Δ.

Wavelet Estimator in the Haar basis

The function $\sigma(t)$ satisfies (A1) or (A2) thus is 0-regular. Recall that the wavelet basis corresponding to an r-regular multiresolution analysis is useful in studying the functional spaces $\mathcal{C}^r$. Therefore, the wavelet estimator is considered in a 0-regular multiresolution analysis associated with the Haar basis. After a change of notation with respect to Genon-Catalot *et al.* (1992), i.e. $\Delta = 2^{-n}, A = 2^{n-j(n)}, N = T\,2^n$, we get:

$$H_{A,\Delta}(t) = \sum_{k=0}^{N/A-1} \left\{ A^{-1} \sum_{i=0}^{A-1} \Delta^{-1} (X_{t_{kA+i+1}} - X_{t_{kA+i}})^2 \right\} \mathbf{1}_{[t_{kA}, t_{(k+1)A})}(t). \quad (3.1)$$

Centred Kernel Estimator

Considering the estimator associated to the kernel $\mathbf{1}_{[-1/2,1/2]}$ and using a change of notation with respect to Florens-Zmirou (1989), i.e., $\Delta = 1/n$ and $A = [nh_n]$, we obtain:

$$K_{A,\Delta}(t) = \sum_{j=A/2}^{N-A/2} \left\{ A^{-1} \sum_{i=-\frac{A}{2}}^{\frac{A}{2}-1} \Delta^{-1} (X_{t_{j+i+1}} - X_{t_{j+i}})^2 \right\} \mathbf{1}_{[t_j, t_{j+1})}(t) \quad (3.2)$$

for $t \in [\Delta A/2, T - \Delta A/2]$.

3.2 Pointwise convergence of the estimators

We first compare the rates of convergence of both estimators at each point of continuity of $\sigma(t)$ in $[0, T]$. This is an asymptotical point of view with the asymptotic $A \to +\infty$ and $A\Delta \to 0$. Theorem 3.1 shows that we obtain, pointwise, the same rate of convergence. In the following subsection, we will compare the functional error (mean integrated square error), and the results will no longer be asymptotic.

Theorem 3.1 *Assume that (A0), (B1) are satisfied, $A\Delta \to 0$ and $A \to \infty$. For both estimators (S=H or K), we have*

(i) *if* (A1) *holds and* $\mathbb{P}(\rho = t) = 0$, *then*

$$S_{A,\Delta}(t) \to \sigma^2(t) \text{in probability and } A^{1/2} \frac{[S_{A,\Delta}(t) - \sigma^2(t)]}{\sigma^2(t)} \Rightarrow \mathcal{N}(0, \sqrt{2}) \quad (3.3)$$

(ii) *If* (A2) *is satisfied then* $S_{A,\Delta}(t) \to \sigma^2(t)$ *in probability and if moreover* $\Delta A^{1+1/2m} \to 0$, *then* (3.3) *holds.*

Proof In order to prove the theorem we decompose the estimator into three terms. We will use the decomposition of the instantaneous quadratic variation, which follows from Itô's formula

$$\Delta^{-1}\left[X_{t_{i+1}} - X_{t_i}\right]^2 = \overline{\sigma}_i^2 + \xi_i + \eta_i \tag{3.4}$$

where

$$\begin{aligned}
\overline{\sigma}_i^2 &= \Delta^{-1}\int_{t_i}^{t_{i+1}} \sigma^2(s)ds \\
\xi_i &= \Delta^{-1}\int_{t_i}^{t_{i+1}} \sigma(s)\left[\int_{t_i}^{s} \sigma(u)\,dW_u\right] dW_s \\
\eta_i &= \Delta^{-1}\left\{\int_{t_i}^{t_{i+1}} b(s,X_s)(X_s - X_{t_i})\,ds + \int_{t_i}^{t_{i+1}} \sigma(s)\left[\int_{t_i}^{s} b(u,X_u)\,du\right] dW_s\right\}.
\end{aligned}$$

We notice that both estimators can be described as linear average operators of the instantaneous quadratic variation

$$H_{A,\Delta}(t) = \mathcal{M}^H(A,\, \Delta^{-1}\left[X_{t_{i+1}} - X_{t_i}\right]^2)(t) \tag{3.5}$$

$$K_{A,\Delta}(t) = \mathcal{M}^{MA}(A,\, \Delta^{-1}\left[X_{t_{i+1}} - X_{t_i}\right]^2)(t). \tag{3.6}$$

From the linearity of the operators, we get for instance

$$H_{A,\Delta}(t) = \mathcal{M}^H(A,\overline{\sigma}_i^2)(t) + \mathcal{M}^H(A,\xi_i)(t) + \mathcal{M}^H(A,\eta_i)(t). \tag{3.7}$$

At each point of continuity, we have $\lim_{A\Delta\to 0}\mathcal{M}^H(A,\overline{\sigma}_j^2)(t) = \sigma^2(t)$. Furthermore, $\mathbb{E}(\eta_i^2) = \mathcal{O}(\Delta)$. The proof is given in Bertrand (1996a), Appendix A. From Jensen's inequality, we deduce $\mathbb{E}\left[\mathcal{M}^H(A,\eta_j)(t)\right]^2 = \mathcal{O}(\Delta)$ which implies the convergence to 0 in $L^2(\Omega)$ of the third term of (3.7) when $\Delta \to 0$.

The second term $\mathcal{M}^H(A,\xi_i)(t)$ provides convergence in law, using a Central Limit Theorem for martingale arrays (Hall & Heyde, 1980). It turns out that the rate of convergence is exactly $A^{-1/2}$. This is another reason to consider the window size as a natural parameter of the problem. □

3.3 Mean integrated square error

Since the two estimators converge pointwise at the same rate we need to compare them more precisely. In order to do so we look at the functional norm of the error between the true and the estimated volatility function. For an estimator $S = H$ or K and a weight function $\gamma(t)$, we define

$$\mathrm{MISE}^S(A,\Delta) := \mathbb{E}\|S_{A,\Delta}(t) - \sigma^2(t)\|^2_{L^2(0,T;\gamma(t)dt}.$$

Using the decompositions (3.5), (3.6) of the estimators we can study precisely the Mean Integrated Square Error and we obtain the following proposition:

Proposition 3.2 *Assume that* (A0) *and* (B1) *are satisfied. Then for both estimators* ($S = H$ *or* K), *we have*

$$\mathrm{MISE}^S(A,\Delta) = \phi^S(A,\Delta) + \mathcal{O}(\Delta^{1/2})\,\phi^S(A,\Delta)^{1/2} \tag{3.8}$$

where

$$\begin{aligned} \phi^H(A,\Delta) &= \sum_{j=0}^{N-1} \gamma(t_j)\Delta V_j^2 + 2A^{-1}\sum_{j=0}^{N-1} \beta_j\,\Delta(\mathbb{E}\xi_j^2) \quad (3.9)\\ &\quad + \sum_{k=0}^{N/A-1}\sum_{j=kA}^{(k+1)A-1} \gamma(t_j)\Delta\mathbb{E}\left[\overline{\sigma}_j^2 - A^{-1}\sum_{i=kA}^{(k+1)A-1}\overline{\sigma}_i^2\right]^2\\ \phi^K(A,\Delta) &= \sum_{j=0}^{N-1} \gamma(t_j)\Delta V_j^2 + 4A^{-1}\sum_{j=0}^{N-1} \alpha_j\,\Delta(\mathbb{E}\xi_j^2) \quad (3.10)\\ &\quad + \sum_{j=0}^{N-1} \gamma(t_j)\Delta\mathbb{E}\left[\overline{\sigma}_j^2 - A^{-1}\sum_{i=-A/2}^{A/2}\overline{\sigma}_{j+i}^2\right]^2 \end{aligned}$$

with $\alpha_j = \mathcal{M}^{MA}(A,\gamma(t_i))(t_j)$, $\beta_j = \mathcal{M}^H(A,\gamma(t_i))(t_j)$ *and* V_j^2 *is the average variance of* $\sigma^2(s)$ *on the interval* $[t_j, t_{j+1}]$, *i.e.*

$$V_j^2 = \Delta^{-1}\int_{t_j}^{t_{j+1}} \left[\sigma^2(s) - \overline{\sigma}_j^2\right]^2 ds.$$

Proof See Bertrand (1996a). □

Formulas (3.9), (3.10) are useful for simulations with a known volatility $\sigma(t)$. Moreover, in the special, but interesting, case of isolated jumps we get an explicit formula from which we infer that $\mathrm{MISE}^H(A,\Delta)$ is an oscillating function of the window size A while $\mathrm{MISE}^K(A,\Delta)$ is a regular function most often smaller than $\mathrm{MISE}^H(A,\Delta)$ (Bertrand 1996b). All the simulations show that this is the case as soon as there is at least a volatility jump. Figure 1 shows a typical representation of $\mathrm{MISE}^K(A,\Delta)$ and $\mathrm{MISE}^H(A,\Delta)$ corresponding to the case of two volatility jumps.

Finally, since $\inf_A \phi^S(A,\Delta)$ is of order $\Delta^{1/2}$ we can disregard the second term on the right hand side of (3.8) and informally rewrite it as

$$\mathrm{MISE}^H(A,\Delta) \simeq \phi^S(A,\Delta).$$

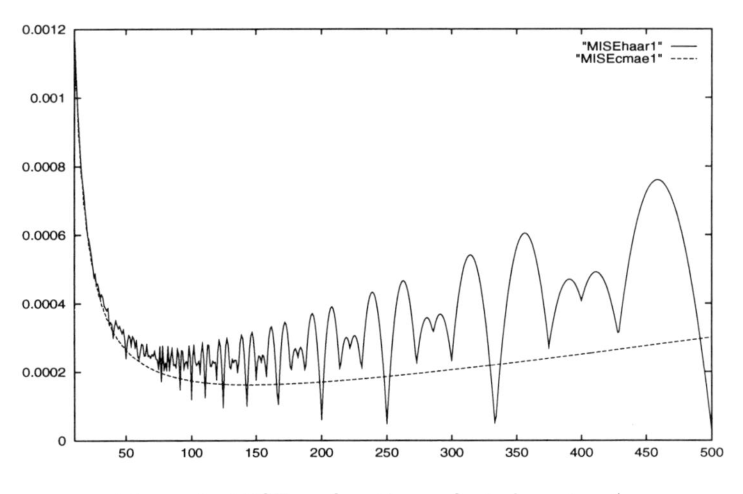

Figure 1: MISE as functions of window size A

3.4 Some numerical simulations

In this section, we consider two examples of known deterministic volatility function. We assume that $T = 1$ and fix $N = 5000$, $\Delta = 1/N$. We simulate a path of the process X_t using the Euler method with step $\Delta/10$. We consider two different specifications of the volatility evolution. For each one we plot two kernel estimators corresponding to two different window sizes, i.e., $A = 105$ and $A' = 3A$.

In the first example, the volatility is piecewise constant with jumps. See Figure 2.

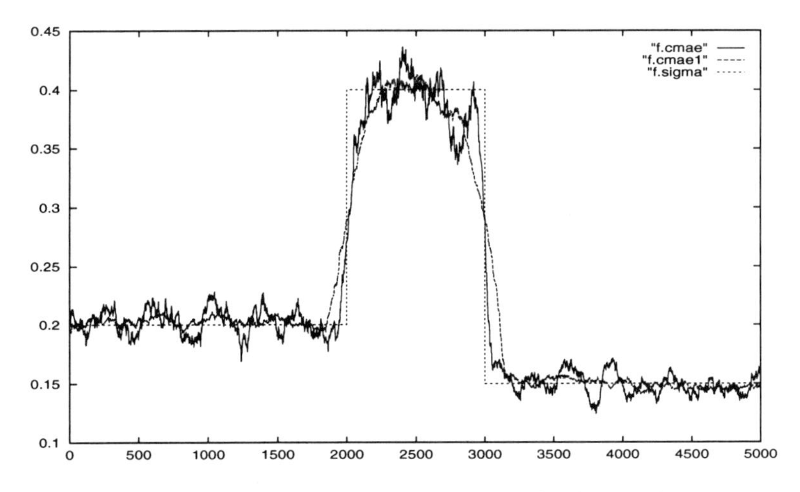

Figure 2: Estimation of piecewise constant volatility with jumps

The second example (Figure 3) is the case of jumps in volatility followed by a continuosly decreasing volatility. We take this as an example of jumps and diffusions in volatility. Notice that, after the second jump, volatility decreases particularly fast so that, *a priori*, it seems difficult to distinguish this from the case of a jump. However our estimator of volatility jump times seems to be able to distinguish this from a volatility jump.

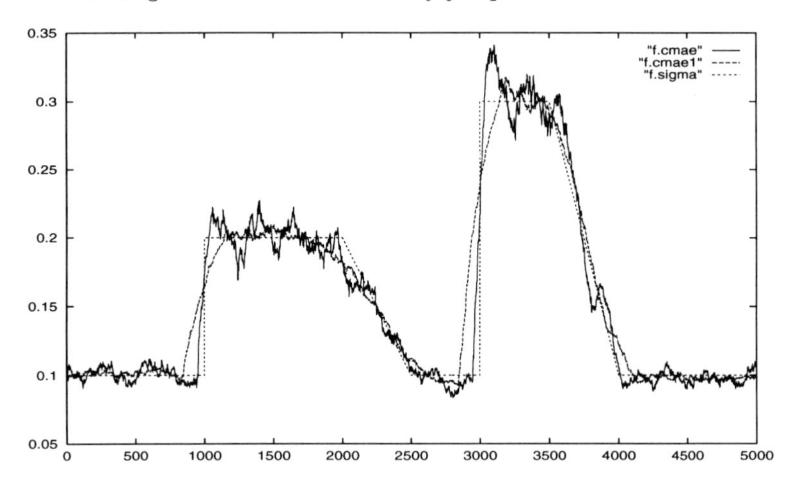

Figure 3: Estimation of volatility jumps and diffusions

4 Volatility Jump Time Estimation

For the two cases contemplated in Figures 2 and 3, the simulations just presented suggest that by varying the size of the windows we obtain two non-parametric estimations of volatility which cross each other in the neighborhood of the volatility jump time. Now we will make this heuristic remark more precise.

We want to built a jump time estimator by using the functional volatility estimator (3.2) since it proved to give better results for the functional norm $L^2(0,T)$. Let λ be a fixed integer $\lambda \geq 2$. To simplify, we assume that t_ρ is an isolated jump time. We denote:

$$\begin{aligned} M_{A,\Delta}(t_j) &:= \mathcal{M}^{MA}(A, \overline{\sigma}_i^2)(t_j) \\ N_{A,\Delta}(t_j) &:= \mathcal{M}^{MA}(A, \xi_i)(t_j) \\ D_{A,\Delta}(t_j) &:= \mathcal{M}^{MA}(A, \eta_i)(t_j). \end{aligned}$$

From decomposition (3.7) we have:

$$\begin{aligned} S_{A,\Delta}(t_j) - S_{\lambda A,\Delta}(t_j) &= M_{A,\Delta}(t_j) - M_{\lambda A,\Delta}(t_j) \\ &+ N_{A,\Delta}(t_j) - N_{\lambda A,\Delta}(t_j) + D_{A,\Delta}(t_j) - D_{\lambda A,\Delta}(t_j) \end{aligned}$$

where

$$\mathbb{E}|N_{A,\Delta}(t_j) - N_{\lambda A,\Delta}(t_j)|^2 \leq KA^{-1}$$

and

$$\mathbb{E}|D_{A,\Delta}(t_j) - D_{\lambda A,\Delta}(t_j)|^2 \leq K\Delta.$$

Therefore these two terms could be disregarded when $A \to \infty$ and $\Delta \to 0$. At this point we are left with the first term which is not small. Let t_ρ be the volatility jump time and assume it is an isolated jump time, i.e., there is no other jump time in the interval $[t_{\rho-\lambda A}, t_{\rho+\lambda A}]$. We have:

$$M_{A,\Delta}(t_j) - M_{\lambda A,\Delta}(t_j) = \phi(A^{-1}(j-\rho))$$

where $\phi(\cdot)$ is an affine function vanishing at zero (Avesani & Bertrand, 1996). This explains why the two estimators cross each other in the neighborhood of the volatility jump time t_ρ. More precisely we have the following result.

Proposition 4.1 *Assume that (A0), (A1) and (B1) are satisfied. Then* $\lim_{A\to\infty,\, A\Delta\to 0} t_c(A) = t_\rho$ *in probability.*

Proof See Avesani & Bertrand (1996). □

When the jump time is deterministic (even with random jumps), we have a more precise result. In this case we have a Central Limit Theorem.

Theorem 4.2 *Assume that* (A0), (A1) *and* (B1) *are satisfied.*

Let $B_{A,\Delta} = \{\omega$ such that $\exists c$ with $S_{A,\Delta}(t_c) = S_{\lambda A,\Delta}(t_c)$ and $|c-\rho| \leq A/2\}$. *If* $A \to \infty$ *and* $A\Delta \to 0$ *then* $\mathbb{P}(B_{A,\Delta}) \to 1$. *Moreover, on the set* $B_{A,\Delta}$ *we have*

$$A^{-1/2}(\rho - c(A)) = U_1(A) + V_1(A) \tag{4.1}$$

with

$$U_1(A) \overset{(\mathcal{L})}{\Rightarrow} \left(\frac{\lambda}{\lambda-1}\right)^{1/2} \nu_\rho^{-1} \mathcal{N}(0,\sqrt{2})$$

where

$$\nu_\rho = \frac{\delta\sigma_\rho^2}{[1/2\sigma_{\rho-}^4 + 1/2\sigma_{\rho+}^4]^{1/2}}$$

and for every $\alpha < 1/4$ *there exist two constants* $K_{\alpha,\lambda}$ *and* $\tilde{K}_{\alpha,\lambda} > 0$ *such that*

$$\mathbb{E}V_1(A)^2 \leq K_{\alpha,\lambda}\Delta + \tilde{K}_{\alpha,\lambda}(\Delta^\alpha + A^{-\alpha}).$$

Proof See Avesani & Bertrand (1996), Appendix D. □

Numerical Application We take $\lambda = 3$ and fix a large enough window size, for example $A = 100$. In this case we have:

$$A^{-1/2}\, U_1(A) \overset{(\mathcal{L})}{\Rightarrow} \sqrt{3}\, \nu_\rho^{-1} \mathcal{N}(0,1)$$

and the parameter ν_ρ^{-1} gives us the variance of the error of the jump time estimation. To give a feeling of the results we present two examples. Suppose we have a volatility jumping from 10% to 13%. In this case $\nu_\rho^{-1} = 2.01$, while if volatility jumps from 10% to 11%, then $\nu_\rho^{-1} = 5.28$. From this it follows that in the 3% jump we have $t_\rho - t_c \simeq \mathcal{N}(0, 35\,\Delta)$, while for the 1% jump $t_\rho - t_c \simeq \mathcal{N}(0, 92\,\Delta)$. Therefore in the case of a small volatility jump the estimator is meaningless and we can detect only jumps bigger than 2%..

5 What do we see in Real Data ?

In order to highlight the differencies in behavior of data with different time resolution we use two sets of data.

The first is the series of daily observations of one month Italian Euro deposit rates from 1984 to 1994. In this period the Italian Euromarket became very liquid and we can observe several shocks. These were caused by the impact on the interest rate of expectations of realignments of the Italian lira.

The second series is the Italian BTP (ten year bond) futures price as a five minutes average of quotations from January 1994 to May 1994. For the BTP futures this period was crucial since the political instability which characterized Italy severely hit, in particular, the bond market. Moreover, at the beginning of 1994 the Federal Reserve suddenly changed the course of its monetary policy and this action had a strong impact on all the major bond markets raising the degree of uncertainty.

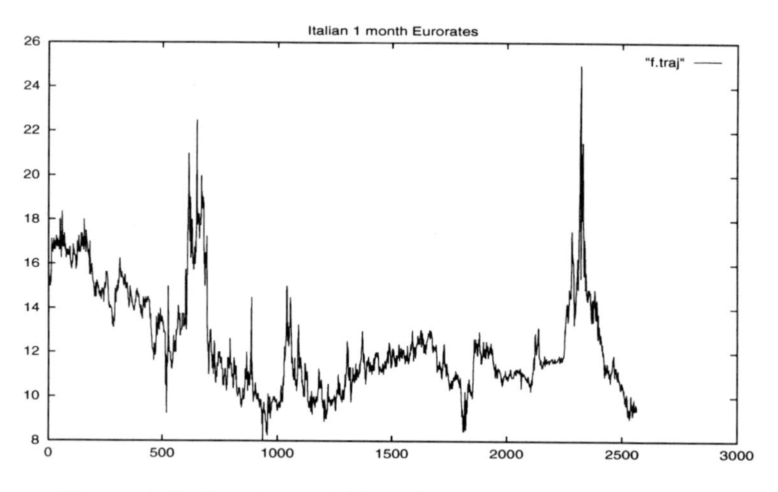

Figure 4: Italian 1 month Euro deposit rates 1984–1994

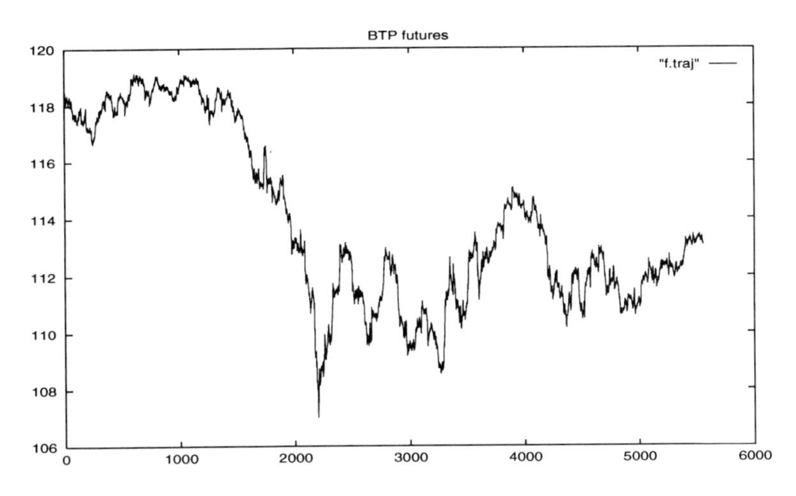

Figure 5: Italian 1 month Euro Italian BTP futures Jan 1994-May 1994

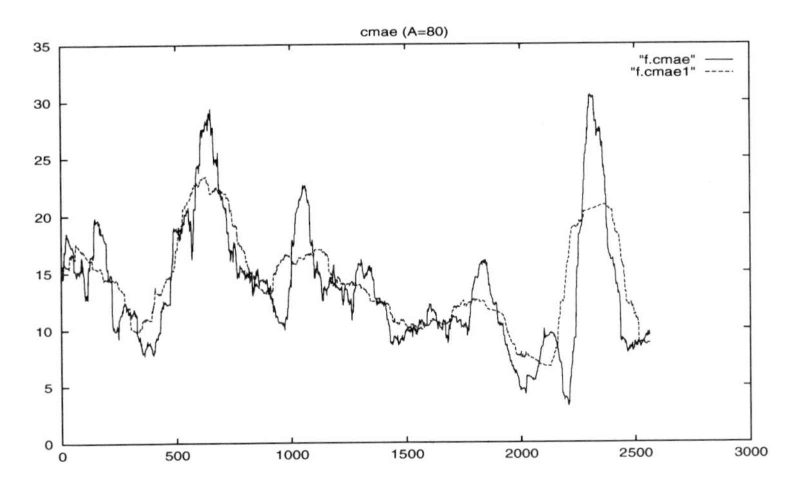

Figure 6: Volatility non-parametric estimation of Euro deposit rates

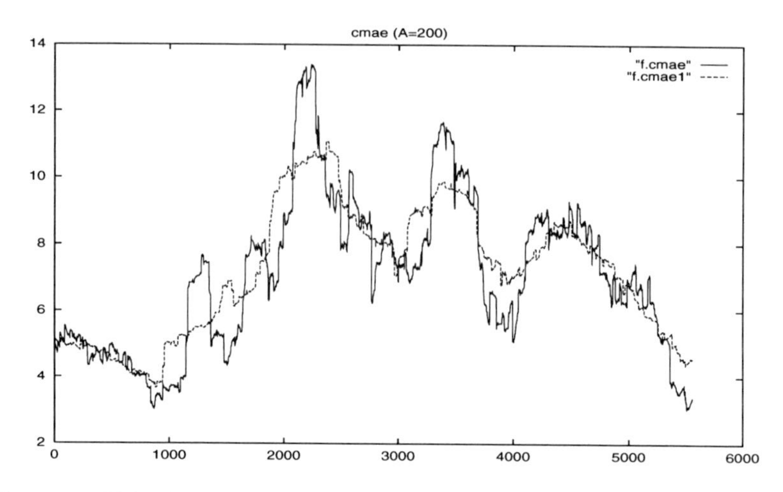

Figure 7: Volatility non-parametric estimation of Italian BTP futures

The observation of these two series seems to suggest a different local behavior of the two data-generating processes. In the interest rate series few clear episodes of changes in volatility are visually detectable.

On the other hand, in the BTP futures series substantial changes in volatility seem to happen more often.

The estimations just proposed suggest several interesting considerations. In both cases the proposed estimator seems to track correctly the expected volatility evolutions. For the Eurorates series we observe episodes of volatility explosion the most notable of which is the one corresponding to September 1992. At that time the Italian lira was forced out of the EMS and we had an enormous increase in uncertainty on interest rate markets. For the BTP futures we can see that in addition to the several sudden jumps we have an initial period of increasing volatility followed by a gradual decline.

What seems to be interesting is that for example in daily data (the Eurorates) if we use a window size of 80 days we see clear abrupt changes. At the same time, as should be expected, with a window which is about three times bigger (270 days) the volatility evolution becomes smoother and a certain degree of autocorrelation seems to emerge. The same behavior is apparent for the BTP futures. Therefore we are brought back to the initial question: when we talk about persistence what is a reasonable time horizon? Furthermore, if the volatility micro structure presents relevant breaks, are we doing the right thing by imposing instead an autoregressive structure?

In any case the crucial question remains that of evaluating the structure of the process followed by the instantaneous volatility.

References

Avesani, R.G. & Bertrand, P. (1996), 'Jumps and diffusion in volatility : it takes two to tango', INRIA Research Report 2848, submitted for publication.

Barone, E., Cuoco, D. & Zautnik, E. (1989) 'La struttura dei rendimenti per scadenza secondo il modello di Cox, Ingersoll e Ross: una verifica empirica', *Temi di discussione* **128** , Banca d'Italia.

Bensoussan, A., Crouhy, M. & Galay, D. (1994) 'Stochastic equity volatility related to the leverage effect I: Equity volatility behavior', *Applied Mathematical Finance* **1**, 63–85.

Bertrand, P. (1996a) 'Estimation of the stochatic volatility of a diffusion process I. Comparison of Haar basis estimator and kernel estimators', Rapport Technique INRIA **2739** (submitted).

Bertrand, P. (1996b) 'Comparaison de l'erreur quadratique moyenne intégrée pour différents estimateurs du coefficient de diffusion d'un processus', note soumise au C.R.A.S.

Black, F., & Scholes, M. (1973) 'The pricing of options and corporate liabilities' *Journal of Political Economy* **81**, 637–659.

Brown, S.J. & Dybvig, P.H. (1986) 'The empirical implications of the Cox–Ingersoll–Ross theory of the term structure of interest rates', *Journal of Finance* **41**, 3.

Brugière, P. (1993) 'Théorème de limite central pour un estimateur non paramétrique de la variance d'un processus de diffusion multidimensionelle', *Annales de l'I.H.P. Probabilités et Statistiques*, **29**, 357–389.

Cox, J., Ingersoll, J. & Ross, S. (1985) 'A theory of the term structure of interest rates.', *Econometrica* **53**, 363–84.

Dacunha-Castelle, D. & Florens-Zmirou, D. (1986) 'Estimation of the coefficient of diffusion from discrete observations', *Stochastic* **19**, 263–284.

Derman, E. & Kani, I. (1994) 'The volatility smile and its implied tree', *Goldman Sachs Quantitative Strategies Research Notes.*

Donhal, G. (1987) 'On estimating the diffusion coefficient', *Journal of Applied Probability* **24**, 105–114.

Dupire, B. (1994) 'Pricing a smile', *RISK* **7**, 18–20.

El Karoui, N., Jeanblanc-Picqué, M. & Shreve, S.E. (1996) 'On the robustness of the Black and Scholes formula', *Mathematical Finance*, to appear.

Florens-Zmirou, D. (1989) 'Estimation de la variance d'une diffusion à partir d'une observation discrétisée', *C.R.A.S.* **309**, Série I, 195–200.

Florens-Zmirou, D. (1993) 'On estimating the variance of diffusion processes', *Journal of Applied Probability* **30**, 790–804.

Forster, D.P. & Nelson, D.B. (1994) 'Continuous record asymptotics for rolling sample variance estimator', *Working Paper, National Bureau of Economic Research* **163**.

Fournié, É. (1993) *Statistiques des Diffusions Ergodiques avec Applications en Finance*, Thèse de l'Université de Nice-Sophia-Antipolis.

Fournié, É. & Talay, D. (1991) 'Application de la statistique des diffusions à un modèle de taux d'intérêt', *Finance* **12**, 79–111.

Genon-Catalot, V., Laredo, C., & Picard, D. (1992) 'Non-parametric estimation of the diffusion coefficient by wavelet methods', *Scandinavian Journal of Statistics* **19**, 317–335.

Genon-Catalot, V. & Jacod, J. (1993) 'On the estimation of the diffusion coefficient for multi-dimensional diffusion processes', *Annales de l'I.H.P. Probabilités et Statistiques*, **29**, 119–151.

Goodhart, C. A., Ito, T., & Payne, R. (1994) 'One day in June, 1993: a study of the working of the Reuters dealing 2000–2 electronic foreign exchange trading system', *NBER Technical Working Paper*.

Gouderioux, C., Monfort, A. & Renault, E. (1993) 'Indirect inference', *Journal of Applied Econometrics* **8**, 85–118.

Hall, P. & Heyde, C.C. (1980) *Martingale Limit Theory and its Applications*, Academic Press, San Diego.

Hall, P. & Patil, P. (1995) 'On wavelet methods for estimating smooth functions', *Bernoulli* **1**, 41–58.

Hamilton, J. & Susmel, R. (1994) 'Autoregressive conditional heteroscedasticity and changes in regime', *Journal of Econometrics* **64**, 307–333.

Harvey, A., Ruiz, E. & Shephard, N. (1994) 'Multivariate stochastic variance models', *Review of Economic Studies* **61**, 247–264.

Hull, J. , & White, A. (1990) 'Pricing interest rate derivative securities', *Review of Financial Studies* **3**, 573–92.

Ibragimov, I.A. & Has'minskii, R.Z. (1981) *Statistical Estimation, Asymptotic Theory*, Springer Verlag.

Ikeda, N. & Watanabe, S. (1989) *Equation and Diffusion Processes*, 2nd ed., North Holland.

Karatzas, I. & Shreve, S.E. (1988) *Brownian Motion and Stochastic Calculus*, Springer Verlag.

Kutoyants, Yu.A. (1984) *Parameter Estimation for Stochastic Processes*, Heldermann Verlag.

Kutoyants, Yu.A. (1994) *Identification of Dynamical Systems with Small Noise*, Kluwer.

Lipster, R. S. & Shiryaev, A.N. (1977) *Statistics of Random Processes I*, Springer Verlag.

Longstaff, F. & Schwartz, E. (1991) 'Volatility and the term structure: a two factor general equilibrium model', *The Journal of Finance* **47**, 1259–1282.

Mandelbrot, B. (1963) 'The variation of certain speculative prices', *Journal of Business* **36**, 394–419.

Müller, U.A., Dacorogna, M., Davé, R.D., Olsen, R.B., Pictet, O.V. & Weizsäcker (1995) 'Volatilities of different time resolutions. Analyzing the dynamics of market components', Olsen and Associate Research Group, Paper presented at the First International Conference On High Frequency Data in Finance.

Nelson, D. B. (1990) 'ARCH models as diffusion approximation' *Journal of Econometrics* **45**, 7–39.

Platen, E. & Schweizer, M. (1994) 'On smile and skewness', *Statistics Research Report* **SRR 027–94**.

Roure, F. (1992) *Stratégie Financière sur le MATIF et le MONEP*, Economica.

Martingale-Based Hedge Error Control

Peter Bossaerts and Bas Werker

1 Introduction

In the seminal paper of Black & Scholes (1973), the payoff on a European option is replicated perfectly by a dynamic strategy in the underlying security and a zero-coupon bond. Hence, the price of the replicating portfolio determines the no-arbitrage value for the option. Since that work of Black and Scholes, option pricing has been following this standard pattern of first specifying a perfect dynamic hedge, followed by an exploration of the pricing implications of the absence of arbitrage.

In Black–Scholes model, markets are complete in the sense that the payoff on the option is attainable through a dynamic trading strategy in the underlying stock and a money account. In practice, however, some of their assumptions can be proven wrong. Foremost, volatility is stochastic. In the absence of an asset that is instantaneously perfectly correlated with the volatility of the stock, market completeness is thereby invalidated. Likewise, actual rebalancing necessarily occurs over discrete intervals of time, in contrast with Black–Scholes continuous rebalancing.

If markets are incomplete, a blind application of complete-markets hedging strategies seems inappropriate. Most importantly, such policies do not self-correct even if a significant deviation from the target payoff is apparent. That immediately raises the question of whether there are hedging strategies that continuously correct in a way that facilitates error control. In other words, are there policies that adjust to past tracking errors such that the stochastic characteristics of the total tracking error become well-specified?

This article introduces one such strategy whose total tracking error can be characterized by martingale theory. In contrast with standard complete-markets hedging policies, the total value to be rebalanced into a new hedge position at any point in time may differ from the value generated by the hedge in the past. Sometimes a surplus is generated, to be invested in risk-free securities; at other times, insufficient funds flow out of past positions, and, hence, money has to be borrowed to reestablish the right hedge. While the hedge portfolio can still be considered self-financing, only part of it is actively engaged in hedging the risk of the option.

By construction, the tracking error of our hedging strategy follows a martingale. Consequently, martingale theory characterizes its stochastic behavior. A stochastic bound on the total tracking error follows immediately. This is useful for *margin analysis*: a particular probability value can be assigned to the likelihood that the total tracking error will exceed a given level.

When restricting the properties of the proposed hedging strategy, we obtain a *central limit result* for the asymptotic behavior of the total tracking error as the frequency of rebalancing increases. This is attractive: the difference between the payoff on the derivative and that of the hedge portfolio will follow the normal distribution, independent of the stochastic properties of the price of the underlying assets or the nature of the derivative. There may be leptokurtosis and the derivative may be an exotic option. Most importantly, two-parameter models (CAPM, APT) can be used to price the risk of the hedging strategy.

One can now see the advantage of our approach. Instead of determining the strategy that minimizes the tracking error for some loss function (as in Föllmer & Sondermann (1986), Duffie & Richardson (1991), Schweizer (1994), Gouriéroux, Laurent & Pham (1995)), we focus on a hedging policy whose error can be characterized by means of martingale theory. In certain cases, this allows one to value the risk of the tracking error using standard asset pricing models.

We subsequently specify a general statistical model of the relationship between the payoffs of the derivative, on the one hand, and those of the underlying securities, on the other. Under the model, the ingredients for our hedging strategy are easily *estimated*, e.g., by means of local parametric estimation (see Bossaerts & Hillion (1996)).

Under our hedging strategies, the hedge portfolio consists of two parts: (i) a portfolio of the underlying securities, effectively used to hedge the derivative, which we refer to as the *active hedge*, (ii) the accumulation of the differences between a newly established active hedge and the proceeds from the past active hedge, which we refer to as the *financing shortfall.* In traditional complete-markets hedges, the latter equals zero. For simplicity, we assume that the financing shortfall (which could be negative) is invested in risk-free bonds. If markets are complete, the strategy that the model generates coincides with traditional Black–Scholes hedging. Hence, within the framework of the posited statistical model, the proposed hedging strategy encompasses standard policies.

Our statistical model of the relation between the derivative's payoffs and those on underlying securities allows there to be a discrepancy between the value of the active hedge and the price of the derivative, to be referred to as the *pricing delta.* (In the empirical literature, differences between a model's price and the actual price are referred to as pricing error; it would be inappropriate to continue this tradition here.) Of course, the presence of a nonzero pricing delta is possible in an incomplete markets context. In complete markets, other justifications for the pricing delta must be advanced. Market microstructure provides one (following Bossaerts & Hillion (1993)). Other interpretations include those based on sampling error in traders' estimation of crucial parameters (as in Lo (1986)), or on randomization of the martingale pricing measure (as in Clément, Gouriéroux & Monfort (1995)).

There have been few attempts to control the error caused by misspecification in a standard hedge because of market incompleteness. El Karoui & Jeanblanc-Picqué (1993) are a prominent exception. They compute bounds on the error from implementing Black–Scholes hedges when stochastic volatility generates market incompleteness. Our approach differs in that we leave Black–Scholes hedging altogether.

The remainder of this article is organized as follows. Section 2 introduces our hedging strategy, designed to facilitate stochastic error control. Section 3 uses martingale theory to characterize the behavior of the tracking error from the proposed strategy. In Section 4, a statistical model of the payoffs of derivatives and underlying securities is specified. Under this model, the ingredients of the proposed hedging strategy can easily be estimated. Section 5 discusses the economic meaning of the statistical model; in particular, its allowing for a discrepancy between the price of the derivative and the value of the active hedge. Section 6 concludes.

2 The Hedging Strategy

Let us consider hedging the payoff on a derivative with expiration date τ. Its payoff at maturity is a function g of the values of a number of underlying assets, collected in the (random) vector V.[1] It is to be hedged with a portfolio of the underlying asset. We do not have to be specific about the nature of the dynamic hedge strategy yet. For the moment, we are mainly interested in the general specification of a duplication strategy.

Let t denote time; $t \in [0, \tau]$. The hedge portfolio is (re)balanced n times, at times $\tau\frac{j}{n}$, for $j = 0, \ldots, n-1$, and liquidated at $t = \tau$. Let R_j^n denote the *return on the hedge portfolio* over period $(\tau\frac{j-1}{n}, \tau\frac{j}{n})$. Let $\overline{R}_j^n$ be the conditional expected return on the hedge portfolio, given information at time $\tau\frac{j-1}{n}$. $\mathcal{F}_{j-1}^n$ denotes the conditioning information at time $\tau\frac{j-1}{n}$. It includes the past values of the underlying asset, as well as any information used in the the determination of the weights of the hedges.

[1]The analysis of the article continues to hold for derivatives whose payoff is path-dependent, in which case g would be a function of past values of the underlying assets as well. We refrain from this level of generality, in order to simplify the exposition.

Recursively define the stochastic process $\{H_j^n\}_{j=0}^n$:

$$\begin{aligned}
H_n^n &= g(V) \\
H_{n-1}^n &= E[\frac{H_n^n}{R_n^n}|\mathcal{F}_{n-1}^n] \\
H_{n-2}^n &= E[\frac{H_{n-1}^n}{R_{n-1}^n}|\mathcal{F}_{n-2}^n] \\
&\dots \\
H_1^n &= E[\frac{H_2^n}{R_2^n}|\mathcal{F}_1^n] \\
H_0^n &= E[\frac{H_1^n}{R_1^n}|\mathcal{F}_0^n].
\end{aligned} \tag{2.1}$$

Consider the following hedge strategy. At time 0, invest H_0^n in the hedge portfolio. Hence, the initial outlay P_0^n equals H_0^n. At time $t\frac{1}{n}$, the portfolio will have grown to

$$P_1^n = H_0^n R_1^n.$$

This is to be rebalanced, with H_1^n to be invested in the hedge portfolio, and the remainder, ϵ_1^n, to be invested in a risk-free asset. For simplicity, we assume that the risk-free rate (r) is zero throughout. Later on, we will comment on the complications introduced by nonzero, deterministic or stochastic interest rates. The hedge position will generate $H_1^n R_2^n$ at time $\tau\frac{2}{n}$, while the remainder generates ϵ_1^n. The total is denoted P_2^n, i.e.,

$$P_2^n = H_1^n R_2^n + \epsilon_1^n,$$

which is to be divided into two parts: H_2^n, to be invested in the hedge portfolio, and $\epsilon_1^n + \epsilon_2^n$, to be invested in the risk-free asset. At time $\tau\frac{3}{n}$,

$$P_3^n = H_2^n R_3^n + \epsilon_1^n + \epsilon_2^n.$$

Rebalance this to an amount H_3^n to be invested in the hedge portfolio, with the remainder,

$$\epsilon_1^n + \epsilon_2^n + \epsilon_3^n,$$

to be put in the risk-free asset.

Continue this strategy until the end, at which time it generates an amount

$$P_n^n = H_{n-1}^n R_n^n + \epsilon_1^n + \epsilon_2^n + \ldots + \epsilon_{n-1}^n,$$

which can be rewritten as

$$P_n^n = H_n^n + \epsilon_1^n + \epsilon_2^n + \ldots + \epsilon_n^n. \tag{2.2}$$

The errors ϵ_j^n define the *cumulative tracking error*:

$$E_j^n = \sum_{i=1}^{j} \epsilon_i^n. \tag{2.3}$$

We will refer to the part of the portfolio that is to be rebalanced into a position with which to hedge the derivative's payoff as the *active hedge*. It has value H_j^n. The remainder, i.e., $P_j^n - H_j^n$, will be referred to as the *financing shortfall*. By construction, the financing shortfall equals the *cumulative tracking error*.

3 Characteristics of the Cumulative Tracking Error

It is straightforward to prove that E_j^n is a *martingale* relative to the information filtration generated by $\mathcal{F}_{j-1}^n$ $(j = 1, \ldots, n)$.

Theorem 3.1 *For all n and j, $E_j^n \in \mathcal{F}_j^n$, and*

$$E[E_j^n | \mathcal{F}_{j-1}^n] = E_{j-1}^n.$$

(All proofs are collected in the Appendix.) This result is important. It implies that the stochastic behavior of the cumulative tracking error can be characterized by a host of martingale theorems. Its simplicity is merely a consequence of the specific formulation of the hedging strategy. Other strategies may not induce martingale behavior. But our strategy is not straightforward to implement, because it requires not only a specification of the composition of the active hedge, but also of the size of the active hedge (H_j^n). In the next section, we will elaborate on techniques meant to recover both inputs.

Here are two examples of how martingale theory can be used to characterize the behavior of the tracking error. First,

Corollary 3.2

$$P\{\max_{j \le n} |E_j^n| \ge M\} \le \frac{1}{M} E[|E_n^n|].$$

This corollary puts a stochastic bound on the tracking error. In other words, it provides an upper bound on the probability that the total (absolute) tracking error exceeds any level M. Corollary 3.2 is obviously useful for the computation of *margins* or *collateral.*

For an even more useful example of the application of martingale theory, notice that the distribution of E_n^n, the total tracking error, can be characterized asymptotically, as the frequency of rebalancing increases.

Corollary 3.3 *Let c^2 be an $\mathcal{F}_0^n$–measurable almost surely positive random variable. Assume:*

$$(i) \quad \max_j |\epsilon_j^n| \to 0 \quad \textit{in probability;} \tag{3.1}$$

$$(ii) \quad \exists \kappa < \infty : E\left(\max_j (\epsilon_j^n)^2\right) \leq \kappa, \quad (\textit{all } n); \tag{3.2}$$

$$(iii) \quad \lim_{n\to\infty} \textstyle\sum_{j=1}^n (\epsilon_j^n)^2 = c^2 \quad \textit{in probability.} \tag{3.3}$$

Then:

$$\frac{E_n^n}{\sqrt{\sum_{j=1}^n (\epsilon_j^n)^2}} \rightsquigarrow N(0,1).$$

To evaluate the scope of this result, notice that c^2 *need not be a constant* (we have not restricted $\mathcal{F}_0^n$ to be the trivial information set!). In other words, c^2 may vary across sample paths. This provides considerable generality.

As an example where (3.3) is violated, assume we use a buy-and-hold portfolio to hedge a call option. (Remember that the composition of the active hedge has been left unspecified; in particular, it may be constant over time, i.e., a buy-and-hold portfolio the size of which varies with H_j^n.) In that case, the limiting sum of squared tracking errors will depend on the option's final payoff.[2] The dependence of the limiting sum of squared tracking errors on the final outcome violates the requirement that c^2 be measurable in $\mathcal{F}_0^n$.

The limiting sum of squared errors can be sample-dependent, but must not depend on the actual sample path of the prices of the underlying assets. It may vary across initial moneyness, or maturity, for instance (both of which are in each $\mathcal{F}_j^n$). The sample-dependence of the limiting sum of squared errors is likely to be of practical relevance: the variance of the hedging error from actual hedging policies may increase with moneyness, for instance. Whether

[2]Take two stock price sample paths which start at the same level (and the call option is in the money). In the first case, the stock price remains at about the same level. If the buy-and-hold portfolio composition happens to be just right to hedge the payoff on the option at that stock price level, the cumulative squared tracking error will be small. Now take the second case, where the stock price decreases, to the point that the option matures out-of-the-money. The buy-and-hold portfolio's hedge becomes worse over time, gradually generating a larger squared tracking error. The cumulative squared tracking error will eventually be much larger than in the first case.

it does is in the first place an empirical question. This is because the characteristics of the tracking error cannot be specified more accurately in the absence of further assumptions on the nature of the prices of options and underlying securities.

If it holds, Corollary 3.3 does state that the asymptotic behavior of the cumulative tracking error is *independent* of either the nature of the derivative or the processes of the values of the underlying assets. The option can be exotic; value changes can be heteroscedastic, skewed, leptokurtotic, etc. This is important from the point of view of asset pricing. Remember that we started from the premise that financial markets are incomplete. Pricing in such an environment necessarily involves identification of a measure of risk and its price. When the cumulative tracking error from a hedging strategy is normally distributed, simple pricing models based on two-parameter models can be appealed to. Moreover, a large methodological literature is available to draw from for the estimation of risk measures and their pricing.

Therefore, Corollary 3.3 is an attractive feature of our hedging policy when it applies. This warrants a closer inspection of the conditions under which it holds, most importantly, whether c^2 is an almost surely positive random variable, measurable in each information set.

The following theorem provides a sufficient condition for Corollary 3.3 to apply.

Theorem 3.4 *Assume that*

$$\lim_{n\to\infty} \sum_{j=1}^{n} (\epsilon_j^n)^2 = c^2,$$

with c^2 an almost surely positive random variable, and that

$$E[\epsilon_j^n | \mathcal{F}_{j-1}^n \vee \sum_{i=1}^{n} (\epsilon_i^n)^2] = 0 \quad \textit{(for all n and j).} \tag{3.4}$$

Furthermore, assume the conditions in (3.1) *and* (3.2). *Then:*

$$\frac{E_n^n}{\sqrt{\sum_{j=1}^{n} (\epsilon_j^n)^2}} \rightsquigarrow N(0,1).$$

The theorem requires the tracking error to be a martingale even after augmenting the information filtration with its quadratic variation. This should provide the right framework within which to evaluate the asymptotic normality of the tracking error in specific cases.

Complete markets provide one case where c^2 is identically equal to zero, which means that Corollary 3.3 does not apply. Define:

$$\tilde{R}_j^n = R_1^n \dots R_j^n.$$

Market completeness is taken to mean that the payoff on the derivative is attainable by a (potentially dynamic) strategy in the underlying securities. If $\tilde{R}^n_j$ denotes the cumulative return on such a perfect replication strategy, this implies that

$$\frac{H^n_j}{\tilde{R}^n_j}$$

is measurable in $\mathcal{F}^n_0$. Assume that

$$\sup_{t\in[0,\tau]} E|\tilde{R}^n_{[tn]} - \tilde{R}_t|^2 \to 0,$$

$$\sup_{t\in[0,\tau]} E|H^n_{[tn]} - H_t|^2 \to 0,$$

as $n \to \infty$. Now consider $E^n_j + H^n_j$, the time-j total value of the hedging strategy:

$$\begin{aligned} E^n_j + H^n_j &= H^n_0 + \sum_{i=1}^{j} H^n_{i-1}(R^n_i - 1) \\ &= H^n_0 + \sum_{i=1}^{j} \frac{H^n_{i-1}}{\tilde{R}^n_{i-1}}[\tilde{R}^n_i - \tilde{R}^n_{i-1}]. \end{aligned}$$

As $n \to \infty$, we obtain convergence to $E_t + H_t$ (where E_t is a continuous martingale):

$$\begin{aligned} E_t + H_t &= H_0 + \int_0^t \frac{H_s}{\tilde{R}_s} d\tilde{R}_s \\ &= H_0 + H_t - H_0 - \int_0^t \tilde{R}_s d\left(\frac{H_s}{\tilde{R}_s}\right) - \left\langle \frac{H}{\tilde{R}}, \tilde{R} \right\rangle_t. \end{aligned}$$

Hence,

$$E_t = \int_0^t \tilde{R}_s d\left(\frac{H_s}{\tilde{R}_s}\right) - \left\langle \frac{H}{\tilde{R}}, \tilde{R} \right\rangle_t$$

and

$$\langle E, E \rangle_t = \int_0^t (\tilde{R}_s)^2 d\left\langle \frac{H}{\tilde{R}}, \frac{H}{\tilde{R}} \right\rangle_s.$$

Consequently,

$$c^2 = \langle E, E \rangle_\tau = \int_0^\tau (\tilde{R}_s)^2 d\left\langle \frac{H}{\tilde{R}}, \frac{H}{\tilde{R}} \right\rangle_s. \tag{3.5}$$

The following obtains immediately.

Theorem 3.5 $c^2 = 0$ *a.s. if and only if the continuous-time market is complete.*

Consequently, c^2 is also a measure of market completeness! Moreover, when $c^2 = 0$, i.e., when markets are complete, $E_t = 0$, which means that the financing shortfall becomes zero. It must then coincide with the usual strategies. In other words, standard hedging obtains as a special case of our approach.

We can also interpret the foregoing in terms of pricing. H_j^n is the *value* of the active part of the portfolio used to hedge the derivative. The value process $\{H_j^n\}_{j=0}^n$ becomes a martingale after deflation by the (predictable) process

$$D_j^n = \overline{R}_1^n \ldots \overline{R}_j^n.$$

In other words, the true probability measure is an equivalent martingale measure with which to price the payoff on the derivative, provided values are deflated appropriately.

Until now, we have assumed that interest rates are zero. Although we will stick to this assumption throughout, let us briefly comment on the alternatives. If interest rates are nonzero and deterministic, the financing shortfall should be borrowed by means of loans maturing at time t. Consequently, the definition of ϵ_j^n changes, as follows. Let b_j^n be the time-$t\frac{j}{n}$ price of a zero-coupon bond with face value of $1 and maturity t. Then:

$$\epsilon_j^n = P_j^n - \sum_{i<j} \epsilon_i^n \frac{b_j^n}{b_i^n} - H_j^n.$$

As before, the ϵ_j^n will constitute a martingale difference sequence relative to the filtration $\mathcal{F}_j^n$. If we then redefine E_j^n as the *discounted* tracking error, i.e.,

$$E_j^n = b_j^n \sum_{i \leq j} \frac{\epsilon_i^n}{b_i^n},$$

then the E_j^ns constitute a martingale sequence up to a discounting factor:

$$E[E_j^n | \mathcal{F}_{j-1}^n] = \frac{b_j^n}{b_{j-1}^n} E_{j-1}^n.$$

If interest rates are stochastic, the financing shortfall should be borrowed *short-term.* Letting b_{j-1}^j denote the time-$t\frac{j-1}{n}$ price of a one-dollar zero-coupon bond maturing at time $t\frac{j}{n}$, and defining E_j^n to be:

$$E_j^n = \sum_{i<j} \epsilon_i^n \left(\frac{1}{b_i^{i+1}} \cdots \frac{1}{b_{j-1}^j} \right) + \epsilon_j^n,$$

one can then show that

$$E[E_j^n | \mathcal{F}_{j-1}^n] = \frac{1}{b_{j-1}^j} E_{j-1}^n.$$

Hence, with nonnegative interest rates, the E_j^n form a submartingale. This also implies that Corollary 3.2 will continue to hold. For the asymptotic normality of Corollary 3.3, mixingale (i.e., asymptotic martingale) central limit theorems would have to be appealed to.

4 Estimation of the Inputs

Let C_j^n denote the option market's clearing price at time $\tau\frac{j}{n}$. Let $V_{i,j}^n$ denote the value of security i in the (active) hedge portfolio at time $\tau\frac{j}{n}$; $i = 1, \ldots, N$. We will *assume* the following statistical model for the relationship between returns on the option and those on the underlying securities. There exists a random vector sequence z_j^n measurable in $\mathcal{F}_j^n$, as well as functions α and β_i ($i = 1, \ldots, N$) mapping z_j^n into some point on the real line, such that the error η_j^n in the following equation,

$$C_j^n - C_{j-1}^n = \alpha(z_{j-1}^n) + \sum_{i=1}^{N} \beta_i(z_{j-1}^n)(V_{i,j}^n - V_{i,j-1}^n) + \eta_j^n, \tag{4.1}$$

satisfies:

$$E[\eta_j^n | \mathcal{F}_{j-1}^n] = 0 \tag{4.2}$$

and

$$E[\eta_j^n (V_{i,j}^n - V_{i,j-1}^n) | \mathcal{F}_{j-1}^n] = 0. \tag{4.3}$$

This model states that the payoff on the option can be replicated with a portfolio composed of β_i units of asset i; apart from a bias term α, the replication error is not predictable, and uncorrelated with the payoff on the hedge portfolio securities. The z_j^n are predetermined *factors* such as the option's moneyness and maturity, or the interest rate level.

In order to avoid identification problems, z_j^n has to be chosen judiciously however; it must not include the interest rate if bond prices are deterministic and a bond is used as one of the underlying assets. Otherwise, α cannot be separately identified from the coefficient to the bond payoff. In this sense, z_j^n has to be 'minimal.'

The functions α and β_i in (4.1) can easily be estimated provided they are smooth. Bossaerts & Hillion (1996) illustrate this: they implement local parametric estimation, whereby the hedge portfolio weighting functions from parametric, complete-markets option pricing models (such as Black–Scholes) are fit locally. For a given factor level z, observations with z_j^n close to z are used to generate least squares estimates of $\alpha(z)$ and $\beta_i(z)$.

The β_is define a hedging strategy. For ease of exposition, we will consider only the case where the first security in the hedge portfolio is a risk-free bond with the same maturity as the option, whereas the other securities ($i = 2, \ldots, N$) are *futures contracts*. In that case, β_i gives the position to be taken in futures contract i ($i > 1$), whereas β_1 gives the number of risk-free bonds to invest in.

Because only the bond position requires initial funds, the value of the hedge portfolio at any time is entirely determined by the position in the risk-free bond, i.e., $\beta_1(z_j^n)V_{1,j}^n$. A question that arises naturally is: what is the relationship between $\beta_1(z_j^n)V_{1,j}^n$, the value of the hedge, and H_j^n, the value of the active hedge in the strategy we proposed in Section 2? Of course, this question makes sense only for H_j^n corresponding to the hedging policy given by β_i $(i = 1, \ldots, N)$. It turns out that there is a simple connection, depending on the nature of α.

To answer the question, define

$$\psi_j^n = C_j^n - \beta_1(z_j^n)V_{1,j}^n. \tag{4.4}$$

ψ_j^n is the difference between the market's call clearing price and the value of the hedge portfolio based on the coefficients in (4.1). We will refer to ψ_j^n as the *pricing delta*. Traditionally, ψ_j^n has been called the pricing error (see, e.g., Lo (1986), Bossaerts & Hillion (1993), Clément, Gouriéroux & Monfort (1993), Jacquier & Jarrow (1995)). This label does not accurately reflect the meaning of ψ_j^n. In an incomplete market, there may be a difference between the value of an option and the portfolio that best hedges its payoff. The pricing delta reflects risk differentials, and may be an equilibrium phenomenon instead of mispricing. Of course, this need not always be the case: from a practical point of view, it is sometimes attractive to allow the option to be mispriced. We will come back to this point.

Theorem 4.1 *Assume:*

$$E[\psi_j^n - \psi_{j-1}^n | \mathcal{F}_{j-1}^n] = \alpha(z_{j-1}^n).$$

Then:

$$H_j^n = \beta_1(z_j^n)V_{1,j}^n.$$

So, if the expected change in the pricing delta can be recovered as the intercept in equation (4.1), then the H_j^n corresponding to the hedge portfolio defined by the β_is in the same equation is given by product of the loading on the risk-free bond payoff, $\beta_1(z_j^n)$, with $V_{1,j}^n$. With these inputs, the hedging policies of Section 2 can be implemented.

5 Economic Interpretation

The model in (4.1) deserves further discussion. In particular, the economic meaning of terms like α and η_j^n must be explored. This is best done by means of examples, thereby also illustrating the diverse set of potential interpretations.

Let us first consider the continuous-time limit case ('$n = \infty$'). The option is a call, written on a futures contract whose price (V_{2t}) follows a geometric Brownian motion; interest rates are constant, i.e., the bond price (V_{1t}) changes deterministically over time; this is the case studied by Black (1976). (We now use t as time subscript; this could be interpreted as emerging from a limit argument, where j is allowed to increase with n: $t = \lim_{n\to\infty} \tau j(n)/n$.) The call price satisfies Black's formula, except for an additive pricing 'error':

$$C_t = C_t^B + v_t \tag{5.1}$$

where v_t is a stochastic process such that:

$$E[dv_t|\mathcal{F}_t] = \overline{v}(z_t), \tag{5.2}$$

$$E[(dv_t - \overline{v}(z_t))dV_{2t}|\mathcal{F}_t] = 0.$$

Under these conditions, it is easy to show that equation (4.1) is satisfied for the following choice of coefficients and factors: z_t consists of the option's moneyness and maturity; β_2, the number of futures contracts to acquire, equals the Black hedge ratio; β_1 is given by the position in the risk-free bond in the same model; and α equals $\overline{v}(z_t)$.

According to Theorem 4.1, H_t in the hedge strategy proposed in Section 2 equals $\beta_1 V_{1t}$, i.e., Black's formula C_t^B. But, since the active hedge is determined by Black's model as well, the financing shortfall becomes zero. In other words, the hedging strategy we proposed encompasses the standard policies in this situation. Also, because $H_t = C_t^B$, $v_t = \psi_t$. In words: the pricing 'error' is identical to the pricing delta defined earlier.

Notice that markets are not complete in this example: the hedge error, η_t, is nonzero (it equals the unpredictable part of dv_t). Yet, the payoff on the option can perfectly be replicated by the Black hedging strategy. Hence, the value of the option must be H_t $(= \beta_1 V_{1t} = C_t^B)$, contradicting (5.1) if v_t is to be nontrivial. So, it seems that nonzero pricing errors are inconsistent with equilibrium.

There are cases, however, where v_t may be nontrivial. This includes the one used in Bossaerts & Hillion (1993). There, the option price is not determined in a Walrasian equilibrium, as is standard in investments and options analysis, but as the clearing price in a batch market. The batch clearing process implies that there is execution price uncertainty. This is reflected in the random variable v_t. If investors are risk neutral, equilibrium restricts the stochastic nature of v_t as in (5.2). (In Bossaerts & Hillion (1993), however, the presence of a representative investor with *logarithmic* preferences induced a different equilibrium restriction on v_t.)

Of course, if $v_t = 0$, we are back in the complete-markets case. Because of this, it should not be surprising that our strategy collapses to the standard one (where the financing shortfall is zero) and that the value of the active hedge (H_t) equals that of traditional hedges (C_t^B in the above example) at each point in time. With respect to the latter, notice that H_t is defined in the same recursive way (see Eqn (2.2)) as hedge portfolio values in traditional continuous-time arguments (where the recursion is expressed as a partial differential equation).

In (5.1), we introduced a 'pricing error' in an additive way. What if we instead model a multiplicative pricing error, as in Jacquier & Jarrow (1995)? This choice would be unfortunate. There would still be functions α, β_1 and β_2 such that (4.1) holds, but the link between these coefficients and the ingredients of the Black model would be obscured by the pricing error. β_2, for instance, would no longer be equal to the Black hedge ratio. This illustrates that even if the underlying assets satisfy the assumptions of a continuous-time model like Black's, the active hedge in the strategy that we propose may not coincide with the complete-markets hedge portfolio.

Turning back to the additive 'pricing error,' it is clear that c^2 in Corollary 3.3 will be zero because the payoff on the call can be spanned by dynamically trading according to the Black policy. As mentioned in the previous Section, c^2 is therefore a measure of market completeness.

In general, c^2 will be nonzero and equal:

$$c^2 = \langle E, E \rangle_\tau = \int_0^\tau (\tilde{R}_s)^2 d \left\langle \frac{H}{\tilde{R}}, \frac{H}{\tilde{R}} \right\rangle_s$$

(see equation (3.5)).

6 Concluding Remarks

This article introduced a new framework in which to analyze hedging error in an incomplete market. Its focus was a hedging strategy whose purpose was not to minimize the tracking error for some loss function, but to generate a tracking error with martingale characteristics. This allows the hedger to evaluate the tracking error by means of martingale analysis. For instance, the tracking error will be asymptotically normal as the rebalancing frequency increases provided it continues to be a martingale after augmenting the information set with the tracking error's quadratic variation. Once normality can be established, the risk and (required) return of the tracking error can be evaluated by means of standard asset pricing models.

We showed how the inputs to our hedging policy can easily be estimated from the data. The estimation procedure does not require one to be very specific about the stochastic processes of the values of the securities in the hedge portfolio. In this sense, our approach differs markedly from standard derivatives analysis. Nevertheless, the standard hedging policies obtain as a special case when markets are complete. In other words, our approach encompasses the usual analysis.

Appendix

Proof of Theorem 3.1

$$
\begin{aligned}
E[E_j^n|\mathcal{F}_{j-1}^n] &= E\left[\sum_{i\leq j}\epsilon_i^n|\mathcal{F}_{j-1}^n\right] = E_{j-1}^n + E[\epsilon_j^n|\mathcal{F}_{j-1}^n] \\
&= E_{j-1}^n + E\left[H_{j-1}^n R_j^n + \sum_{i<j}\epsilon_i^n - H_j^n + \sum_{i<j}\epsilon_i^n|\mathcal{F}_{j-1}^n\right] \\
&= E_{j-1}^n + E[H_{j-1}^n R_j^n - H_j^n|\mathcal{F}_{j-1}^n] \\
&= E_{j-1}^n + H_{j-1}^n \overline{R}_j^n - E[H_j^n|\mathcal{F}_{j-1}^n] \\
&= E_{j-1}^n.
\end{aligned}
$$

Proof of Corollary 3.2

Follows immediately from Theorem 2.1 in Hall & Heyde (1980).

Proof of Corollary 3.3

See Theorem 3.2 in Hall & Heyde (1980).

Proof of Theorem 3.4

Simply augment the filtration $\{\mathcal{F}_j^n\}$ with the quadratic variation of E_n^n and apply Theorem 3.2 of Hall & Heyde (1980).

Proof of Theorem 3.5

$c = 0 \Leftrightarrow \langle E, E\rangle_\tau = 0 \Leftrightarrow E = 0$ (i.e., perfect replication is possible).

Proof of Theorem 4.1

Proof by induction. Set: $\beta_1(z_n^n)V_{1,n}^n = C_n^n$. Hence, $\beta_1(z_n^n)V_{1,n}^n = H_n^n$. Assume that $\beta_1(z_j^n)V_{1,j}^n = H_j^n$. Then derive:

$$
\begin{aligned}
\beta_1(z_{j-1}^n)V_{1,j-1}^n &= C_{j-1}^n - \psi_{j-1}^n \\
&= C_{j-1}^n + E[\psi_j^n - \psi_{j-1}^n|\mathcal{F}_{j-1}^n] - E[\psi_j^n|\mathcal{F}_{j-1}^n] \\
&= C_{j-1}^n + \alpha(z_{j-1}^n) - E[C_j^n - \beta_1(z_j^n)V_{1,j}^n|\mathcal{F}_{j-1}^n] \\
&= \alpha(z_{j-1}^n) - E[C_j^n - C_{j-1}^n|\mathcal{F}_{j-1}^n] + E[H_j^n|\mathcal{F}_{j-1}^n] \\
&= -\sum_{i=1}^N \beta_i(z_{j-1}^n)E[(V_{i,j}^n - V_{i,j-1}^n)|\mathcal{F}_{j-1}^n] + E[H_j^n|\mathcal{F}_{j-1}^n] \\
&= -\beta_1(z_{j-1}^n)V_{1,j-1}^n(\overline{R}_j^n - 1) + E[H_j^n|\mathcal{F}_{j-1}^n].
\end{aligned}
$$

Hence,

$$
\beta_1(z_{j-1}^n)V_{1,j-1}^n = E\left[\frac{H_j^n}{\overline{R}_j^n}|\mathcal{F}_{j-1}^n\right],
$$

i.e., $\beta_1(z_{j-1}^n)V_{1,j-1}^n = H_{j-1}^n$.

References

Black, F. (1976) 'The pricing of commodity contracts', *Journal of Financial Economics* **3**, 167–179.

Black, F. & M. Scholes (1973) 'The pricing of options and corporate liabilities', *Journal of Political Economy* **81**, 637–659.

Bossaerts, P. & P. Hillion (1993) 'A test of a general equilibrium stock option pricing model', *Mathematical Finance* **3**, 311–348.

Bossaerts, P. & P. Hillion (1996) 'Local parametric analysis of hedging in discrete time', *Journal of Econometrics*, to appear.

Clément E., C. Gouriéroux & A. Monfort (1993) 'Linear Factor Models and the term structure of interest rates', *Annales d' Economie et de Statistique* **40**, 93–124.

Duffie, D. & H.R. Richardson (1991) 'Mean-variance hedging in continuous time,' *Annals of Applied Probability* **1**, 1–15.

El Karoui, N. & M. Jeanblanc-Picqué (1993) 'Robustness of the Black and Scholes formula', Laboratoire de Probabilités, Université Pierre et Marie Curie, working paper.

Föllmer, H. & D. Sondermann (1986) 'Hedging of contingent claims under incomplete information', in *Contributions to Mathematical Economics*, A. Mas-Colell & W. Hildenbrand (eds.), North-Holland.

Gouriéroux, C., J.P. Laurent & H. Pham (1995) 'Quadratic hedging and numéraire', CREST working paper.

Hall, P. & C.C. Heyde (1980) *Martingale Limit Theory and its Application*, Academic Press.

Jacquier, E. & Jarrow, R. (1995) 'A new framework for evaluating error in option pricing models', *Risk* **8**, 2–4.

Lo, A. (1986) 'Statistical tests of contingent-claims asset-pricing models: a new methodology', *Journal of Financial Economics* **17**, 143–174.

Schweizer, M. (1995) 'Variance optimal hedging in discrete time', *Mathematics of Operations Research* **20**, 1–32.

The Use of Second-Order Stochastic Dominance To Bound European Call Prices: Theory and Results

Claude Henin and Nathalie Pistre

1 Introduction

Since the seminal article of Black and Scholes (1973) on option pricing, a vast amount of literature has been written concerning options. Most of these articles consider perfect and complete markets, where options can be replicated by self-financing strategies involving continual revisions in a portfolio containing the underlying security. The price is then shown to be the 'risk-neutral' expected cash-flows discounted at the risk-free rate: see for instance the seminal articles of Harrison & Kreps (1979), Harrison & Pliska (1981, 1983), Karatzas (1988) and El Karoui & Rochet (1989).

However, very broadly speaking, in any financial model, the equilibrium price of any asset is found to aggregate the risk-aversions of the individuals and their demands for this asset taking into account its correlation with the state variables of the economy (see for instance the general equilibrium model of Cox, Ingersoll & Ross (1985)) or with a 'well-defined' variable (Breeden 1989). But expected utility is, in general, a function of all the moments of the distribution: rules and pricing involving only two moments or duplication are valid only for a limited class of utility functions, or for specific distributions. For several years, finance research has got rid of utility functions in option pricing through the inspired idea of duplication. If the option is simply, at each instant, a portfolio composed of two others assets, then of course its price is the simple sum of the prices of the two assets (the underlying and the risk-free assets), these two prices being determined elsewhere. Given the two asset prices, everybody agrees to give this unique price to the option. This methodology is thus intrinsically linked to the completeness of the market (which implies the uniqueness of the equivalent probability under which asset returns are martingales).

The existence of perfectly hedged portfolios yields single-valued equilibrium formulas for option values, but is achieved by making some strong assumtions regarding the distribution process and by assuming no transaction

[1]The authors thank Claire Gauthier and Katell Savidan for their very useful computing assistance.

costs. A precise value for the premium of the option could only be obtained by supposing some properties for the distribution of the underlying security. This being the case, however, the financial instrument being valued does not enrich the existing financial markets. On the contrary, the impossibility of continuous trading is one economic explanation for the opening of secondary markets in options. At the present time, many of the classical assumptions such as the lognormality for example, are under scrutiny. More and more papers study what would happen if volatility were to be stochastic or if a jump process were to affect the evolution of the underlying security. But if the markets are not complete there are many equivalent probabilities (and thus many possible prices) and the choice for one of them is a little rash, because it should be determined at (a Pareto efficient) equilibrium.

Moreover when continuous trading is excluded, or with transaction costs, Black-Scholes option prices can be generated only with very restrictive preferences (Rubinstein 1976, Brennan 1979).

Rather than dealing with an exact pricing formula some papers have dealt with bounding the option premium in order to take into account market friction or uncompleteness and risk-aversion – for example, the articles of Leland (1985), Henin & Rentz (1984, 1985), Levy (1985), Perrakis & Ryan (1984), Perrakis (1984), Ritchken (1985), Ritchken & Kuo (1988), Jacka (1992) and El Karoui & Quenez (1995). The disadvantage of these frameworks is that they obtain a range – an upper and a lower bound – rather than one value. But the uniqueness of the price is totally linked to the completeness hypothesis.

Essentially, three different approaches have been developed: the stochastic dominance approach, the discrete time replication approach and the state price approach.

Bounds derived with a stochastic dominance approach

The stochastic dominance approach makes no assumption whatsoever on the distribution of the underlying security, in order to obtain the minimum necessary conditions for no stochastic dominance between simple strategies. These considerations do not require complete markets. They simply mean that, whatever the risk on the underlying security, the options capture that risk and that even if this risk were not hedgable, its price would still be limited. The notions of first and second degree stochastic dominance introduced by Hadar & Russel (1969, 1971, 1974) are adapted to situations where it is desirable to make a prediction about an agent's preference without having much information on the decision maker's utility function, and without making any simplifying hypothesis, allowing for a mean-variance approach.

We now recall some definitions:

Definition 1.1 *Let X and Y be two variables representing the gain of two assets. Let F and G represent respectively the cumulative distribution func-*

tions of X and Y. We say F stochastically dominates G to the first order (F FSD G) iff for all t:

$$F(t) \leq G(t), \tag{1.1}$$

with strict inequality for at least one t.

In this case it can be shown that every nondecreasing utility function prefers X to Y. Using FSD, and allowing for riskless assets, Levy (1985) finds simple upper and lower bounds on call and put prices, by considering any combination of stock and riskless asset and comparing it with any combination of call and riskless asset of the same cost. The lower bound is exactly the same as Merton's (1973), derived without imposing any restrictions on the stock price behavior or the investor characteristics.

Definition 1.2 *Let X and Y be two variables representing the gain of two assets. Let F and G represent respectively the cumulative distribution functions of X and Y. We say F stochastically dominates G to the second order (F SSD G) iff for all t:*

$$\int_{-\infty}^{t} F(x)dx \leq \int_{-\infty}^{t} G(x)dx, \tag{1.2}$$

with strict inequality for at least one t.

As a matter of fact, this implies that the 'truncated' expected value of X is always equal or superior to the 'truncated' expected value of Y for any 'cutting' value t. Second-order stochastic dominance is a very useful notion, because it defines an order relation which is universal among risk-averse individuals. Any person with a concave utility function prefers X to Y (Fishburn 1964, Hadar & Russel 1969, Rotschild & Stiglitz 1970, 1971). The notion of second-order stochastic dominance takes into account the risk and return of the asset; it is not only a measure of risk as the mean preserving spread, which is a particular case.

Using subjective SSD, Henin & Rentz (1984, 1985) show how changing expectations with respect to the subjective density function for stock returns affects the choice between the call purchase (put writing) strategy and the stock purchase startegy. Levy (1985) finds significant improvements on Merton's bounds for the call and put prices. One can extend the usefulness of stochastic dominance by including riskless asset in the strategies (Levy & Kroll 1978, Kroll & Levy 1979). Under the assumption that the investors display decreasing absolute risk-aversion (DARA), Ritchken & Kuo (1988) derive bounds by using third stochastic dominance or higher order.

This method is particularly useful if neither the contingent claim nor the underlying asset is traded, which can be the case for real assets, for instance.

Discrete time replication approach

Transaction costs invalidate the Black–Scholes arbitrage argument, since continuous revision implies infinite trading. Leland (1985) shows that (when transaction costs are included) discrete revisions using Black–Scholes deltas generate errors that are correlated with the market, and do not approach zero, with more frequent revision. Leland develops a modified replicating strategy dependent on the size of transaction costs and the frequency of revision, in addition to the standard variables of option pricing. This permits calculation of the transaction costs of option replication and provides bounds on option prices. The additional parameters enter in a simple way, through adjustment of the volatility in the Black–Scholes formula. Inclusive of all transaction costs, the net purchasing (selling) price of the stock is higher (lower). This accentuation of movements can be modelled as if the volatility of the actual price was higher. Leland's dynamic portfolio strategy is not really self-financing, because he uses a continuous model with discrete revision times.

Conversely, Boyle & Vorst (1992) propose a discrete time model with proportional transaction costs, where the long call price corresponds to the current value of a portfolio which exactly replicates the long call. The frequency of transactions is nevertheless specified exogenously.

State price and linear programming approach

Considering that the Black–Scholes formula is only an approximation to the discrete-time solution, while the closeness of that approximation has not be ascertained in general, Perrakis & Ryan (1984) use the Rubinstein (1976) approach, to derive upper and lower bounds for option prices with both a general price distribution and discrete trading opportunities. Assuming the existence of an average investor with concave increasing utility function and the existence of a Pareto efficient equilibrium, they found bounds forming a range of prices in equilibrium derivable for any general distribution. They use the equilibrium formula derived by Rubinstein to evaluate any asset and comparing three strategies involving calls, riskless assets and stocks, they determine minimum necessary bounds. The bounds exist as equilibrium values given a consensus on stock price distributions. Perrakis (1984) tightens up significantly the bounds, and develops similar bounds on American put options. Ritchken (1985) develops bounds for option prices based on primitive prices in incomplete markets, where state prices are not unique, and compares these bounds to the single-period bounds derived by Perrakis & Ryan. The linear programming approach of Ritchken places constraints on time preferences, risk-aversion, and state probabilities. The optimized objective function is the call price. Minimization and maximization of the objective function gives lower and upper bounds on the call. Perrakis & Ryan have shown that bounds developed by their approach are identical to the stochastic dominance bounds of Levy and the linear programming bounds of Ritchken (1985). Ritchken & Kuo (1988) extend the single-period linear programming option-bounds model to allow for a finite number of revision opportunities. In an incomplete market, the bounds are derived using a modified binomial option-pricing model. When the bounds are developed under more restrictive assumptions on probabilities and risk-aversion, the upper bounds are shown to be identical to those of Perrakis, while the lower bounds are tighter.

The present article aims at presenting general conditions that European options must satisfy under very broad assumptions, in particular concerning the distribution of the underlying asset. We do not suppose that the markets are complete, which means that we do not use duplication strategies to price options. If the market has homogeneous anticipations on the distribution of the underlying asset, we provide bounds in which all risk-averse investors agree to transact, each of them with a personal price that depends on his/her risk-aversion. These bounds on the options price are defined in order to prevent stochastic dominance of strategies with the same initial cost. It does not pretend to obtain necessarily the best possible bounds as we could possibly obtain more precise bounds by creating other comparisons between different strategies. Bounds on put prices can be achieved in the same way. In this article, since the distribution is general, the market is not supposed complete either. We do not try to find an equilibrium price for the option, as we do not suppose the present price for the stock to be an equilibrium price. It can be a transaction price. We are simply able to assert: according to the price of the underlying asset today, and the expectations for its distribution at time T, the price at which any risk-averse individual agrees to transact is between two bounds. The approach we use is to obtain the minimum necessary conditions for the absence of second-order stochastic dominance between simple strategies. These considerations do not require complete markets. We make the hypothesis that investors are risk-averse, they have assumptions about future distributions of a stock, possibly subjective, and we define bounds on the options price in order to prevent stochastic dominance of strategies with the same initial cost.

The article is organized as follows. Section 2 presents definitions and preliminary lemmas. Section 3 provides theoretical bounds on calls when the market is bullish and Section 4 when it is bearish. Section 5 provides results and comparisons with Black–Scholes calculations. Section 6 concludes.

2 Some Useful Lemmas and Main Hypotheses

In this section we recall some lemmas necessary to the proofs of the theorems and present our main hypotheses.

2.1 Useful lemmas on stochastic dominance

Lemma 2.1 (Hammond (1974)) *We suppose that X and Y have a finite first moment. When $\exists(x_A, F(x_A))$ such as*

$$F(x) < G(x), \forall x < x_A, \tag{2.1}$$
$$F(x) > G(x), \forall x > x_A, \tag{2.2}$$
$$F(x_A) = G(x_A), \tag{2.3}$$

F SSD G if and only if

$$E(X) \geq E(Y). \tag{2.4}$$

For two intersection points we obtain the following lemma:

Lemma 2.2 *We still suppose that X and Y have a finite first moment. When $\exists(x_B, F(x_B))$ and $(x_C, F(x_C))$, with $0 < x_B < x_C$ such as*

$$F(x) < G(x), \forall x < x_B \text{ or } x > x_C, \tag{2.5}$$
$$F(x) > G(x), \forall x_B < x < x_C, \tag{2.6}$$
$$F(x_B) = G(x_B) \text{ and } F(x_C) = G(x_C), \tag{2.7}$$

F SSD G if and only if

$$\int_{-\infty}^{x_C} F(x)dx \leq \int_{-\infty}^{x_C} G(x)dx, \tag{2.8}$$
$$or$$
$$\int_{-\infty}^{x_C} xdF(x) \geq \int_{-\infty}^{x_C} xdG(x). \tag{2.9}$$

For more than two intersection points, we have the following result.

Lemma 2.3 *When F intersects G more than twice, the first time being from below, F SSD G if (2.8) is satisfied for all even-numbered intersection points, infinity being considered as such a point (i.e. Condition (2.8) applies to the mean if there is an odd number of intersection points).*

2.2 Definitions and hypotheses

A generalized option on a given financial instrument (security, index, future, etc., ...) is an instrument that will give a reward at a given date (expiry date), which is a function of the price of the underlying security vis a vis a certain price called the exercise price. In what follows, we shall limit ourselves to the study of a classical type of option, European linear calls, but nothing prevents the application of the techniques presented here to generalized convex or concave calls and puts; see Henin & Pistre (1996a, b, c).

Our hypotheses are the following:

H1. The options are European options, i.e. they can only be exercised at expiry.

H2. There exists for the underlying security a probabilistic distribution of values at the expiration of the options; this distribution has a finite mean but otherwise is unconstrained. In particular the distribution may not be symmetrical at all.

H3. There exists a reward (or final value) function defining each option. This reward function has also a final expected value.

H4. There exists a constant risk-free rate of interest during the life of the option and borrowing and lending can be made at this rate.

These conditions are very general and do not impose a given behavior on stock prices. They just ensure the existence of finite parameters for bounding the options. They do not preclude the existence of dividends – discrete or continuous – on the underlying security.

The price of the underlying security is S_0 at time 0 and $\tilde{S}_T$ at time T, the expiry date of the call. There is no particular restriction on the underlying distribution of the security price at time T which is only defined between 0 and infinity by a cumulative distribution function F and its density function f. In particular, the distribution of the price is not necessarily stable between 0 and T, nor Gaussian. We now state the hypothesis that leads all our proofs.

H5. (non-dominance hypothesis) If two non-losing strategies have the same cost at time 0, there cannot exist a relation of stochastic-dominance of the first or second order between the laws of the terminal wealths of these two strategies or *a fortiori*, a relation of strict dominance.

This last hypothesis, on which we base our proofs, plays a role similar to the no-arbitrage hypothesis, but witout any idea of duplication. The completeness of the market plays no role.

What we say is that if two strategies have the same cost, neither can dominate the other (or nobody will pay the same price for the two strategies). There is no direct link with the notion of arbitrage, because the notion of risk is integrated in the notion of stochastic dominance. We do not try to eliminate the risk as in the Black–Scholes derivation. We consider it impossible. This means that we are never making our calculus in the so-called risk-neutral probability, but in the 'real', objective or subjective probability. Notice that the order defined by second-order stochastic dominance is universal, if the expectations are homogeneous. If this is the case, the bounds we found are universal; if investors have different expectations on the distributions of the stock price, these bounds will be subjective.

3 Bounds on Call Price in a Bullish Market

With no particular assumptions on the shape of the distribution and the shape of the reward function of the options, it is impossible to derive exact formulas from hedging strategies, even if these strategies remain feasible conceptually.

This section aims at presenting upper and lower bounds for the premiums at time 0, C_0 of a European call, from comparisons between non-dominating strategies, using the following notation for the underlying security. The expected final value of the stock, and this same expectation stopped at the y value of the stock, are respectively defined as:

$$E(\tilde{S}_T) = \int_0^\infty x f(x) dx \tag{3.1}$$

and

$$E_y(\tilde{S}_T) = \int_0^y x f(x) dx. \tag{3.2}$$

We need to define two other items of notation for values that appear in our results. The first one is 'stopped' at the y expected value of the stock divided by the distribution function at this point. This can be viewed as the mean value of the distribution until point y. The second one is the complementary data; it can be viewed as the mean value of the distribution after point y.

$$E_y^*(\tilde{S}_T) = \frac{\int_0^y x f(x) dx}{F(y)} \tag{3.3}$$

$$E^{y^*}(\tilde{S}_T) = \frac{[E(\tilde{S}_T) - E_y(\tilde{S}_T)]}{[1 - F(y)]}. \tag{3.4}$$

The call is defined by a payoff function at expiry of the type:

$$\max(0, \tilde{S}_T - K).$$

We set:

$$E(\tilde{C}_T) = \int_K^\infty (x - K) f(x) dx \tag{3.5}$$

$$E_y(\tilde{C}_T) = \int_K^y (x - K) f(x) dx \tag{3.6}$$

$$E_y^*(\tilde{C}_T) = \frac{\int_K^y (x - K) f(x) dx}{F(y)} \tag{3.7}$$

$$E^{y^*}(\tilde{C}_T) = \frac{[E(\tilde{C}_T) - E_y(\tilde{C}_T])}{[1 - F(y)]}. \tag{3.8}$$

We now compare strategies between them, by defining strategies of the same cost and explicit bounds which allows H5 to be satisfied. In order to be coherent, we compare strategies which are not trivially dominated by a risk-free strategy, i.e. bullish strategies. Note that it is always possible to define 'losing' strategies; a losing strategy is defined as a bearish (bullish) strategy if the expected return on the underlying security is larger (smaller) than the risk-free rate. In case of a bullish market, selling the asset is dominated by buying the asset and borrowing at the risk-free rate; moreover Arrow (1970) showed that all investors have a positive proportion of their wealth invested in the risky asset. To solve this problem Levy (1985) simply supposes that in the Capital Asset Pricing Model (CAPM) framework, the asset has a non-negative beta. We prefer not to be linked with this framework and make the assumption that stock buyers make their bullish hypothesis reflected in the present stock price and the expected return (greater than the risk-free rate), whereas sellers has the inverse expectation.

If we have the following condition:

$$E(\tilde{S}_T) > (1+r)S_0 \tag{3.9}$$

then we can compare bullish strategies.

Three equal-cost strategies will be defined in order to make some comparisons and derive bounds on the call premium. At time 0:

Strategy 1 (S1): To buy one unit of the underlying security at price S_0.

Strategy 2 (S2): To buy calls of premium C_0, for the same amount, i.e. a quantity S_0/C_0.

Strategy 3 (S3): To buy α calls with $\alpha < S_0/C_0$ and to invest the remaining amount $(S_0 - \alpha.C_0)$ in the risk-free asset with a yield of r until expiry.

The three strategies have the same initial cost S_0.

At time T, we have:

Final value of Strategy 1: $\tilde{S}_T$

Final value of Strategy 2: $S_0/C_0.(\tilde{S}_T - K)^+$. The cumulative distribution function of this final value will be called G.

Final value of Strategy 3: $\alpha.(\tilde{S}_T - K)^+ + (S_0 - \alpha.C_0)(1+r)$. H represents the corresponding distribution function.

We now compare the above strategies pairwise in order to derive conditions of no-dominance by using H5.

3.1 Comparison 1: between S1 and S2

In order to prevent strict dominance from Strategy 1 over Strategy 2, F and G must have at least one intersection point (see Figure 2). As the reward is convex, this implies that

$$\frac{S_0}{C_0} > 1 \tag{3.10}$$

which gives the trivial bound:

$$C_0 < S_0. \tag{3.11}$$

But then there is only one value of the underlying security where both strategies give the same final value, i.e., $S_0/C_0.(\tilde{S}_T - K) = \tilde{S}_T$. This point corresponds to A, the only intersection point of both distribution functions, of coordinates $(x_A, F(x_A))$. x_A is the fixed point of the function $S_0/C_0.(x - K)$ as shown by the following figures:

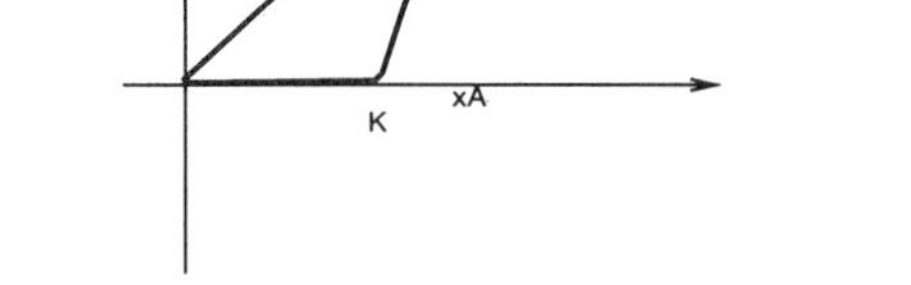

Figure 1: Final values of S1 and S2

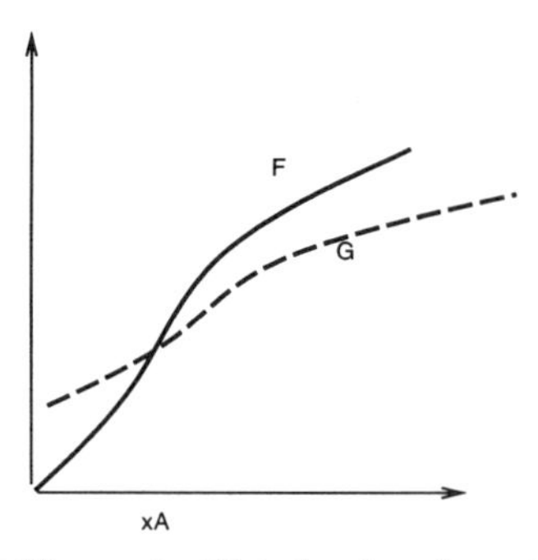

Figure 2: Distribution functions

By using the hypothesis of no-dominance H5 and Lemma 2.1, we will be able to define conditions which prevent F from dominating G. This condition will create bounds on the call price. Intuitively, since F dominates G for small final values, the expected value of $\tilde{S}_T$ must not be too high, unless S1 dominates stochastically S2 at the second order. This will give an upper bound on the call price. S1 and S3 can now be compared in the same manner.

3.2 Comparison 2: between S1 and S3

To avoid strict dominance between the two strategies, the existence of at least one intersection point is necessary. As both strategies must intersect, one must have for any value of $\alpha < S_0/C_0$ at least one value x for which the final value given by Strategy 3 is below the final value given by Strategy 1, i.e. there exists a value x such as

$$\alpha(x-K)^{+}+(S_0-\alpha C_0)(1+r)<x. \tag{3.12}$$

One gets therefore the necessary condition for the non-dominance of S3:

$$C_0>\max_{\alpha}\min_{x}((\alpha(x-K)^{+}+S_0(1+r)-x)/(\alpha(1+r))). \tag{3.13}$$

There should be one or two intersection points in the final values of strategies $S1$ and $S3$, depending on the value of α, $B(x_B, F(x_B))$, and $C(x_C, F(x_C))$ such that x_B and x_C are the two fixed points of the function

$$\alpha.(x-K)^{+}+(S_0-\alpha.C_0)(1+r).$$

(If there is only one intersection point, it implies that x_C is infinite and it does not in any way affect the validity of the results obtained when there are two intersection points.) These are the graphs of S1 and S3:

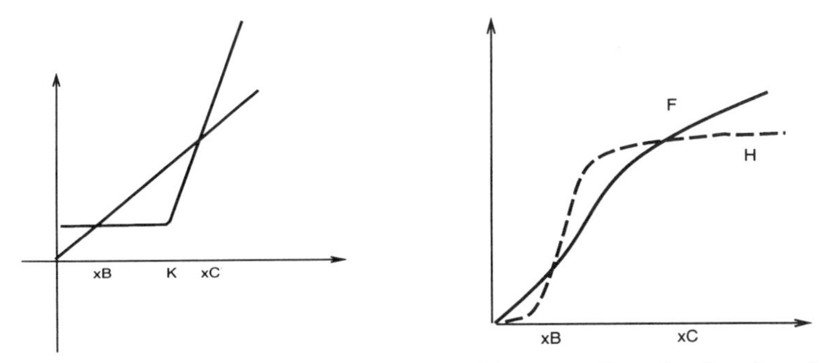

Figure 3: Final values of S1 and S3 Figure 4: Distribution functions

We can see graphically that this time the option strategy dominates the stock strategy, for small final values. Thus, the expected value of S3 must not be too high, unless S3 dominates stochastically S1. The constraint is thus the opposite of the one obtained in the former comparison. This is the reason why we obtain bounds on the two sides of the call price.

Theorem 3.1 *For a European call, if* $E(S_T)>(1+r)S_0$ *then*

$$\max\left[\max_{\alpha\in[0;\frac{S_0}{C_0}]}\left[\frac{S_0(1+r)+\alpha E^*_{x_C}(\tilde{C}_T)-E^*_{x_C}(\tilde{S}_T)}{\alpha(1+r)}\right], S_0\frac{E_{x_A}(\tilde{C}_T)}{E_{x_A}(\tilde{S}_T)}\right]<C_0. \tag{3.14}$$

and

$$C_0<S_0\frac{E(\tilde{C}_T)}{E(\tilde{S}_T)}. \tag{3.15}$$

Proof

Comparison 1

For $x < x_A$, $S_0/C_0.(x-K)^+ < x$; this implies that $F(x) < G(x)$. In addition, for $x > x_A$, $S_0/C_0.(x-K)^+ > x$; hence, $F(x) > G(x)$. This implies that the conditions of Lemma 2.1 are satisfied (F intersects G once from below).

Thus, if $E(S1) \geq E(S2)$, then S1 DS2 S2: hence a contradiction with H5. Therefore,

$$E(S1) < E(S2)$$

and

$$E(\tilde{S}_T) \quad < \quad E\left(\frac{S_0}{C_0}(\tilde{S}_T - K)^+\right).$$

This gives:

$$E(\tilde{S}_T) \quad < \quad \frac{S_0}{C_0}E((\tilde{S}_T - K)^+).$$

From this we obtain (3.15).

Comparison 2

When $x < x_B$ then Final Value (S3) > Final Value (S1) and $H(x) < F(x)$ and again for $x > x_C$, $H(x) < F(x)$. We have chosen α in order to obtain two intersection points; using Lemma 2.2 to prevent stochastic dominance of S3, we obtain

$$\int_{-\infty}^{x_C} F(x)dx \quad < \quad \int_{-\infty}^{x_C} H(x)dx,$$

from which

$$\int_{-\infty}^{x_C} xf(x)dx \quad > \quad (S_0 - \alpha C_0)(1+r)F(x_C) + \alpha \int_K^{x_C} (x-K)^+ f(x)dx$$

therefore:

$$(S_0 - \alpha C_0)(1+r)F(x_C) + \alpha E_{x_C}(\tilde{C}_T) \quad < \quad E_{x_C}(\tilde{S}_T)$$

$$\frac{S_0(1+r)F(x_C) + \alpha E_{x_C}(\tilde{C}_T) - E_{x_C}(\tilde{S}_T)}{\alpha(1+r)F(x_C)} \quad < \quad C_0$$

$$\frac{S_0(1+r) + \alpha E^*_{x_C}(\tilde{C}_T) - E^*_{x_C}(\tilde{S}_T)}{\alpha(1+r)} \quad < \quad C_0.$$

Moreover, by having α going toward S_0/C_0, we have

$$(S_0 - \alpha C_0)(1+r) \to 0.$$

Since B tends towards the origin, and C towards A, the gain tends towards $(S_0/C_0)(\tilde{S}_T - K)^+$, thus we obtain:

$$S_0 \frac{E_{x_A}(\tilde{C}_T)}{E_{x_A}(\tilde{S}_T)} < C_0$$

which gives (3.14). □

Notice the lower bound of the call premium makes the stochastic dominance of the stock upon the option impossible, while the upper bound of the call premium makes the stochastic dominance of the option upon the stock impossible.

4 Bounds on Call Price in a Bearish Market

When the market (or the investor's expectation) is relatively bearish on the security, i.e., if:

$$E(S_T) < S_0(1+r) \tag{4.1}$$

the three former strategies are dominated by an investment in the risk-free rate and must be replaced by the reverse (selling) strategies to obtain bounds on the calls. Notice the flat market is not directly treated as if the expected return of the stock is the risk-free return, any risk-averse individual will not invest in this asset nor sell short this asset either. This remark applies also for the call. This asset is not traded except by risk-neutral investors, who simply price the option by its expected value, actualised by the risk-free rate (in the real probability), because they do not ask for a risk premium.

At time 0:

Strategy $1'$: to sell one unit of the underlying security at price S_0.

Strategy $2'$: to sell S_0/C_0 options to get initially the same amount S_0.

Strategy $3'$: to sell α options with $\alpha < S_0/C_0$, (initial amount $\alpha.C_0$), and to borrow $(S_0 - \alpha.C_0)$ at the risk-free rate to start with the same initial amount as the two other strategies S_0.

At time T:

Final value of Strategy $1'$: $\tilde{S}_T$ with a distribution function F', and a density function f'.

Final value of Strategy $2'$: $-S_0/C_0.(\tilde{S}_T - K)^+$, with a distribution function G'.

Final value of Strategy $3'$:$-\alpha.(\tilde{S}_T - K)^+ - (S_0 - \alpha.C_0)(1+r)$, with a distribution function H'.

4.1 Comparison 3: between S1′ and S2′

If one notices that $F'(x) = P(-\tilde{S}_T < x) = P(\tilde{S}_T > -x) = 1 - F(x)$ and that $G'(x) = 1 - G(x)$, the sole intersection point has the same first coordinate x_A as in the comparison of the corresponding bullish strategies, i.e., the fixed point of the function $S_0/C_0.(x-K)^+$.

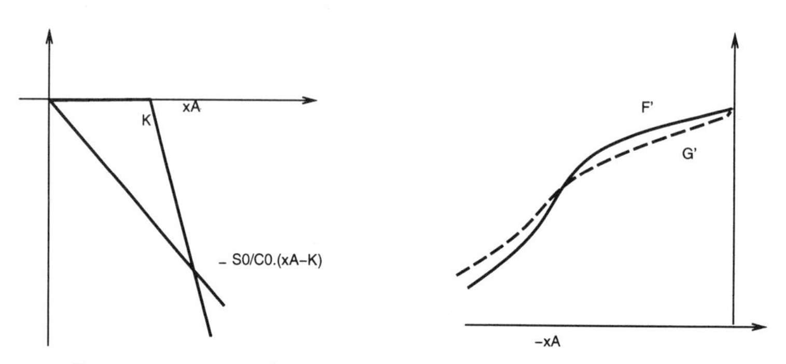

Figure 5: Final values of S1′ and S2′ Figure 6: Distribution functions

Once again we use Lemma 2.1 in order to obtain bounds on the call that will prevent S1′ from dominating S2′. Notice the stock strategy still dominates for small final values (which happens this time for high values of the stock).

4.2 Comparison 4: between S1′ and S3′

Here again, to get an intersection between the two strategies, in order to prevent strict dominance, one gets the same lower bound as for Comparison 2. Two cases may happen. We can choose α such as there are two intersection points which happens exactly for the same value of α and $\tilde{S}_T$ as in Comparison 2. Similarly the two intersection points have the same first coordinates x_B and x_C as in Comparison 2. They obey the same conditions and $F'(x) = P(\tilde{S}_T < x) = P(\tilde{S}_T > -x) = 1 - F(x)$ and $H'(x) = 1 - H(x)$.

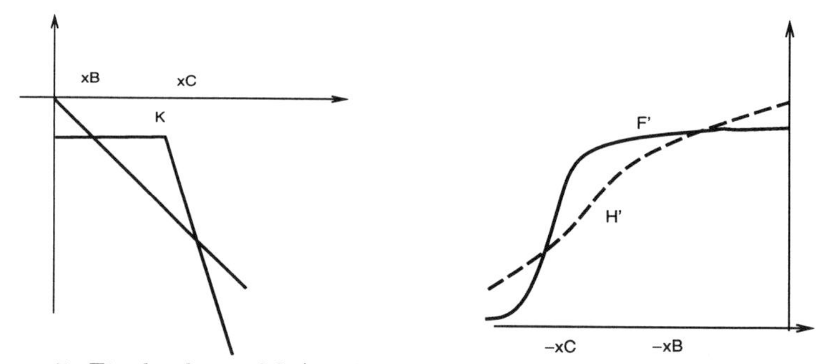

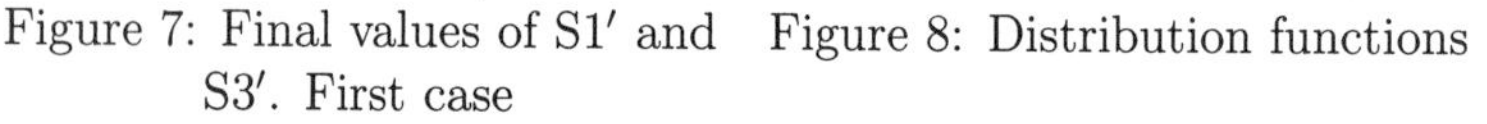

Figure 7: Final values of S1′ and S3′. First case Figure 8: Distribution functions

Notice that, surprisingly, stock strategies still dominate for small final values, which was not the case in bullish inverse strategies. Comparison 4 is thus insufficient to obtain an inverse bound on the call.

We obtain a second case by taking $\alpha = \alpha'$ in order to have only one intersection point for a final value x_B (x_C becomes infinite in this case), which is simply obtained for

$$\alpha' < 1. \tag{4.2}$$

These are the graphs of $\mathrm{S1}'$ and $\mathrm{S3}'$ in the second case:

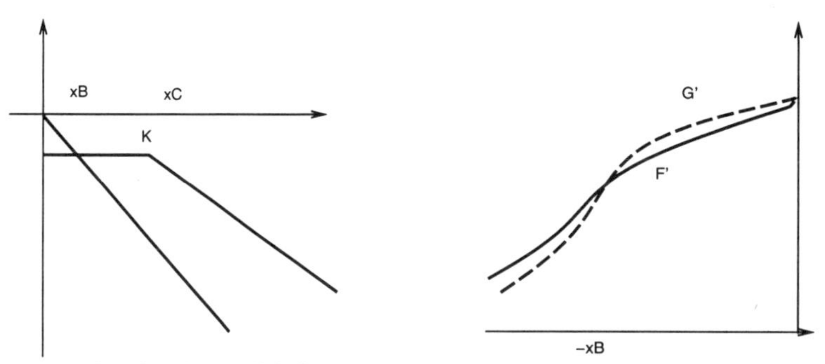

Figure 9: Final values of $\mathrm{S1}'$ and $\mathrm{S3}'$. Second case

Figure 10: Distribution functions

In this case, stock strategy $\mathrm{S1}'$ is dominated for small values by $\mathrm{S3}'$. We are thus able to obtain an upper bound for the call in the bearish case.

Theorem 4.1 *For a European call, if* $E(S_T) < (1+r)S_0$ *then*

$$\max\left[\max_{\alpha\in[1;\frac{S_0}{C_0}]}\left[\frac{S_0(1+r)+\alpha E^{x_B^*}(\tilde{C}_T)-E^{x_B^*}(\tilde{S}_T)}{\alpha(1+r)}\right],\ S_0\frac{E(\tilde{C}_T)}{E(\tilde{S}_T)}\right] < C_0 \tag{4.3}$$

and

$$C_0 < \min_{\alpha'\in[0,1]}\left[\frac{S_0(1+r)+\alpha' E(\tilde{C}_T)-E(\tilde{S}_T)}{\alpha'(1+r)}\right]. \tag{4.4}$$

Proof

Comparison 3

Here, Lemma 2.1 can be used again for the value of x_A. Notice that because there is only one intersection point, even by inverting the strategies, F' lies under G' for the most negative values of the stategies at T, and then above it, starting at point A. Thus if $E(\mathrm{S1}') > E(\mathrm{S2}')$ then $\mathrm{S1}'$ SSD $\mathrm{S2}'$: hence a contradiction. We must have $E(\mathrm{S1}') < E(\mathrm{S2}')$. This yields

$$E(\tilde{S}_T) > \frac{S_0}{C_0}E(\tilde{C}_T).$$

Comparison 4

In Comparison 4, we have two cases. The first is when α is such that there are two intersection points which happens exactly for the same values of α and S_T as in Comparison 2 (only the sign of the final values of the strategies are inverted). Notice however that the sign of the possible final values of the strategies are inverted. When $x > -x_B$ or $< -x_C$ then Final Value(S3′) $<$ Final Value(S1′) and $H'(x) > F'(x)$. Moreover $-x_B > -x_C$ which is important for the application of the lemma. When $-x_C < x < -x_B$ then $H'(x) < F'(x)$. Because there are two intersection points we must prevent again stochastic dominance of S1′. Lemma 2.2 can be used for the common final values $-x_C$ and $-x_B$. But the less negative gain is for $-x_B$. We have necessarily:

$$\int_{-\infty}^{-x_B} F'(x)dx > \int_{-\infty}^{-x_B} H'(x)dx.$$

Easy computations (see Henin & Pistre (1996a) for details) provide (4.3).

Second case: We can use the same strategies but in taking α' in order to have only one intersection point for a final value x_B of S_T which permits us to use Lemma 2.1 and to get a different result. Effectively in this case when $x < -x_B$ then Final Value(S3′) $>$ Final Value(S1′) and $H'(x) < F'(x)$. When $-x > -x_B$ then $H'(x) > F'(x)$. This time we must prevent stochastic dominance of S3′. Again, easy computations lead to (4.4). □

In this bearish situation, the upper bound of the call premium makes the stochastic dominance of the latter upon the stock impossible (the call must not be sold too high), while the lower bound of the call premium makes the stochastic dominance of the stock impossible (the stock must not be sold too high).

Notice something somewhat surprising concerning the proofs in the bullish and bearish cases. While the 'primed' strategies seem symmetrical (buy/sell), the proofs are not symmetrical. This is due to the number of intersection points. When there is only one intersection point there exists a symmetry in the results, because the stock always dominates for the smallest final values of the strategies (compare Figures 1 and 5). But when there are two intersection points, there is no symmetry in the results, because the stock and the call alternately dominate for the smallest final values of the strategies (compare Figures 3 and 7). This is the reason why, in the case of the primed strategy, we need to consider α and α' to obtain bounds on either side of the premium.

On the other hand, it may seem surprising that all the conditions obtained depend solely on the first moment of the distributions of the options and the underlying security, without even including the variance. Actually, the whole distribution is included from the asymmetry in the reward function of the option and from the fact that its expected value is an intrinsic part of the constraints. When the yields are normally distributed, the Black–Scholes formula which is the expected value of the actualized flows of the option under risk-neutral probability, makes the variance appear explicitly. In a more general fashion, computing the expected value of the call under real probability and comparing it with the distribution and expected value of the underlying security, really captures the whole distribution, any eventual asymmetry being made evident by their comparison.

5 Bounds on Call Price: Results Compared with Black–Scholes Price

In this section, we simply compute bounds by making the hypothesis that the underlying security follows a geometric Brownian motion. Of course, in this case, the market is complete and Black–Scholes can be applied. These calculations are made to compare our bounds with Black–Scholes prices, and to have an idea of the tightness of the bounds. We make the calculus for different values of μ, the instantaneous expected return of the stock, in order to see if the, possibly heterogeneous, expectations of the investors have a consequence on the bounds. On the other hand, we take one single value for the volatility by supposing it can be observed on the market. The stock price is 100, the risk-free interest rate is 0.08, the volatility is 0.15 and the maturity of the option is 90 days. We compute bounds for out, at and in the money calls. Some of the bounds obtained from the conditions of non stochastic dominance include implicitly the value of the option itself as a parameter; this is the case of any constraint derived at a finite point of intersection such A, B or C. They can be used however to preclude the existence of options values for which the corresponding bounding inequalities do not hold, and give results by an iteration process.

Here are the bounds for an out of the money call:

$\sigma = 0.15$; $t = 90$ days; $r = 0.08$; **K = 110**; **Black–Scholes value=0.63**				
Bullish bounds(buyers)	$\mu = 0.15$	$\mu = 0.125$	$\mu = 0.10$	$\mu = 0.085$
Lower Bounds	0.23	0.23	0.32	0.39
Upper Bounds	0.94	0.82	0.71	0.646
Bearish bounds(sellers)	$\mu = 0.075$	$\mu = 0.07$	$\mu = 0.06$	$\mu = 0.05$
Lower Bounds	0.61	0.59	0.55	0.52
Upper Bounds	6.25	6.35	6.54	6.74

They are not very good, but always include the Black–Scholes value, which is not surprising, because this value corresponds to a specific utility function. Notice the bounds on the selling price (bearish) are very bad; this means that very risk-averse individuals ask for a very high price to be the seller on a call. The buying and selling bounds are not at all symmetric, because selling a call is of course much riskier than buying a call (even if you are bearish). In this case the behavior of the investors is more heterogeneous. Looking at the same time at bullish and bearish bounds lets us understand what type of transactions can really be performed. No transactions can be realized near the lower bound of the buyers, nor the upper bound of the sellers. If we suppose that all the expectations exist on the market (from $\mu = 0.05$ to $\mu = 0.15$), the transactions will take place between 0.52 (for the best seller) and 0.94 (for the best buyer). And the less the expectations are heterogeneous between sellers and buyers, the tighter will be the spread.

These results can be interpreted also from a market maker point of view. Once again only the upper bounds for the buying prices and the lower bounds for the selling prices are interesting for him/her. He/she knows that by quoting 0.52/0.94 he/she will be able to find an investor willing to transact.

Here now are the results for a call at the money and calls in the money.

$\sigma = 0.15$; $t = 90$ days; $r = 0.08$; **K = 100; Black–Scholes value=4.02**				
Bullish bounds(buyers)	$\mu = 0.15$	$\mu = 0.125$	$\mu = 0.10$	$\mu = 0.085$
Lower Bounds	3.41	3.61	3.82	3.96
Upper Bounds	5.06	4.67	4.29	4.07
Bearish bounds(sellers)	$\mu = 0.075$	$\mu = 0.07$	$\mu = 0.06$	$\mu = 0.05$
Lower Bounds	3.93	3.86	3.72	3.59
Upper Bounds	9.39	9.43	9.52	9.62

$\sigma = 0.15$; $t = 90$ days; $r = 0.08$; **K = 90; Black–Scholes value=11.89**				
Bullish bounds(buyers)	$\mu = 0.15$	$\mu = 0.125$	$\mu = 0.10$	$\mu = 0.085$
Lower Bounds	11.82	11.84	11.86	11.87
Upper Bounds	13.32	12.81	12.29	11.98
Bearish bounds(sellers)	$\mu = 0.075$	$\mu = 0.07$	$\mu = 0.06$	$\mu = 0.05$
Lower Bounds	11.77	11.66	11.46	11.25
Upper Bounds	16.78	16.78	16.80	16.81

$\sigma = 0.15$; $t = 90$ days; $r = 0.08$; **K = 80; Black–Scholes value=21.56**				
Bullish bounds(buyers)	$\mu = 0.15$	$\mu = 0.125$	$\mu = 0.10$	$\mu = 0.085$
Lower Bounds	21.55	21.55	21.55	21.55
Upper Bounds	22.88	22.41	21.93	21.64
Bearish bounds(sellers)	$\mu = 0.075$	$\mu = 0.07$	$\mu = 0.06$	$\mu = 0.05$
Lower Bounds	21.45	21.35	21.16	20.96
Upper Bounds	25.91	25.91	25.91	25.91

$\sigma = 0.15$; $t = 90$ days; $r = 0.08$; **K = 70; Black–Scholes value=31.37**				
Bullish bounds(buyers)	$\mu = 0.15$	$\mu = 0.125$	$\mu = 0.10$	$\mu = 0.085$
Lower Bounds	31.35	31.35	31.35	31.35
Upper Bounds	32.52	32.11	31.686	31.43
Bearish bounds(sellers)	$\mu = 0.075$	$\mu = 0.07$	$\mu = 0.06$	$\mu = 0.05$
Upper Bounds	31.26	31.17	31.01	30.84
Higher Bounds	35.17	35.16	35.17	35.17

Of course, the higher the value of μ, the more the less risk-averse individual is ready to buy (sell) at a high price. The more surprising is the behavior of the lower buying bound when μ is increasing; this one is decreasing whereas investors are more sure to exercice. As they are less risky, the lower bound should be higher. Actually this bound is found with non-dominance arguments. This bound diminishes when μ increases or the stock would dominate the call (which has good chances to be exercised). This decreasing bound expresses the comparison between the option strategy and the stock strategy, but this is more accurate as the option is in the money.

Moreover, the more the option is in the money, the more the bounds are similar. The lower buying bound no longer depends on the expected μ, or the upper selling bounds. When the option is in the money, the value of μ has more impact on the less risk-averse individuals, than on the more risk-averse individuals. Only the volatility has an impact on the lower bound of the buying price, and the upper bound on the selling price. The upper bound for the buying price (lower for the selling price) still depends on μ, but less than for an out-of-the-money call. Moreover investors are more homogeneous in their behavior (because being long in a in-the-money call is less risky) and the spread between lower and upper bounds significantly tightens. This is also true for the sellers (the exercise is almost sure).

Finally the spread in which the transactions can take place becomes:

Transactions spread		
K=100	3.59	5.06
K=90	11.25	13.32
K=80	20.96	22.88
K=70	30.84	32.52

Conclusion

The present article presents very general bounds that do not depend in their derivation on the process followed by the underlying asset. For example, they will allow for the payment of dividends as they are based on expected values at expiry of the options. The payment of dividends simply affects the distribution of the stock prices at expiry, thus resulting simply in a modification of the gain to be used in the results. This is an important extension which comes from the fact that only the final distribution of the stock price is used in the bounding conditions. The inclusion of transaction costs is also trivial, because there are no reallocations of a portfolio in the chosen methodology. The transaction costs can be introduced in the buying/selling price of the asset and the option, at time 0 and T. The remark applies also for taking into account the bid/ask spread of the stock. The computation of bounds give an interesting insight into investors' behavior.

Further research could be performed on the generalization of the method to American options.

Computational results applied to a variety of distribution and options functions, as well as research on tightening the bounds obtained so far, also offer wide scope for research.

Concerning applications, this method can be applied by market makers who have an idea of the expectation of their customers concerning the underlying asset process, with possibly stochastic volatility processes, dividends and transaction costs. In particular, the method can be extended to individuals who already have a position in assets or options. The bounds will surely depend on this original position.

References

Arrow, R. (1970) *Essay in the Theory of Risk Bearing*, North-Holland.

Beiner, N. (1994) 'The rationale for limit order trading', *Working paper*, University of Geneva, Department of Economics.

Black, F. & Scholes, M. (1973) 'The pricing of options and corporate liabilities', *Journal of Political Economy* **81** , 637–54.

Boyle, P.P. & Vorst, T. (1992) 'Option replication in discrete time with transaction costs', *Journal of Finance* **XLVII**, 271–293.

Breedon, D. (1979) 'An intertemporal asset pricing model with stochastic consumption and investment opportunities', *Journal of Financial Economics* **7**, 265–296.

Brennan, M. (1979) 'The pricing of contingent claims in discrete time methods', *Journal of Finance* **34**, 53–68.

Cox, J., Ingersoll, J. & Ross S. (1985) 'An intertemporal general equilibrium model of asset prices', *Econometrica* **53**, 363–384.

El Karoui, N. & Rochet, J.C. (1989) 'A pricing formula for options on coupon-bonds', *Cahier de Recherche du GREMAQ-CRES* **8925**.

El Karoui, N. & Quenez, M.C. (1995) 'Dynamic programming and pricing of contingent claims in incomplete markets', *SIAM J. Control and Opt.* **33**, 29–66.

Fishburn, P.C. (1964) *Decision and Value Theory*, Wiley.

Föllmer, H. & Sondermann, D. (1986) 'Hedging of non-redundant contingent claims', in *Contributions to Mathematical Economics in Honor of Gerard Debreu*, W. Hildebrand & A. Mas-Colell (eds.), North-Holland.

Gollier, C. (1993) 'Portfolio dominance, lower conditional expectation and the monotone likelihood ratio order', *Working paper*, IDEI and GREMAQ.

Hadar, J. & Russel, W. (1969) 'Rules for ordering uncertain prospects', *American Economic Review* **59**, 25–34.

Hadar, J. & Russel, W. (1971) 'Stochastic dominance and diversification', *Journal of Economic Theory* **3**, 288–305.

Hadar, J. & Russel, W. (1974) 'Diversification of interdependent prospects', *Journal of Economic Theory* **7**, 231–240.

Hammond, J. S. (1974) 'Simplifying the choice between uncertain prospects where preference is non-linear', *Management Science* **20**, 1047–1072.

Harrison, M.J. & Kreps, D.M. (1979) 'Martingales and arbitrage in multiperiod securities markets', *Journal of Economic Theory* **29**, 381–408.

Harrison, M.J. & Pliska, S.R. (1981) 'Martingales and stochastic integrals in the theory of continuous trading', *Stochastic Processes and their Applications* **11**, 215–260.

Harrison, M.J. & Pliska, S.R. (1983) 'A stochastic calculus model of continuous trading; complete markets,' *Stochastic Processes and their Applications* **15**, 313–316.

Henin, C. & Pistre, N. (1996a) 'Bounding the generalized convex option price', *European Journal of Finance* **2**.

Henin, C. & Pistre, N. (1996b) 'Bounding the generalized concave option price', *Working paper*, Groupe Ceram.

Henin, C. & Pistre, N. (1996c) 'Using second-order stochastic dominance and linear options to bound nonlinear options premia', *Working paper*, Groupe Ceram.

Henin, C. & Rentz, W. (1984) 'Call purchases, stock purchases and stochastic dominance', *Journal of Business Finance and Accounting* **11**.

Henin, C. & Rentz, W. (1985) 'Subjective stochastic dominance, put writing and stock purchases with extensions to options pricing and portfolio composition', *Management Science* **31**.

Jacka, Saul D. (1992) 'A martingale representation result and an application to incomplete financial markets', *Mathematical Finance* **2**, 239–250.

Karatzas, I., (1988) 'On the pricing of American options', *Applied Math. Opt.* **17**, 37–60.

Kroll, Y. & Levy, H. (1979) 'Stochastic dominance with a riskless asset: an imperfect market,' *Journal of Financial and Quantitative Analysis* **XIV**, 179–204.

Landsberger, M. & Meilijson, I. (1993) 'Mean-preserving portfolio dominance', *Review of Economic Studies* **60**, 479–485.

Leland, H.E. (1985) 'Option pricing and replication with transaction costs', *Journal of Finance* **XL** 1283–1301.

Levy, H. & Kroll, Y. (1978) 'Ordering uncertain options with borrowing and lending', *The Journal of Finance* **XXXIII**, 553–574.

Levy, H. (1985) 'Upper and lower bounds of put and call option value: stochastic dominance approach', *Journal of Finance* **X**, 1197–1217.

Merton, R. (1973) 'The theory of rational option pricing', *Bell Journal of Economics and Management Science* **4**, 141–183.

Perrakis, S. (1984) 'Option pricing bounds in discrete time: extensions and the pricing of the American put', *Journal of Business* **59**, 119–41.

Perrakis, S. & Ryan P. (1984) 'Option pricing bounds in discrete time', *Journal of Finance* **39**, 519–525.

Ritchken, P.H. (1985) 'On options pricing bounds', *Journal of Finance* **XL**, 1219–1233.

Ritchken, P.H. & Kuo, S. (1988) 'Option bounds with finite revision opportunities', *Journal of Finance* **43**.

Rothschild, M. & Stiglitz, J.E. (1970) 'Increasing risk I: a definition', *Journal of Economic Theory* **2**, 225–243.

Rothschild, M. & Stiglitz, J.E. (1971) 'Increasing Risk II: its economic consequences', *Journal of Economic Theory* **3**, 66–84.

Rubinstein, M. (1976) 'The valuation of uncertain income streams and the pricing of options', *Bell Journal of Economics* **7**, 407–425.

Ryan, P. (1995) 'A duality and tightening of option pricing bounds', *Working paper*, University of Ottawa.